The Geochemistry of Manganese and
Manganese Nodules in the Ocean

SEDIMENTOLOGY AND PETROLEUM GEOLOGY

The Geochemistry of Manganese and Manganese Nodules in the Ocean

G. N. Baturin

Institute of Oceanology
Academy of Sciences of the U.S.S.R.

D. Reidel Publishing Company

A MEMBER OF THE KLUWER ACADEMIC PUBLISHERS GROUP

Dordrecht / Boston / Lancaster / Tokyo

Library of Congress Cataloging in Publication Data

Baturin, G. N. (Gleb Nikolaevich)
    The geochemistry of manganese nodules
in the ocean.

    (Sedimentology and petroleum geology)
    Bibliography: p.
    Includes index.
    1. Manganese. 2. Manganese nodules.
3. Geochemistry. I. Title. II. Series.
QE 516.M6B38   1987   553.4'629   87–16309
ISBN-13:978-94-010-8167-2   e-ISBN-13:978-94-009-3731-4
DOI: 10.1007/978-94-009-3731-4

Published by D. Reidel Publishing Company,
P.O. Box 17, 3300 AA Dordrecht, Holland.

Sold and distributed in the U.S.A. and Canada
by Kluwer Academic Publishers,
101 Philip Drive, Norwell, MA 02061, U.S.A.

In all other countries, sold and distributed
by Kluwer Academic Publishers Group,
P.O. Box 322, 3300 AH Dordrecht, Holland.

TABLE OF CONTENTS

INTRODUCTION

Manganese nodules were first discovered on the ocean floor 160 miles south-west of the Canary Islands on February 18, 1803, during the first complex oceanological cruise of the <u>Challenger</u>. They surprised researchers by their unusual shape and also by their unusual chemical composition; nevertheless for many years afterwards, they were considered merely as one of Nature's exotic marine tricks. After the Second World War, a comprehensive investigation of the World Ocean started, and new data were obtained on a wide distribution of manganese nodules and their polymetallic composition, that made scientists consider nodules as one of the major characteristics of the deep oceanic zone. Recently, meaning since the 1960's, nodules have been recognized as a potential ore source, investigation of which is stimulated by the progressive depletion of land-based mineral resources.

Several generations of scientists from various countries have contributed to the problem of exploration of manganese nodules on the ocean floor. Though the problem has been posed, it has not been solved yet because it required, in its turn, a scrutiny of some fundamental aspects such as composition, nature, accretion rate of nodules and retrieval of nodule fields. These problems have been discussed in thousands of papers and larger publications; see, in particulare, Mero, 1965; Horn, 1972; Morgenstein, 1973; Bezrukov, 1976; Glasby, 1977; Bischoff and Piper, 1979; Lalou, 1979; Manganese nodules, 1979; Varentsov, 1980; Cronan, 1980; Manganese nodules..., 1984, 1986. However, many problems of nodule composition and origin are far from being solved and it is not surprising that further research and interpretation is stimulated on the basis of recent observational material, analytical techniques and systematic approach.

Current oceanological thought considers the Ocean as a single system, the geochemistry of which, to a great extent, is controlled by the interaction of the ocean with the lithosphere and the atmosphere. Arising from this assumption, the composition of nodules should be analysed with due respect to the geochemistry of sea water, sediments and pore water and also to the ways and forms of material supply to the ocean from various endogenous and exogenic sources.

The present book has two aims. First, to give a general review of the marine geochemistry of manganese as being the most representative element in nodules. Manganese is characterized by complex migration forms and is a host element for major ore and microelements of nodules. Our second purpose is to draw together all available individual data on the geochemistry of nodules proper.

To cope with these two problems, representative data from numerous publications were used on content and forms of manganese migration in river discharge, eolian dust and precipitate, sea and oceanic water and suspension, marine organisms, bottom sediments and pore water. Published and some original data on the geochemistry of macro- and microelements in nodules, partly, on microstructure of nodules, their mineralogy and accretion rates were also used.

The last chapter is devoted to the problems of nodule genesis.

When revising the available material for the book we learned that the data on some microelements were scarce or not reliable, sometimes no data were available at all. To fill in the gaps in our knowledge we carried out special analytical research to determine cadmium, tungsten, mercury, rhenium, gold, silver, platinum and elements of the platinum group in representative collections of nodules and ore crusts gathered in various regions of the ocean, mainly in the Pacific, during the cruises organized by P.P. Shirshov Oceanology Institute (R/V <u>Vityaz</u>, <u>Akademik Kurchatov</u>, <u>Dmitry Mendeleev</u>, <u>Akademik Mstislav Keldysh</u>).

The material was analysed by various techniques, namely, chemical, spectrochemical, atomic-absorption, x-ray fluorescence, neutron activation, fire assay, laser photo-ionization of atoms, microprobing and others.

The author would like to thank his colleagues for invaluable assistance in preparing the material for the book, namely, Kochenov, Dubinchuk, Boíko, Isaeva, Kurskyi, Oreshkin, Savenko, Skornyakova, Fisher. The staff of the Analytical laboratory of P.P. Shirshov Oceanology Institute, in particular Zavadskaya, Kazakova, Tolmacheva and Shevchenko, were very helpful in analytical work. Vasilieva and Chaikina spent a lot of time and effort preparing this book for publication.

PART I

MARINE GEOCHEMISTRY OF DISPERSED MANGANESE

According to one of the fundamental laws in geochemistry, dispersion and concentration of elements in the Earth's crust and, in particular, within the sedimentary cover co-exist in time and space, dispersion predominating considerably over concentration though the same geochemical rules govern both groups of processes (Clarke, 1924; Vernadsky, 1954; Goldschmidt, 1954; Vinogradov, 1959).

So, to learn about the regularities of ore processes one should have an understanding of the laws that govern the distribution of background concentration of elements in geological objects. This is especially true for a manganese nodule formation process in the ocean which, in some cases, was analysed in ignorance of manganese behaviour in the sea.

The general geochemical background of ore formation process in the sea is considered in the present part of the monograph, in particular: manganese supply to the ocean; its behaviour in sea water; suspension; sea organisms; the processes of Mn deposition to the bottom; its distribution in a surface layer and throughout a sedimentary cover; its mobility in interstitial water; and its geochemical balance.

CHAPTER I

MANGANESE SUPPLY TO THE OCEAN

Manganese and other elements are supplied to the ocean from various sources and
by various routes. Continental massifs and the Earth's interior under the oceanic
floor are believed traditionally to be the main sources for the supply of material,
and river discharge volcanic eruptions and hydrothermal fluids are the main
routes for this material transportation. Besides this, some material can arrive
as a part of an air suspension and as aerosols, as cosmic dust, ground water discharge, and due to coastal erosion. We shall show further that these sources are
of different significance and value.

1.1. MANGANESE IN RIVER DISCHARGE

The World ocean sedimentation depends considerably upon river discharge. The
World rivers drain an area of about 100 million km^2 and their annual water discharge to the ocean is 35 to 44 thausand km^3, 13 to 22 billion tonnes of suspension
and 3.3 to 4.8 billion tonnes of dissolved material (Lopatin, 1950; Alekin, 1966;
Holeman, 1968; Alekseev and Lisitsin, 1974; Lisitsin, 1974; L'vovich, 1974;
Gordeev and Lisitsin, 1978; Martin and Meybeck, 1979; Milliman, 1981; Milliman and
Meade, 1983).
    River discharge intensity and composition depend upon the climate, topo-
graphy, rock and soil composition within a drainage area (Strakhov, 1960). Manganese in river water migrates in a solution and suspension, in mineral and
organic forms.

Dissolved and suspended manganese: concentration and their ratio

Table 1 presents a general behaviour pattern of suspended and dissolved manganese
in a river discharge. It combines data on 34 lowland and mountain rivers in the
northern humid, arid and tropical areas of Europe, Asia, Africa, North and South
America. Data on average water, ion and suspension discharge of rivers were taken
from Lopatin, 1952; Shamov, 1954; Strakhov, 1962; Livingstone, 1963; Alekin and
Brazhnikova, 1964; Alekin, 1966; Gibbs, 1967; Zalogin and Rodionov, 1968;
Depetris and Griffin, 1968; Holeman, 1968; Turekian, 1971; Lisitsin, 1974; Carbonnel
and Meybeck, 1975; World Register, 1978; Discharge..., 1979; Martin and Meade,
1983; Martin and Gordeev, 1984. Data on suspended and dissolved manganese
concentration were taken from Glagoleva, 1959; Nesterova, 1960; Konovalov et al.,
1968; Kontorovich, 1968; Trefrey and Presley, 1976; Gibbs, 1977; Chudaeva, 1978;
Martin and Meybeck, 1978, 1979. The author's data on manganese concentration in
suspension of the Northern Dvine, Kuban, Don, Volga, Ural, Amu Darya, Rion were
also used.
    Despite some discrepancy in the results obtained by different authors for
various years, the Table displays some regularities in manganese behaviour in
river discharge.
    The water discharge of the rivers considered varies from 3 (the Southern Bug)
to 5500 km^3/yr (the Amazon), mean mineralization varies from 50 (the Mezen) to 760
mg/l (the Colorado), and mean turbidity from 22 (the Yenisei) to 6660 mg/l (the
Colorado).

Dissolved manganese concentration in river water varies from several micrograms
per litre (the Ob, Lena, Kuban) to 45 µg/l (the Dniester). Anomalously
high concentrations of dissolved manganese (to 160-690 µg) were
recorded in the Kura water and especially in the Rion, both of which drain
manganese-bearing sedimentary layers. The average manganese concentration in the
water of the World rivers is 8 µg/l (from recent estimates), whereas the total
river discharge is 330000 t/yr (Gordeev, 1984; Martin and Gordeev, 1984).

The manganese concentration in river particulate suspension varies from 0.027
in the Parana to 0.1-0.2% in some other major World rivers, in particular, in the
Hwang Ho and Yangtze Rivers (Li et al., 1984). A higher percentage (0.6%) of
manganese was detected in the suspension from the Dnieper River and its maximum
was registered in individual suspension samples from Rion (to 5.9%) that is caused
by washout and supply of a suspended ore material in the river system (in the area
of the Rion River discharge, the Chiatura manganese field is located).

From Table 1 and some other papers (Turekian and Scott, 1967; Turekian,
1971; Windom et al., 1971; Martin et al., 1973; Duinker and Nolting, 1976;
Borole et al., 1977; Evans et al., 1977), the average manganese concentration in
river suspension is assumed to be 0.11% (Gordeev, 1983, 1984; Martin and Gordeev,
1984), which is consistent with its mean concentration in the Earth's crust, i.e.
0.125% (Ronov and Yaroshevsky, 1967). However, it is noticeably higher by comparison with its mean concentration in soils (0.09% (from Vinogradov, 1957)) and
sedimentary rocks (0.07% (from Vinogradov, 1962)).

Since the study of manganese concentration in rivers was initiated not so long
ago, i.e., during the period of active intrusion of a man in the natural geochemical
regime of river discharge, relatively high mean concentrations of manganese in river
suspensions may probably be the result of anthropogenic pollution (Teraoka and
Kobayashi, 1980).

The general concentration of suspended manganese in river water ranges from
10 (the Pechora, Lena) to hundreds and even thousands of micrograms per litre
(the Syr Darya, Amu Darya, Colorado), being principally a function of the
turbidity of river water, as has been shown by Volkov (1975). In cases where a
suspension proper is enriched by manganese, concentrations of suspended manganese
reach even higher anomalous values (in the Rioni, it is up to 28200 µg/l). Total
suspended manganese supply from the World rivers is estimated as 20.4 million tonnes
per year (Gordeev, 1984).

Suspended manganese concentration in river water is always higher than the
solution concentration. In some lowland rivers of the northern humid zone, the water
of which is characterized by low turbidity, a relative share of suspended manganese
can sometimes be 60 to 80% (the Pechora, Ob, Irtysh, Lena), while in all mountain
rivers of the arid and humid zones and in the majority of lowland rivers with water
turbidity over 200 mg/l, a relative share of suspended manganese is 90 to 99% (see
Table 1), being 98.4% on average for all rivers (Gordeev, 1984).

The amount of an element in a solution relative to general mineralization ("dry
residue") is one of the indicators of element mobility. In the rivers considered,
manganese concentration in a "dry residue" is within the range from 0.0003% (the
Dnieper) to 0.4% (the Kura), i.e., it is always lower than its mean concentration in
crustal sedimentary rocks.

Another indicator of an element's mobility in hypergenetic processes is its
"coefficient of migration in water", that is, the ratio of its concentration in a "dry
residue" to its mean concentration in the rocks within a drainage area (Perelman,
1961, 1968). In our case, since we lack the corresponding data, we can use
manganese concentration in suspension (last column in Table 1) which also gives a
distincitive picture: the ratio is always less than 1; for the majority of rivers it is
about 0.1 or lower. Such an estimate permits us to rate manganese as a moderate or
slightly mobile component in a river discharge.

Earlier, the migration potential of elements was assessed many times by the
"mobility sequence" derived from the ratio of solution to suspension (Strakhov, 1962).
That leads us to conclude that there is no uniform behaviour of elements in various
rivers of the same climate zone. In the rivers within the Black Sea basin this series
is as follows (in order of increasing mobility): V-(Cr, Ni, Be, Ga, Zn)-Fe-Mn-P-
(Pb, Sn)-Ba-Cu-Sr (Strakhov, 1962), in the Ob River: V-Mn-Cr-Fe-P-Ni-Cu-Co
(Nesterova, 1960). Thus, in the first case, manganese turned out to be a relatively

Dissolved and suspended manganese in river water*

| River | River discharge (km³/yr) | Minerali-zation (mg/l) | Turbidity (mg/l) | MANGANESE | | | | Relative share of suspended Mn (%) | Ratio of Mn in dry residue to Mn in suspension |
				Dissolved (μg/l)	Suspended (μg/l)	In dry residue (%)	In suspension (%)		
Northern European part of the USSR and Siberia									
Northern Dvina	110	115	53	(8)	42	0.005	0.077-0.095	84	0.06
Mezen	26	49	30	(8)	50	0.016	0.17	86	0.1
Pechora	106	83	50	8.0-13.2	9.0-24	0.01-0.016	0.11	53-64	0.1
Ob	400	120	36	3.4-25	17-790	0.003-0.02	0.13	77-99	0.1
Irtysh	68	200	163	63-67	43-430	60.03	(0.11)	39-87	0.3
Yenisei	560	140	22	3.7-13	286-650	0.003-0.009	(0.11)	98-99	0.05
Lena	510	143	300	2.9-3.5	12-24	0.002-0.004	(0.11)	77-89	0.03
Kolyma	120	-	45	3.1-6.3	117-440	-	(0.11)	95-99	
Amur	347	70	150	3.5-62	19-25	0.05-0.09	0.10	76-88	0.4
The Drainage Basin of the Black and Azov Seas									
Dniester	4	300	230-380	8.4-45.5	196-540	0.003-0.015	0.024.0.225	87-97	0.1
Dnieper	43	287	30	1.0-19.8	290	0.0003-0.007	0.079-0.60	95	0.1
Southern Bug	2.8	400	200	3.9-10.4	200-500	0.001-0.002	0.09-0.26	95-99	0.02
Danube	203	300	300-	10-70	156-3110	0.003-0.02	0.09-0.274	94-98	0.1
Kuban	11	195	700	2.5-8	49-405	0.001-0.003	0.03-0.05	92-99	0.05
Don	28	400	200	(8)	60-400	0.002	0.03-0.21	88-98	0.02
Rioni	12.8	167	620	31-690	2140-28200	0.02-04	0.12-5.90	97-99	0.1
Chorokh	8.5	205	885	13.4-18	800-1200	0.006-0.009	0.08-0.12	98-99	0.08
Ingur	6	100	330	(8)	370	0.008	0.072	98	0.1
Kura	157	382	19	4.6-160	360-11600	0.001-0.04	0.08	86-99	0.2
The Caspian Sea Drainage Basin									
Volga	255	182	102	3.0-37.5	67-126	0.002-0.02	0.030-0.055	94-98	0.2
Ural	6.2	440	252	3.2-3.6	1100	0.0008	0.050	99.7	0.02
Terek	11	280	2400	6.8-23.2	442-2720	0.003-0.008	0.07	97-99.7	0.1

Table 1 (continued)

Middle, South and Southeastern Asia

Syr Darya	14	430	1160	10.4–14.8	630–5030	0.002–0.003	0.06	98–99.6	0.05
Amu Darya	42	420	2510	7.4–9.1	394–2980	0.002	0.055	98.99.6	0.04
Ganges	366	152	1200	(8)	1200	0.005	0.106	99.4	0.05
Mekong	470	105	438	–	400		0.094	98	0.05

Africa

Congo	1250	80	48	(8)	67	0.01	0.14	89	0.07
Niger	192	98	229	(8)	183	0.008	0.08	96	0.1

North and South America

Mississippi	560	223	833	(8)	425–1900	(0.004)	0.051–0.23	98–99.6	0.04
Colorado	20	760	6660	(8)	2800–5200	(0.001)	0.043–0.078	99.7	0.02
Mackenzie	306	214	34	(8)	22	(0.004)	0.065	73	0.06
Amazon	5519	57	156	5.5–16.5	125	0.01–0.03	0.08	89–96	0.2
Orinoco	1070	54	196	(8)	145	(0.015)	0.074	95	0.5
Parana	470	114	215	(8)	58	(0.008)	0.027	88	0.3

* In brackets we give supposed values on the basis of average manganese concentration in solution (8 µg/l) and in suspension (0.11%).

mobile element, while in the second it is less mobile.

In the World river discharge the ratio of suspension element concentration to dissolved concentration varies from 0.02 (chlorine) to 770 (aluminium) (Gordeev, 1984). From these data, the sequence of 15 elements in the World river discharge will be:

Al - Ti - Fe - Mn - Si - P - K - F - Mg - Na - Ga - C_{org} - S - I - Cl
770 - 617 - 590 - 62 - 24 - 21 - 5 - 2.3 - 1.6 - 0.9 - 0.8 - 0.8 - 0.4 - 0.06 - 0.02

Manganese forms in river discharge

The migration intensity of elements in surface water is connected with the forms of their occurrence in solutions and suspensions.

It is a difficult problem to determine the forms of dissolved manganese (like many other microelements) in river water. This problem is approached by both analytical and numerical methods.

From experimental data by Glagoleva (1959), up to 75% of dissolved manganese in the rivers running in the south of the european part of the USSR migrates as a component of a humic complex. It was determined that 73% of dissolved manganese in the rivers of the Black Sea basin is in a colloidal form (Garanzha and Konovalov, 1977).

From data by Eremenko (1964, 1966), dissolved manganese in the water of the Volga and Don Rivers occurs in the following forms: 61.9 to 85.4% are simple positive ions; 10.8 to 28.6% are negative ions and complex molecules; 3.8 to 9.5% are colloids and pseudocolloids with particle size less than 0.35 μm.

In the Dnieper River, downstream of Kiev Dam, free cations of bivalent manganese co-exist with complex manganese-organic compounds with molecular mass of three classes: (120-150) x 10^3, about 70 x 10^3 and (0.5-50) x 10^3. Free cations predominate during winter and spring, whereas complex high molecular compounds predominate in summer during the vegetative season of water plants (Linnik and Nabivanetz, 1978).

In the water of two rivers in England, manganese was detected by an ultra-filtering technique as being predominant in the fraction with molecular weight less than 30000 (Moore et al., 1979).

Analysis of dissolved manganese from the rivers of the Black Sea, Azov and Caspian basins by a complex system of successive extractions revealed three forms of its occurrence, namely, labile organic (complexes with lipoids), stable organic (complexes with humic and fulvicacids and non-organic (ions, molecules, non-organic complexes) forms (Table 2). Relative concentration of labile complexes in lowland and mountain rivers turned out to be equal (23-24%), whereas concentration of stable complexes decreases sharply in lowland rivers as compared to the mountain ones (3.5 and 24.6%, respectively). The non-organic form is the predominating one: 72.5% in lowland and 52.2% in mountain rivers. Iron forms are remarkably different - stable organic complexes prevail in lowland rivers (57.7%), whereas in mountain rivers iron shows relatively regular distribution among all three fractions (Demina et al., 1978).

Many researchers believe that modern fractioning techniques distort the natural ratio between natural water components and therefore they are not reliable. So, alongside experimental study, one uses numerical methods to determine various forms of metals in river water. They are based on the analysis of total content of elements and on the data on stability of compounds (Semenov et al., 1968). However, absence of sufficient data on many compounds hinders its widespread application.

Forms of manganese occurrence in river suspension are as variable as they are in solution. On the whole, they can be subdivided into labile and inert. The first group embraces sorbed complexes, hydrooxides, readily soluble carbonate and organic compounds; the second embraces elements in a crystal lattice of stable clastic minerals and stable organic assemblages such as kerogen. To distinguish the two forms samples are successively processed by various solvents; however, as no unified technique exists, this system is very ad hoc (Gordeev, 1983).

It is clear from Table 3 that relative concentration of superficially sorbed manganese in different rivers varies from 0.6-0.9% (the Amazon and Yukon Rivers) to 46.60% (the Ganges and Brahmaputra, and rivers of the Sea of Japan basin).

TABLE 2

Forms of Mn and Fe dissolved in the rivers
of the Black, Azov and Caspian Sea areas (Demina et al., 1978)*

Forms	Manganese				Iron			
	Concentration				Concentration			
	Absolute (μg/l)		Relative (%)		Absolute (μg/l)		Relative (%)	
	Lowland	Mountain	Lowland	Mountain	Lowland	Mountain	Lowland	Mountain
Labile organic	0.38	0.82	24	23.2	17.5	11.4	24	30
Stable organic	0.06	0.87	3.5	24.6	42.2	11.4	57.7	30
Non-organic	1.16	1.85	72.5	52.2	13.3	15.2	18.3	40
Total dissolved	1.6	3.54	100	100	73	33	100	100

* Lowland rivers: Danube, Dnieper, Don, Southern Bug, Volga, Ural.
  Mountain rivers: Rioni, Ingur, Codor, Chorokh, Bzyb, Kura, Samur, Sulak, Mzymta, Kacha, Supsa, Sochi, Kuban, Terek.

TABLE 3

Manganese forms in river suspension
(% of the total content)

River	Manganese forms					References
	Sur-ficial sorbed	In compo-sition of amorphous hydrooxides	In compo-sition of crystallized hydrooxides	Organic	Silicate	
Amazon	0.9	59.4		6.0	33.7	Gibbs (1973)
Yukon	0.6	51.3		6.8	41.3	"
Rivers of the Black Sea drainage basin	15.1	62.2		–	–	Glagoleva (1959)
Same, lowland	–	84.1	5.8	1.5	17	Demina et al. (1978)
Same, mountain	–	65.4	12.2	1.2	21.2	"
Rivers of the Sea of Japan drainage basin	60.3	9.0		12.9	16.8	Chudaeva et al. (1982)
Ganges-Brahmaputra:						
bulk sample	46.4	12.7		13.2	27.7	Gordeev et al. (1983)
fraction 0.01 mm	42.8	15.1		15.2	26.9	"
fraction 0.001 mm	48.2	14.3		15.5	22	"

In the rivers of the Black Sea, Azov and Caspian basins this form is 15% of the total Mn content of suspended material.

The concentration of hydrooxide forms of manganese varies from 9-15% in the Ganges, Brahmaputra, rivers of the Sea of Japan basin to 51-90% in other rivers. From the data obtained for the rivers in the Black Sea area it follows that manganese is associated with the hydrooxide form 5 to 13 times more often than with the crystallized one because the former types are considerably more active geochemically.

The organic form is of lesser importance: from 1.2-1.5% in the rivers of the Black Sea and Azov basins to 6-7% in the Amazon and Yukon rivers and up to 13-15% in the rivers of India and Far East (USSR).

The share of the silicate form of suspended manganese is a minimum in the rivers of the Black Sea, Azov, Caspian and Sea of Japan basins (17-21%), a bit higher in the Ganges and Brahmaputra (22-28%) and a maximum in the Amazon and Yukon rivers (34-41%).

Thus, about 20.4 million tonnes of suspended and 0.33 million tonnes of dissolved manganese are the annual supply to the World Ocean through river discharge. Suspended manganese is mainly in mineral sorbed and hydrooxide forms, whereas dissolved manganese is in mineral and organic forms.

1.2. MANGANESE IN THE ATMOSPHERE

The atmosphere plays a remarkable role in the energy and mass exchange between the ocean and continents, as it supplies dust particles, aerosols and atmospheric precipitation to the ocean various chemical elements, heavy metals in particular, being incorporated in them. The content of some elements (Zn, Cu, Cd, Pb, Hg, Se, Sb, As) with respect to iron and aluminium is anomalously high in the atmosphere in contrast to that in the crustal rocks (Lantzy and Mackenzie, 1979; Prospero, 1981; Miklishansky, 1983).

These elements arrive from various sources by different ways; for instance, due to weathering of continental rocks and soils (Litsitsin, 1978; Winchester et al., 1981; Schutz and Rahn, 1982), as gas and solid products of volcanism (Peterson and Rotschi, 1952; Petterson, 1959; Zelenov, 1972; Mroz and Zoller, 1975; Sobotovich, 1976, 1982; Miklishansky et al., 1979; Golenetzky et al., 1981, 1982), due to anthropogenic pollution (Bertine and Goldberg, 1971; Nriagu, 1979), as crustal sublimation (Goldberg, 1976; Blimbcombe and Hunter, 1977), as a result of low-temperature processes of biological methylization (Wood, 1974; Ridley et al., 1977), by emission from the vegetation surface (Beauford et al., 1975, 1977).

The main manganese supply to the atmosphere comes from the continental aeolian material – the role it plays in oceanic sedimentation has been studied for over hundred years since the expedition of the Challenger.

In some pelagic oceanic deposits, components of obviously aeolian origin are widely spread, namely, quartz grains transported from deserts, mica particles, freshwater diatoms, etc. (Lisitsin, 1978; Schutz et al., 1980). In the Northern Pacific, the share of aeolian components in the fine-grained bottom sediments reaches 30% to 50%, whereas in the southern arid climate belt it seems to be over 50% in some places (Windom, 1969, 1970; Duce et al., 1980).

Up to 12 million tonnes of dust is transported by wind yearly to the pelagic zone of the North Pacific from the arid Asian regions (Uematsu, 1983) and about 250 million tonnes from the Sahara Desert to the Atlantic (Graham and Duce, 1979). In total, from 2 to 3 billion tonnes of solid terrigeneous material is supplied yearly to the atmosphere as a part of aeolian material (Goldberg, 1971; Petterson and Junge, 1971; Robinson and Robbins, 1971; Graham and Duce, 1979).

From Lisitsin's estimates (1974, 1978), the aeolian dust annual supply to the World Ocean is up to 1.6 billion tonnes, whereas the amount of terrigenous material in river discharge that reaches pelagic zone does not exceed 1.3-1.6 billion tonnes.

Concentrations and compositions of aerosol material in the lower atmosphere, troposphere and stratosphere vary within a wide range. Aerosol concentration in the atmosphere over the ocean varies from 0.003 to 30 and, on average, is about 0.5 $\mu g/m^3$ (Savenko et al., 1975; Lisitsin, 1978; Prospero, 1979).

Terrigenous material predominates in aerosol composition in the lower atmosphere over near-continental oceanic zones, whereas in the stratosphere and troposphere, the role of atmospheric components, such as sulphate and other salts, increases. According to their solubility potential the following fractions can be distinguished in aerosols: water-soluble, exchangeable, acid-soluble and residual (Miklishansky et al., 1977; Miklishansky, 1983).

Several reviews have been devoted to the data on composition and distribution of aerosols over the oceans (Chester, 1972; Duce et al., 1976; Lisitsin, 1978; Winchester, 1978; Lantzy and Mackenzie, 1979; Prospero, 1981; Miklishansky, 1983).

The manganese content in the atmosphere over the ocean, in atmospheric precipitation and aerosols has been assessed by many researchers (Tables 4 to 6).

Total manganese content in the air depends upon meteorological conditions and upon the area of sampling. It varies from 0.004 to 650 ng/m³. The most clean atmosphere was registered over the South Pole, the manganese concentration in the air there being 0.0044 to 0.019 ng/m³, and, on average, it is 0.0103 ng/m³ (Zoller et al., 1974).

Manganese concentration in the air over the parts of the ocean at great distance from continents is several tenths of a nanogram or several nanograms per cubic meter, however, it increases by hundreds or even a thousand times, reaching several micrograms per cubic meter in the Atlantic near the Sahara Desert shore and in the north-western Pacific, east off China, in the areas affected by dust storms (Lisitsin, 1978; Tsunogai and Kondo, 1982).

A remarkable contrast in manganese concentration in the air is observed also in the northern and southern parts of the Indian ocean, and in some pelagic and near-continental parts of the Pacific.

The average manganese concentration in the air over pelagic oceanic zones calculated from Table 4 is 2.6 in the Atlantic, 1.3 in the Indian ocean, 0.4 in the Pacific, the total concentration over the oceans being 1.4 ng/m³. Manganese concentration in the air over continents is usually higher and is some tens or hundreds of nanograms per cubic meter, with a distinct maximum in arid zones.

Manganese concentration in precipitations (rain, snow, fog) varies from 0.015 to 94 µg/l. Its minimum was registered in humid climate belts, in regions far from industrial centres; namely, the snow cover and glaciers of Antarctica and Greenland (0.015 to 0.557), Bermuda Islands region (0.1 to 0.83), along the Puerto Rican coast (0.67 to 1.7 µg/l).

Over continents (Antarctica excluded), the manganese concentration in precipitations is higher and reaches tens of micrograms per litre, especially in industrial areas of Europe and America. Its average concentration in precipitations over the ocean is about 0.1 to 0.2 µg/l from the available data sources (Martin and Harriss, 1973; Galloway et al., 1982).

Manganese concentration in aerosols, considered over both oceans and continents, is within the range of 0.003 to 1.45%. In the aerosols sampled over the Atlantic it is 0.029-0.68%, over the Norwegian and Greenland Seas - 0.004-0.026%. In the aerosols over the Indian ocean, manganese concentration varies from 0.014 to 0.15%, and, over the Pacific, from 0.0054 to 0.067%.

Manganese concentration is 0.026-0.055% in the aerosols transferred by western winds from Australia to New Zealand and precipitating on the Tasman and Franz Joseph Glaciers.

In the coastal oceanic regions bordering the continental arid and industrial regions, manganese concentration in aerosols usually goes up. Its maximum in sea aerosols (0.25-0.68%) is registered in the northern part of the Atlantic and China Sea. Even higher manganese concentrations (to 1.45%) were determined in some industrial continental regions. Average manganese concentration in aerosols is 0.075% over the Atlantic, 0.06% over the Indian ocean, 0.03% over the Pacific, 0.026% over the Ross Sea (Antarctica), 0.13% over the seas of the South-East Asia. Average manganese concentration in the total mass of oceanic aerosols is 0.018% from 119 samples (Lisitsin and Gordeev, 1974), and about 0.05% in their mineral fraction.

The rate of aerosol precipitation varies considerably but falls mainly within the range of 0.2 to 2 cm/s and is 0.8 cm/s, on average, for the northern hemisphere, as has been detected from [7]Be and ^{210}Pb isotopes (Turekian et al., 1977; Young and Silker, 1980; Turekian and Cochran, 1981a, b).

The total amount of manganese supply to the ocean through aerosols can be calculated from the absolute mass of aeolian material (1.6 billions tonnes per year,

TABLE 4

Manganese concentration in the air (ng/m³)

Region	Concentration			Number of samples	References
	from	to	avarage		
The Arctic Ocean					
Novaya Zemlya, Dickson, Salekhard	0.25	1.73	1.03	3	Egorov et al. (1970)
The Atlantic					
Northern part	1.06	25.8	9.2	3	Savenko et al. (1975)
"	0.055	1.25	1.52	3	Chester et al. (1974)
"	0.03	2.86	1.97	10	Hoang and Servant (1974)
"	0.16	5.3	1.25	24	Hoffman et al. (1974)
"	0.05	5.4	2.5	25	Duce et al. (1975)
Bermuda Islands	0.03	27	1.5	134	Duce et al. (1976)
Northern tropical zone	–	–	2.2	53	Olet (1979)
Same, near Sahara Desert	–	–	217	1	Chester et al. (1974)
	37	650	230	4	Hoffman et al. (1974)
Southern part	0.47	1.6	1.1	3	Savenko et al. (1975)
The Indian Ocean					
Northern part	–	–	7.86	–	Egorov et al. (1970)
Southern part	–	–	0.24	–	"
"	0.3	5	2.6	2	Chester et al. (1974)
"	0.42	1.47	1.13	5	Savenko et al. (1975)
The Pacific					
Region of Hawaii Islands	0.02	0.6	0.2	87	Hoffman et al. (1972)
Northern part, south of Japan	–	–	0.87±0.58	4	Yano et al. (1974)
"	0.20	8.29	2.0	13	Tsunogai and Kondo (1982)
Equatorial zone	–	–	0.07	30	Maenhaut et al. (1983)
Eniwetok Atoll	0.025	1.7	0.29	55	Duce et al. (1983)
Near continental zones	0.34	31.4	11	6	Chester et al. (1974)

Table 4 (continued)

South Pole					
South Pole	0.0044	0.019	0.0103	–	Zoller et al. (1974)
Continental areas					
Eastern and southern coastal regions: U.S.S.R.	11.76	20.73	15.3	3	Egorov et al. (1970)
Intracontinental regions: U.S.S.R.	141.9	434.4	267.4	6	"
Region of Tokyo	–	–	38 ± 2	4	Yano et al. (1974)
Tennessee state, U.S.A.	2	18	10	10	Lindberg and Harriss (1983)

TABLE 5

Concentration of manganese in the atmospheric precipitation (µg/1)

Region	Mn concentration			Number of samples	References
	from	to	average		
U.S.S.R.;					
Northern part	1	32	8.7	8	Drozdova and Mahon'ko (1970)
North-western part	3	43	10.9	35	"
Central part	2.3	2.9	12.7	32	"
Moscow region	14	94	50	3	Savenko et al. (1978)
Crimea coast	-	-	10	1	"
"	0.35	20.50	14.8	11	Belyaev and Ovsyanyi (1969)
Washington state, U.S.A.	2.3	2.9	2.6	2	Rancitelli and Perkins (1970)
United Kingdom	5	60	8	7	Peirson et al. (1973, 1974)
Puerto Rico	0.67	1.7	0.89	10	Martens and Harriss (1973)
Bombay	0.2	26.9	8.6	40	Sadasivan et al. (1974)
Antarctic ice sheet	3.0	3.3	3.1	2	Vilensky and Miklishansky (1976)

TABLE 6

Manganese concentration in aerosols (%)

Region	Mn concentration			Number of samples	References
	from	to	average		
Northern Atlantic	0.036	0.47	0.156	18	Chester and Johnson (1971)
"	0.10	0.19	0.13	50	Chester and Stoner (1974)
"	0.029	0.68	0.15	8	Savenko et al. (1975)
"	-	-	0.04	-	Miklishansky (1983)
Norwegian Sea	-	-	0.004*	-	Wilkniss et al. (1974)
Greenland Sea	-	-	0.026*	-	"
Indian Ocean	0.014	0.059	-	-	Egorov et al. (1970)
"	0.06	0.15	0.11	2	Chester et al. (1974)
"	0.014	0.10	0.049	5	Savenko et al. (1975)
Pacific Ocean	0.042	0.067	0.050	3	Prospero and Bonatti (1969)
"	0.0054	0.0360	0.0143	25	Savenko et al. (1975)
"	-	-	0.0124	-	Miklishansky (1983)
China, Japan and Java Seas	0.07	0.25	0.13	6	Chester et al. (1974)
New Zealand, glaciers	0.026	0.055	0.04	6	Windom (1970)
Tennessee state, U.S.A.	0.0031	0.058	0.034	10	Lindberg and Harriss (1983)
U.S.S.R., coastal regions	0.0303	0.0435	0.0342	5	Egorov et al. (1970)
U.S.S.R., intracontinental regions	0.18	1.45	0.56	6	"

* In fractions over 2 µm, free of salts.

Lisitsin, 1978) and from the average manganese concentration in it (0.05%). It is 800 thousand tonnes per year.

According to the research carried out in the North Sea, the rate at which atmospheric manganese is supplied to this sea is 1.2×10^{-12} mol/m^2/s (Cambray et al., 1975). If we assume this as an average for the entire oceanic area, which equals 361 mkm^2, the total manganese supply through aerosols will be 730 thousand tonnes per year. However, such an amount is probably overestimated since, in the high latitudes (the Arctic and Antarctic oceans), the rate of manganese supply by aerosols is minimum and varies within the range from 2.2 to 8.5 ng/cm^2/yr (Herron et al., 1977; Boutron, 1979; Davidson et al., 1981).

From the data on manganese concentration in the atmosphere over the Bermuda Islands, atmospheric flows of this element were calculated by three different theoretical models. The results are 0.87, 1.9 and 5.5×10^{-5}/g/cm^2/s (Duce et al., 1976). For the entire ocean it will be 93000, 200000 and 585000 tonnes per year.

From another calculation, the total manganese supply to the ocean from the atmosphere is 70 ng/cm^2/yr, on average (Buat-Menard and Chesselet, 1979), that is, 250000 tonnes per year for the entire ocean.

Some scientists believe that a certain ratio does exist between the concentration of chemical elements in the air and rain. It is:

$$\frac{\text{concentration in the air, ng/m}^3}{\text{concentration in rain water, } \mu g/l} \approx 1.$$

(Zhigalovskaya et al., 1973; Pierson et al., 1973; Cambray et al., 1975). If this ratio holds, then the average manganese concentration in rain water over the ocean should be about 1.4 μg/l in accordance with its average concentration in the air (1.4 ng/m^3). However, it was noted earlier that direct estimates provide another average, about 0.2 μg/l.

The total precipitation over the ocean is 324000 km^3/yr, on average (Defant, 1961). Consequently, from various estimates, 64000 to 450000 tonnes of manganese, washed out of aerosols, are supplied yearly to the ocean by these precipitations; this is probably 50% of its total supply to the ocean through the atmosphere.

According to the earlier determination, water-soluble manganese in aerosols varies within the range 10 to 80%, being 50% on average (Martens and Harriss, 1973; Duce et al., 1976; Miklishansky et al., 1977; Hodge et al., 1978); this does not contradict the earlier estimate.

The problem of the origin of aerosol material can be solved by mineralogical, geochemical and isotopic techniques.

The terrigenous component in aerosols can be identified by quartz, feldspar, clay minerals, oxygen-isotope composition of quartz, by [228]Th in submicron fractions (Lisitsin, 1978; Hirose and Sugimura, 1984). To define the sources of some elements one can use correlations and ratio with characteristic "reference" elements for terrigenous material (aluminium, iron) and for sea water (sodium). In the aerosols near the Australian coast, the coefficient of Mn to Al correlation is 0.87, on average (Andreae, 1982).

Over the ocean, the (Mn/Al in the air) to (Mn/Al in the Earth's crust) ratio varies from 0.03 to 5 and, on average, is 1.2 to 2.6 for the North Atlantic, 1.5 for Bermuda Islands, 2.6 for Hawaii, 1.7 for the southern Indian ocean and 5 for its northern part, about 5 near Norwegian and Scotland Islands (Table 7).

In aerosols, the average coefficient varies from 1 to 6.

A similar coefficient with iron as an index was determined in the air, precipitations and aerosols. In the air its average is 1.4 for the tropical zone of the North Atlantic and over Hawaii, 1.7 to 2 in the Indian ocean and 2.6 over Great Britain.

This coefficient is 0.7 to 3.9 in the snow and glaciers of the Antarctica continent. In aerosols, as in the air, it is from 1 to 2.5 on average, which verifies the terrigenous origin of the main manganese mass in the atmosphere material. A coefficient for the correlation between manganese and definitely terrigenous aluminium was 0.99 at the 0.01 level of confidence determined in 53 aerosol samples from the tropical zone of the Northern Atlantic (Buat-Menard and Chesselet, 1979).

One obtains a different result when comparing the Mn/Al and Mn/Fe rations in the air and sea water, the average concentration of manganese, iron and aluminium being 0.027, 0.056 and 0.54 μg/l, respectively (Bruland, 1983). The average of the coefficient is 0.2 for aluminium and 0.04 for iron, which differs by one or two orders from the atmosphere/crust coefficient.

TABLE 7

Coefficients of aerosol enrichment with manganese

Region	Enrichment coefficient*	References
Relative to aluminium		
Arctic coast region	(2.3-4)	Egorov et al. (1970)
United Kingdom	6(4-8)	Peirson et al. (1973)
Northern Atlantic	1.2(0.03-3)	Hoang and Servant (1971)
"	2.6	Duce et al. (1975)
Bermuda Islands	1.5	Duce et al. (1976)
Shetland Islands	4.6	Peirson et al. (1974)
Norwegian coast	5	Rahn (1975)
Indian Ocean	4(2-7)	Egorov et al. (1970)
Northern tropical zone	1.3(0.8-2.2)	Tsunogai and Kondo (1982)
of the Pacific	1.3(0.8-2.2)	Tsunogai and Kondo (1982)
Hawaii Islands	2.6	Hoffman (1972)
Eniwetok Atoll	0.96(0.86-1.1)	Duce et al. (1983)
Relative to iron		
Nothern Atlantic	1.4(0.9-2.1)	Hoffman et al. (1974)
United Kingdom	2.6(2.0-3.5)	Peirson et al. (1973)
Indian Ocean	1.8(1.7-2.5)	Egorov et al. (1970)
Hawaii Islands	1.4	Hoffman et al. (1972)
Northern tropical zone, Pacific	1.6(1.0-3.7)	Tsunogai and Kondo (1982)
South Pole	(0.69-1.1)	Zoller et al. (1974)
Atlantic,Indian,Pacific Oceans	1.5(0.9-2.5)	Maenhaut et al. (1977)

* Extrema are given in brackets.

One can have a more distinct picture when comparing concentrations of manganese and sodium, which is the main cation of sea water. The ratio: (Mn/Na in the air, precipitations, sea aerosols) to (Mn/Na in sea water) varies from 4600 to 40000 (Rancitelli and Perkins, 1970; Hoffman et al., 1972; Duce et al., 1983). This testifies to a continental, rather than a marine origin of manganese in the atmosphere and aerosols over the ocean.

The recirculation of manganese from the ocean to atmosphere has not been studied thoroughly yet. Some specialists believe that the supply of chemical elements to the atmosphere from the ocean results mainly from the bursting of air bubbles as they go upward and reach the water surface (Horn, 1972).

To verify this supposition special tests were made. Nitrogen bubbles were barbotted through sea water sampled in the Atlantic shelf (near Rhode Island) and then the foam was collected and investigated for heavy metals (Wallace and Duce, 1975). It turned out that gas bubbles extract a considerable part of the metals dissolved in the water.

Numerically, the process was estimated as follows (in picogr/m^2/s) : 2.1 for manganese, 13 for aluminium, 260 for iron. When summing up the total manganese extraction from the entire ocean one obtains 23000 tonnes per year.

From other estimates, based on a hypothetical mass exchange model between the ocean and atmosphere, manganese extraction from the ocean surface is about 31000 tonnes per year (Miklishansky, 1983).

Both these estimates show that manganese supply to the ocean from the atmosphere is at least one order higher than its extraction from the ocean to the atmosphere. Thus, the atmosphere, like river discharge, is an additional route for manganese transfer from continent to the ocean.

1.3. COSMOGENIC MANGANESE

Cosmogenic components in the form of magnetic and silicate spherules are widespread both over the continents and in sediments of the World Ocean where they were first discovered during the Challenger expedition (Murray and Renard, 1891). The majority of the spherules are products of meteorite ablation in the atmosphere (Papanastassiou et al., 1982; Nishiizumi, 1983; Thiel et al., 1983).

In the pelagic zone of the Pacific, the concentration of cosmogenic spherules over 30 microns in diameter is 330 units per kilogram of sediments, on average (Brownlow et al., 1966). Sometimes in manganese nodules, the number of magnetic spherules of 100 microns in diameter only reaches several thousands of units per kilogram (Finkelman, 1970). According to the first calculations based on the data of one deep-sea core from the Equatorial Pacific, the total outfall of magnetic spherules to the Earth's surface is 125 tonnes per year (Laevastu and Mellis, 1955). Later, however, it was supposed that their total mass is two to three orders greater (Sobotovich, 1976). One of the first estimates for the total fall of cosmic material to the Earth's surface was 35 to 70 thousand tonnes per year (Buddhue, 1950). Recent estimates are 16 to 18 thousand tonnes per year (Bhandary et al., 1968; Kondratiev et al., 1983) and 10 to 100 million tonnes per year (Sobotovich, 1976; Golenetskt et al., 1981, 1982). Maximum values result from the attempt to consider particles of submicron size of cometary origin which were not considered earlier, though they might form the bulk of the cosmic material on the earth (Fesenkov, 1965; Bronshten, 1975; Nazarova et al., 1975).

The composition of meteorite and cosmic material is analysed in detail in some reviews (Holweger, 1979; Mason, 1979; Meyer, 1979; Palme et al., 1981; Anders and Ebihara, 1982; Cameron, 1982). Some data on manganese content in stony meteorites are given in Table 8. From these data, one can assess the average concentration of manganese in cosmic material of meteoritic origin as 0.19%. It can be slightly lower, about 0.16% for some varieties of carboniferous chondrites (Kallemeyn and Wasson, 1982). On the whole, manganese concentration in cometary material seems to be the same as in meteorites (Golenetsky et al., 1977), in lunar rocks it is 0.17 to 0.25% (Green and Ringwood, 1973), whereas in cosmic spherules it varies within the range from 0.01 to 0.72%, thus being about 0.3% on average (Marvin and Einaudi, 1967; Papanastassiou et al., 1982).

If one assumes a mean absolute mass of cosmic material fall as being 10 million tonnes per year and the mean concentration of manganese in it as being 0.2%, the yearly supply of cosmogenic manganese will be 20 thousand tonnes. This is at least one order lower, as compared to the total supply of manganese by aerosols. If one assumes the maximum estimate of cosmogenic material supply (100 million tonnes per year), the supply of cosmogenic manganese will be close to that of terrigenous aerosols.

1.4. VOLCANOGENIC AND HYDROTHERMAL MANGANESE

Determining the role of endogenous processes in sedimentation and ore formation is traditional in geology. As for marine geology, lately, rich new observational data were obtained on this problem and they were subjected to generalization in various aspects (Böstrom, 1973, 1980; Lisitsin, 1974, 1978, 1983; Strakhov, 1976; Bonatti, 1981; Fife and Losdale, 1981; Honnorez, 1981; Rona et al., 1983; Thompson, 1983; Rona, 1984). The geochemical aspect of volcanic processes concerning marine sedimentation was considered in papers by Markhinin (1967), Zelenov (1972), Trukhin and Shuvalov (1979), Naboko (1980), Kononov (1983).

The main sources of endogenous material supply to the ocean are volcanic pyroclastic material, exhalation products and hydrothermal solutions.

Manganese concentration in solid volcanic material varies from some hundredths

TABLE 8

Manganese in stony meteorites (Schmitt et al., 1972)

Class	Number of samples	Mn concentration (%)			In atoms per 10^6 Si
		from	to	average	
Chondrites					
C1	3	0.174	0.230	0.188	9300
C2	7	0.154	0.174	0.163	6200
C3	9	0.128	0.165	0.149	4900
H	40	0.186	0.280	0.226	6700
L	47	0.221	0.278	0.246	6700
LL	17	0.233	0.283	0.256	7000
E4	3	0.180	0.250	0.220	6800
E5, 6	5	0.148	0.254	0.179	5000
Achondrites depleted of Ca					
Ae	5	0.083	0.255	0.140	2800
Ah	5	0.357	0.439	0.399	8300
Ac	1	–	–	0.41	12000
Au	2	0.286	0.291	0.289	7700
Achondrites enriched with Ca					
Aa	1	–	–	0.07	1700
An	2	0.352	0.386	0.37	8200
Aho	8	0.346	0.40	0.38	8200
Aeu	20	0.244	0.581	0.40	9000

TABLE 9

Manganese concentration in volcanogenic material

Material	Region	Mn (%)	Number of samples	References
Volcanic glass and lava ashes	North America	0.025-0.04	–	Smith and Westgate (1969)
"	Kamchatka and Kuril Is.	0.13	67	Markhinin and Sapozhnikova (1962)
Pumice	South Sandwich Is.	0.065	1	Lisitsin (1978)
"	Indian Ocean	0.09	4	"
"	Antarctic zone	0.14		"

to some tenths of one percent, but, on the whole, the mean composition of volcanic material is close to that of the crustal rocks (Table 9).

The total mass of volcanic products that comes to the surface from the Earth's interior reaches 3 billion tonnes per year (Markhinin, 1966). Only 20 to 75 million tonnes are produced by gas during explosive volcanic activity (Markhinin, 1966; Cadle, 1975). The amount of fine dispersed volcanic products arriving at the upper atmosphere is 150 million tonnes per year (Goldberg, 1971).

The share of labile (soluble) manganese in this material does not exceed several percent (Sung et al., 1982).

The manganese content in a condensate of Tolbachik basalt eruption of 1975/ 1976 is 0.27 mg/l, whereas the manganese content in precipitations from ash cloud is 0.0011 mg/l (Miklishansky et al., 1979; Menyailov et al., 1982).

In exhalative products of Etna Volcano, the manganese content was 1.3 to 13 $\mu g/m^3$, the enrichment coefficient relative to iron being 2-4 (Buat-Menard and Arnold, 1978).

During eruptions of Heimaey Volcano in Iceland, the manganese concentration in aerosol was determined as being 79±12 ng/m³, in fumarole sediments 0.12±0.006% (Mroz and Zoller, 1975).

About 150-200 thousand tonnes of manganese are supplied to the ocean daily with solid volcanic products. However, its geochemical activity is relatively low; so, it is deposited as a component of volcanic material together with materials of terrigenic and biogenic origin and does not provide for any ore concentrations, as has been verified by results of the analysis of sea and oceanic sediment compositions (Strakhov, 1960, 1963, 1976; Lisitsin, 1966, 1974, 1978; Butuzova, 1969).

As has been calculated, only 200 tonnes of manganese are supplied yearly to the ocean through volcanic exhalative products (Lantzy and Mackenzie, 1979).

Sea hydrothermal activity, no doubt, plays a more significant role in the geochemical processes in the ocean. This was noted by Zelenov (1964) who discovered from 0.89 to 2.56% of manganese in a coagulating suspension of Banu Wahu sea volcano (Indonesia). Later, it turned out that the majority of sea hydrothermal springs at the oceanic bottom are confined to spreading zones, not to sea volcanoes.

By contrast to sea water, hydrothermal solutions within continents, islands and at the oceanic bottom are usually considerably enriched in manganese (Table 10). Manganese concentration in hydrothermal springs of the Kuril islands, Kamchatka, Indonesia, and Santorini volcano is 0.001 to 12 mg/l, whereas in hydrothermals of Matupi Harbour (New Britain) and ore brines in the Red Sea (Atlantis II Deep), it reaches 100 mg/l. The maximum concentration of manganese, about 1 g/l, was registered in the brine of the Salton Sea in California, in the zone of continental extension of the East Pacific Rift system covered by thick sediments (Helgeson, 1968). A high concentration of metals was also observed in hydrothermal brines of Cheleken (near the eastern Caspian coast) draining sedimentary cover (Lebedev, 1975, 1977).

Hydrothermal solutions discharging to the open sea bottom are also rich in manganese; however, since they are mixed with sea water, the concentration of metals in them is usually less than 1 mg/l, and only in some cases higher concentrations (to 33-41 mg/l) were recorded (East Pacific Rise, 21°N); such concentration was predicted earlier by extrapolation of the initial data on the Galapagos Rift (Edmond et al., 1979). The "torches" of manganese rich suspension detected in the region of the Galapagos spreading centre (Figure 1), were the first evidence for a considerably active influence of hydrothermal processes on the manganese geochemistry in the ocean. Hydrothermal solutions in oceanic spreading zones are believed to originate as the result of sea water penetration to the interior of hot fissured basaltic rocks down to a depth of several hundred meters. The water, thus, is being transformed to an ore fluid rich in metals. It is then pushed under high pressure to discharge zones (Corliss, 1971). As the mid-oceanic ridges extend for about 60000 km (in total), the oceanic crust is 6.5 km thick on average, and the mean spreading rate is 5 cm/yr, the total mass of hot basaltoids contacting sea water seems to reach 50 billion tonnes per year (Seyfried and Mottle, 1982), which reveals the global scale of the process. Most of the specialists believe that basic rocks (basaltoids) are the source of a metal supply in hydrothermal solutions of rift zones in the open ocean; however, following from theoretical considerations, one cannot ignore the possibility of juvenile component input in their formation (Bostrom, 1973).

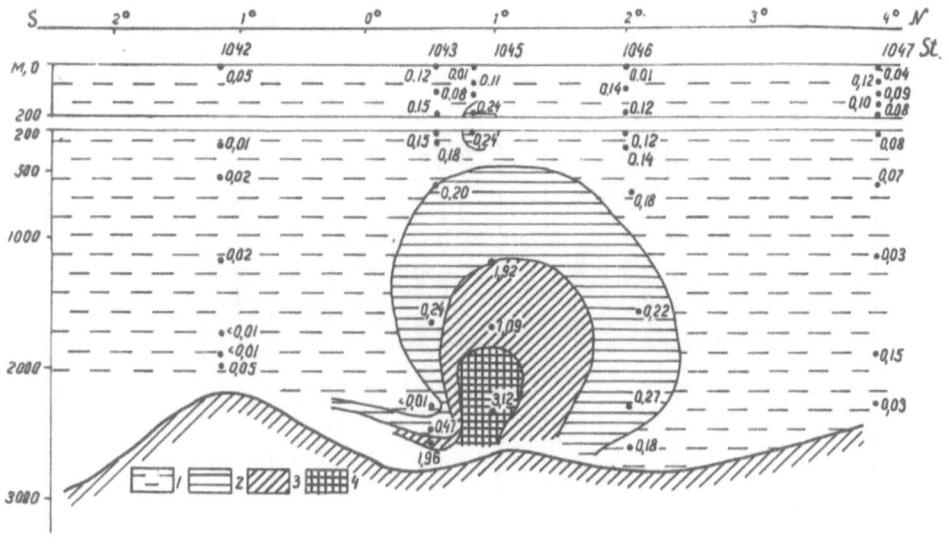

Fig. 1. Distribution of suspended manganese along the latitudinal profile (86°W) in
the Galapagos zone (Gordeev et al., 1979).
Content of manganese, in µg/l: (1) less 0.2; (2) 0.2 to 0.5; (3) 0.5 to 2.0;
(4) over 2.0.

A great number of experiments consider the interaction of basic rocks with water at
various pressures, temperatures, durations and solid/liquid phase ratios to reveal
the actual role of the oceanic crust as a primary source of metals of oceanic hydrothermal solution (Ellis, 1968; Bischoff and Dickson, 1975; Hajash, 1975; Bischoff and
Seyfried, 1977, 1978; Seyfried and Bischoff, 1977, 1979, 1981; Dickson, 1978; Seyfried and Mottle, 1978, 1982; Thompson and Hemphris, 1978; Hajash and Archer,
1980; Kholodkevich and Geptner, 1982).
    The range of manganese concentration in a solution detected in the course of
these experiments is from 0.26 (Hajash, 1975) to 290 mg/l (Mottle et al., 1979).
Table 11 shows the results of one of these experiments. At temperature 200°C,
pressure 500 bar and water-to-rock ratio 10:1, the concentration of manganese and
iron in a solution first increases gradually to 32-34 mg/l and then decreases gradually to 5 mg/l. The relative amount of manganese that remains in solution is several
percent (Kholodkevich and Geptner, 1982).
    When, in the process of the experiment, the hydrothermal fluid obtained was
mixed with cold sea water, manganese remained in solution, which demonstrates
its potential ability to be transferred in a sea medium (Seyfried and Bischoff, 1977).
Thus, experimental results agree with natural observations which detected peculiar
"torches" of water rich in manganese and other metals extending for tens and even
hundreds of kilometers off the source (Galapagos Rift) (Bolger et al., 1978;
Gordeev and Demina, 1979).
    Numerous investigations were also made to analyse the composition of oceanic
basic rocks contacting sea water in natural conditions (Baragar et al., 1977;
Thompson and Humphris, 1977; Humphris and Thompson, 1978; Humphris, 1979;
Pritchard, 1979; Frolova et al., 1979; Staudigel and Hart, 1983; Kurnosov, 1984).
    The results of these investigations are not uniform: in most cases, when under-

TABLE 10

Manganese concentration in hydrothermal solutions

Region	Mn (mg/l)	Concentration of salts (g/l)	pH	References
Biryuzovoe Lake, Shimushiru Is., Kuril Is.	0.05	3.77	6.8	Zelenov (1972)
Uzon Caldera, Kamchatka	to 2.0	0.67-3.68	2.12-6.35	Lebedev (1975)
Chiater Spring, Central Java	1.4-2.0	0.85-1.02	2.3-2.8	Bemmelen (1949)
Santorini Caldera, Aegean Sea	0.05-2.05	37	6-8	Pushkina (1967)
Matupi Harbour, New Britain	2.7-111	11.8-46.0	3.4-6.1	Ferguson and Lambert (1972)
Atlantis II Deep, Red Sea	70-98	256	5.6	Brewer et al. (1969, 1971); Brewer and Spencer (1969); Hartmann (1969)
East Pacific Rise, 21°N	0.5-0.9	35	8	Klinkhamer (1980)
Mixture with sea water				
East Pacific Rise, 21°N	0.005-0.31			Lupton et al. (1980)
"	33.5			Edmond et al. (1982)
", 13°N	41			Michard et al. (1984)
Galapagos Rift	to 0.024			Klinkhamer et al. (1977)
"	to 0.3	(suspended Mn)		Bolger et al. (1978), Gordeev and Demina (1979)
Deception Is., Antarctica	to 2.4			Elderfield et al. (1977)
Reykjanes, Iceland	4-5			"
Salton Sea	950-1370	219-259	-	Helgeson (1968)

TABLE 11

Content of Mn and other elements in sea water (300 g) mixture with basalt powder (30 g) heated up to 200°C at pressure of 500 bar (Bischoff and Dickson, 1975)

Time (hrs)	(g/kg)					(mg/kg)		
	Mg	Ca	K	SO_4	SiO_2	Fe	Mn	pH
0	1.18	0.41	0.40	2.61	0.2	0.05	0.05	5.2
24	0.96	0.10	0.44	1.26	0.3	15	10	3.2
71	0.83	0.15	0.49	0.92	0.5	31	10	3.9
90	0.76	0.15	0.47	0.78	–	31	13	4.0
142	0.71	0.21	0.52	0.58	–	37	20	4.2
212	0.64	0.28	0.53	0.39	–	35	25	4.2
262	0.60	0.33	0.53	0.33	–	29	25	4.2
335	0.50	0.36	0.153	0.29	0.5	23	28	4.2
426	0.48	0.41	0.55	0.25	–	32	33	4.5
552	0.43	0.47	0.55	0.22	–	29	35	4.4
738	0.36	0.52	0.55	0.20	0.78	34	32	4.4
785	0.35	0.55	0.55	0.17	0.79	33	32	4.7
1576	0.19	0.81	0.55	0.12	0.67	24	23	4.7
2328	0.12	0.95	0.58	0.10	0.88	17	23	4.7
2928	0.08	0.95	0.58	0.10	0.87	16	17	4.9
3576	0.06	0.95	0.55	0.06	0.80	12	11	4.9
4176	0.05	0.90	0.54	0.06	–	8	8	–
4752	0.03	0.90	0.53	0.06	0.80	5	5	4.9

Reference: sea water at

25°C	1.29	0.41	0.40	2.70	0.01	0.05	0.05	7.9

water weathering and hydrothermal changes in basalts occur, manganese turned out to be labile, though it remains in altered rocks in the composition of secondary minerals which are apt to extract additional manganese from sea water. These observations are verified by the data of spectral and microprobe analyses (Table 12, Figure 2).

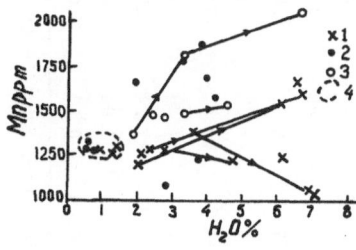

Fig. 2. Mn to H_2O relation in fresh and hydrothermally altered basalts (Hamphris and Thompson, 1978). (1) sample from 4°S; (2) sample from 22°N; (3) sample from 22°S; (4) fresh basalts.

TABLE 12

Content of Mn and other elements in oceanic basalt glass*
and its alteration products by microprobe data (Staudigel and Hart, 1983)

Oxide	Fresh glass	Varioles	Yellow palagonite	K-rich zone along the fault	Dark yellow palagonite	Dark yellow palagonite near variolated zone	Smectite, 18 samples (Pritchard, 1979)	Carbonate, 3 samples (Hamphris et al., 1979)
MnO	0.16	0.18	0.18	0.07	0.05	0.01	0.00	0.44
FeO	10.98	11.40	11.04	7.03	10.10	17.38	17.04	1.32
SiO_2	51.02	50.32'	50.56	58.19	48.89	50.27	48.32	0.00
TiO_2	1.29	1.32	1.28	1.36	3.73	4.15	0.02	0.00
Al_2O_3	14.52	14.47	14.55	17.42	15.69	15.43	6.55	0.00
MgO	7.68	7.90	7.75	4.44	5.44	5.93	9.69	2.50
CaO	12.09	12.24	12.36	6.30	5.79	5.46	1.04	54.42
Na_2O	2.17	1.04	2.19	1.58	0.40	0.44	0.24	0.00
K_2O	0.09	0.13	0.08	3.61	0.92	0.92	4.87	0.00

* Basalt material from Site 418 A, western Atlantic.

TABLE 13

Estimates of absolute hydrothermal manganese supply to the ocean

Hydrothermal Mn (m t/yr)	Estimation technique	References
0.7	From global balance of sediment accumulation	Horn and Adams (1966)
3.5	From Mn supply as a function of spreading rate (Fe+Mn/Al=0.230+SRx0.233)	Bostrom (1973)
0.7-7.2 av. 1.4	From geochemical balances (several modifications)	Elderfield (1976)
0.7	From hydrothermal fluxes	Lyle (1976), Bender et al. (1977) Weiss (1977)
10	From ratio $Mn/^3He=4 \times 10^5$ g/cm^3	Corliss et al. (1979)
4-10	From ratio Mn/heat flow = $(1.5-3.9) \times 10^{-9}$ mol/cal	Staudigel and Hart (1983)
0.51	From submarine weathering of basalts	Lisitsin (1983)
1.6-1.9	From composition and absolute mass of metalliferous sediments	

* SR = spreading rate.

Various approaches were attempted to estimate numerically the absolute mass of hydrothermal manganese being supplied to the ocean: they used several varieties of sedimentary material balance; dependence of metal supply upon spreading rate; manganese relations to heat flow and endogenic helium; composition and supposed discharge of hydrothermal solutions; amount of basalt material reacted with water; composition and absolute mass of metalliferous sediments. The results of these estimates, summarized in Table 13, are within the range of 0.5 to 10 million tonnes per year.

A part of hydrothermal manganese most obviously participates in the oceanic sedimentations and ore formation, whereas the rest of it is involved in a closed cycle as a part of the products of basalt transformation. However it should be noted that the maxima of the presented estimates are not consistent with the data on manganese accumulation in sediments. We shall consider this problem later, in the chapter concerning manganese balance in the ocean.

CHAPTER II

MANGANESE IN SEA WATER AND SUSPENSION

The manganese content, distribution and forms of occurrence in sea and oceanic water depend not only on the properties of the former, but also on sea water parameters and on the impact of mineral and organic suspension, living matter, gas regime etc. Therefore, to provide a distinct description, we shall speak in turn about manganese in solution, suspension and in living matter.

Attempts to detect dissolved manganese in sea water have been made since the 1890's. A review of the first results in this field are given in the book by Vinogradov (1967) who arrived at the conclusion that average concentration of manganese in sea water is 2 µg/l. Similar estimates were given in other reviews of that period (Goldberg, 1965; Bowen, 1966).

Later, as the techniques improved, new lower values, about some tenths of a microgram per litre, were derived (Slowey and Hood, 1966, 1971; Turekian, 1969; Knauer and Martin, 1973; Chester and Stoner, 1974). Mean manganese concentration in sea water was assumed to be 0.4 µg/l (Turekian, 1969; Gordeev and Lisitsin, 1979).

Since the development of ultraclean sampling techniques and most sensitive techniques for analysis, the manganese concentration in sea water has been going down (Table 14). From recent data, dissolved manganese concentration in the open ocean is only some hundredths of a microgram per litre. In surface and shallow waters it might be considerably higher due to discharge of rivers and atmospheric dust and due to diffusion out of shelf sediments (Kremling, 1983a). On the whole, fluctuations of manganese concentration in water of pelagic oceanic zone are within 0.01 to 0.16 µg/l, being 0.027 µg/l on average (Bruland, 1983; Quinby-Hunt and Turekian, 1983).

The surface film of sea water (100 to 300 microns) is often rich in manganese as compared to the near-surface layer (Piotrowicz et al., 1972; Hoffman et al., 1974). In a vertical water column, manganese reaches its maximum concentration in the intermediate layer of oxygen minimum (Figure 3). As a result of reduction and

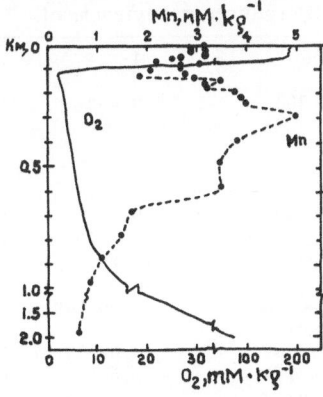

Fig. 3. Distribution of dissolved manganese and oxygen in a vertical cross-section in the Pacific, 18°N, 108°W (Martin and Knauer, 1984).

TABLE 14

Concentration of Mn dissolved in oceanic and sea waters

Region	Layer (m)	Mn (µg/l)	References
Atlantic Ocean			
Sargassa Sea	Surface	0.10-0.12	Bender et al. (1977)
	1800-4000	0.03	
	Bottom water (5400)	0.15	
North-western region	600-3100	0.1	
", shelf	Surface	1.15	Bruland and Franks (1983)
", open sea	"	0.13	
North Sea, shelf	"	1	Kremling (1983a)
", open sea	"	0.04	
Bermuda Basin	1750	0.012-0.064	Bewers and Windom (1983)
Cape Basin	0-50	0.16	Burton et al. (1983)
	120	0.09	
	263-4960	0.014-0.036	
Pacific Ocean			
North-eastern region	Surface	0.08	Bender et al. (1977)
	Deep water	0.03	
Entire deep sea region	Surface	0.05-0.16	Klinkhamer and Bender
(26 sites)	40-100	0.02-0.12	(1980)
	Layer of O_2 min	0.05-02	
	Deep water	0.03	
North-eastern region	Surface	0.21	Martin and Knauer (1980,
	150-300	0.05	1982)
	600	0.08	
	2500	0.06	
	Deep water	0.05-0.08	
California Current region,			
", shelf	Surface	0.1-0.3	Landing and Bruland (1980)
", open sea	Same	0.03-0.05	
	Bottom water	0.1	
Guatemala Basin	Surface	0.49	Murray et al. (1983)
	Layer of O_2 min	0.30	
	Bottom water	0.49	
Shelf of Mexico	0-590	0.17	Martin and Knauer (1984,
	690-1955	0.06	1985a)
28°N, 155°W	Surface	0.044	Martin and Knauer (1985b)
	200-450	0.011	
	450-700	0.029	
	1000	0.019	
	1000-2000	0.015	
Basins with reducing environment			
Black Sea, oxic zone	Surface	1- 4	Ovsyanyi and Eremeeva (1980)
", H_2S zone	Below 200	300-400	Mokievskaya (1961), Skopintsev and Popova (1963), Spencer and Brewer (1971), Brewer and Spencer (1974)
Baltic Sea, Bornholm Deep	Bottom water to	50	Kremling and Peterson (1978)
", Gotland Depression	", (152-230m)	643-1150	Kremling (1983b)
Atlantic, Kariako Depression	485-1396	17-25	
Canada, Jervis Inlet	Bottom water	0.94-12.3	Grill (1978)
", Saanich Inlet	From surface to bottom	0.09-748	Emerson et al. (1979), Grill (1982)
", St. Lawrence estuary	1-362	0.07-20,27	Yeats et al. (1979)
Harima Sea, Japan	From surface to bottom	1.5-12.2	Shiozawa et al. (1982)
Narragansett Bay, Rhode Is., eastern coast of U.S.A.	"	0.5-60	Graham et al. (1976)

desorption from suspended particles (Landing and Bruland, 1980, 1982; Bruland, 1983; Martin and Knauer, 1984). The same process is responsible for a sharp increase of manganese content in sea water poor in oxygen and, in particular, rich in hydrogen sulphide. In hydrogen sulphide waters of the Black Sea, manganese concentration reaches 300-440 $\mu g/l$, in the bottom waters of the Baltic basin it is 50 to 1150, in Kariako trench 17 to 25, whereas in fjords, gulfs and estuaries it is 1 to 748 $\mu g/l$.

Some attempts have been made to find the forms of manganese occurrence in oceanic water by thermodynamic calculations (Krylov et al., 1979; Gramm-Osipov and Shulga, 1980; Savenko, 1981; Savenko and Baturin, 1981; Tikhomirov and Gromov, 1983; Tischenko and Gramm-Osipov, 1985) and also by experiments; in particular, using ^{54}Mn and ^{57}Mn isotopes, by sorption and desorption, ultrafiltering, ion exchange, and other techniques (Khitrov and Kholina, 1972; Gromov, 1974; Legin et al., 1974; Gromov and Spitsin, 1975; Gromov and Surikov, 1976; Tikhomirov et al., 1978, 1982; Tikhomirov and Gromov, 1971, 1983; Shumilin and Tikhomirov, 1982; Tikhomirov, 1983-1985; Tikhomirov and Lukashin, 1983; Amdurer, 1983; Carpenter, 1983).

The majority of researchers believe that the predominant form of dissolved manganese in sea water is the ion Mn^{2+} (58-77%), then follows the $MnCl^+$ ion (14-30%). Such forms as $MnSO^\circ_4$ (to 7-12%), $MnCl^\circ_2$, $MnCO^\circ_3$, $MnHCO^+_3$, $MnOH^+$, $Mn(OH)^\circ_2$ can be found in a subordinate amount (Table 15).

In the bottom waters of the Pacific radiolarian belt a colloidal form of manganese was detected. It comprises to 23-37% of the total Mn content in water (Tikhomirov, 1985).

Both theoretical and experimental data obtained suppose the occurrence of dissolved tetravalent manganese such as MnO_2 x H_2O in sea water (Raevsky et al., 1981; Savenko, 1984).

The occurrence of manganese cation assemblages with organic ligands (amino-acid, protein and humic) dissolved in sea water was also posited (Desai and Ganguli, 1980). This supposition was supported by test results: up to 36% of dissolved manganese can be transformed to the organic form as sea water passes through the biogenic cycle (Shumilin and Tikhomirov, 1982).

Forms of dissolved manganese were also determined in the samples taken from a vertical water column in the Sea of Okhotsk (Demina, 1982). There, in contrast to less-productive oceanic areas, the process of metal-organic complexation is believed to be more probable due to high biological productivity. In surface water, about 50% of dissolved manganese turns out to be in the composition of organic lipid hydrophobic compounds, about 2% in the composition of organic polar hydrophilic high-molecular compounds and less than 50% in a non-organic form. As the depth increases, concentration of the former decreases gradually to 36%, on average; occurrence of the second form increases to 24%, on average, whereas non-organic dissolved manganese remains stable, i.e. 40-50% of its total content (Table 16).

Suspended manganese in the sea and oceanic water has been the focus of analysis since the 1960's as the interest in suspension composition grew; the latter was considered to be the initial phase of bottom deposits (Lisitsin, 1961, 1964). At that time, total suspended manganese content in oceanic water - from less 0.01 to over 0.2 $\mu g/l$ - was first estimated from the analysis of samples obtained by separation (Lisitsin, 1964). Later, these estimates were verified.

From recent data, suspended manganese content in open-sea and ocean water may vary within six orders, i.e., from 0.0002 to 20 $\mu g/l$ (Table 17). In turbid continental shelf surface water, the suspended manganese content reaches its maximum and decreases sharply as the total depth increases off the coastal zone. So, at the latitudinal cross-section from New England (U.S.A.) coast to the Bermuda Islands, the suspended manganese content decreases from 0.330 to 0.001-0.006 $\mu g/l$ (Wallace et al., 1977). The same seems to be typical of the Pacific, namely, the suspension manganese content is 0.02 to 0.13 $\mu g/l$ n its northern part and in the Sea of Okhotsk, near the continents (Demina, 1982; Tsunogai et al., 1982), whereas in the pelagic zone it decreases to 0.003-0.008 $\mu g/l$ (Baker et al., 1979). It should be noted, however, that higher concentration of suspended manganese was also registered in the central parts of the ocean (Gordeev and Khandros, 1976).

In the bottom oceanic water, suspended manganese content tends to increase as

TABLE 15

Forms of manganese dissolved in sea water (%)

Mn^{2+}	MCl^+	$MnClO_2$	$MnClO_3$	$MnHCO^o_3$	$MnSO^o_4$	$MnOH^+$	$Mn(OH)^o_2$	Techniques	References
62	16	4	8	3	7	0.1	3×10^{-9}	Thermodynamic estimates	Tikhomirov and Gromov (1983)
58	30	2	–	2	7	–	–	Experiment with ^{57}Mn	Amdurer et al. (1984)
72–77	14–16	–	–	–	10–12	–	–	Determination of electron spin resonance	Carpenter (1983)

TABLE 16

Forms of manganese dissolved in the Sea of Okhotsk water (µg/l and %)
(Demina, 1980)

Layer	Number of samples	Organic lipid hydrophobic compounds		Organic polar hydrophylic high-molecular compounds		Non-organic forms		Total content	
			average		average		average		average
0–100	26	0.10–2.1	0.62 (53.9%)	0.00–0.08	0.02 (1.8%)	0.10–1.0	0.51 (44.3%)	0.3–2.8	1.15
100–1500	14	0.08–1.1	0.33 (41.2%)	0.05–0.1	0.06 (7.6%)	0.18–1.1	0.41 (51.2%)	0.3–1.6	0.80
1500	14	0.14–0.90	0.29 (36.2%)	0.08–0.3	0.19 (23.8%)	0.15–0.8	0.32 (40%)	0.4–1.4	0.80
Average	54	0.08–2.1	0.47 (42.7%)	0.00–0.3	0.15 (14.2%)	0.10–1.1	0.43 (39.1%)	0.3–2.8	1.05

the nepheloid layer of high turbidity occurs there. The suspended manganese
minimum in surface, intermediate and deep water of the pelagic part in the Atlantic
is 0.0033-0.0045 μg/l, whereas in the bottom water, it is 0.007 μg/l; its maxima
are 0.0155-0.0261 and 0.234 μg/l, respectively (Lambert et al., 1984). In the
Pacific, suspended manganese content in the bottom water is two to three times
higher by contrast to the rest of the water column (Baker et al., 1979; Tsunogai
et al., 1982).

In anaerobic basins, the suspended manganese content varies over a wide
range, from 0.02 to 687 μg/l (Table 17). Such a high content seems to be
caused by Mn^{2+} ion oxidation and is accompanied by sorption of the hydrooxide
formed on suspension particles or by the appearance of colloidal particles of manganese hydrooxide in the result of the interaction between hydrogen sulphide and
oxygenated waters, which happens at the contact between water masses or is caused by
fluctuations in the water circulation regime.

In the lower water layers of the St. Lawrence estuary, a high content of suspended manganese, up to 7.3 μg/l, was registered; in the Jervis Inlet it reaches
39, in the Black Sea (in the interface layer at depth of 150-200 m) it is 18-58, in
Saanich Inlet it is up to 687 μg/l (Table 17).

Our knowledge of the suspended-to-dissolved manganese ratio in oceanic water
has been advanced as new, more reliable, analytical data were obtained, mainly
concerning dissolved manganese. When the content of dissolved manganese in sea
water was assumed as being 2 μg/l on average, the relative suspended manganese
concentration did not exceed 1 to 3%. From recent data (Tables 16, 17),
dissolved manganese, in general, still predominates in sea and oceanic waters, though
its relative concentration is considerably lower. In the pelagic oceanic zone it is
about 90% for the entire water mass and about 80% for the nepheloid layer (Bender
et al., 1977). In some cases, however, in shelf parts in particular, the dissolved-
suspended manganese ratio can be inverse or, at least, equal (Table 17).

Manganese concentration in oceanic suspension varies in the range from 0.0005
to 0.18%, whereas in sea suspension it varies from 0.004 to 1.8% (Table 18).
In the surface suspension of the open ocean, manganese concentration is at a minimum because they contain predominant amounts of siliceous, calcareous and organic
biogenous components. As suspension sinks to the bottom and biogenous components
dissolve, the manganese concentration usually increases and reaches its maximum
(0.07-0.18%) in the nepheloidal layer.

The highest concentration of manganese in suspension is observed in anaerobic
basins at the oxygen-hydrogen sulphide-water contact where the suspension becomes
rich in manganese up to tenths of one percent and, sometimes, to 1% and over, as
in the St. Lawrence Bay (Yeats et al., 1979).

The forms of manganese occurrence in oceanic suspension are defined using
various solvents, which permits one to distinguish between several fractions. When
analysing the material collected by sediment traps, three fractions are clearly recognized, namely: a) exchangeable manganese in the trap salt solution (which is added
to the collection vessels to suppress bacterial activity of the sampled material);
b) manganese removed by weak leach, and c) refractory manganese. The first
fraction is associated with the most labile organic and calcareous components, the
second with mainly sorbed and hydrooxide forms, and the third with alumosilicate
terrigenous material. The water column shows that relative concentration of the first
fraction decreases downwards from 30-70% to 2-15%, the second increases from 18-23%
to 53-70% and the third also increases from 4-14% to 23-36% (Martin and Knauer,
1983).

Other modifications of the method of successive solution applied to study the
surface separational suspension (suspension obtained by separation method) allow
one to determine that organic suspended manganese takes up about 50% of the total
in the Pacific. It comprises organic complexes extracted by pyrophosphate (2.4-
32.4%), and albumen and carbohydrate extracted by chlorinated spirit (14.8-54.9%),
these being in approximately inverse relations to each other (Table 19). From
15 to 42% of suspended manganese comprises adsorbed forms and amorphous hydrooxides, and from 6.1 to 27.7% is a residual alumosilicate fraction resistant to weak
solvents (Demina, 1982).

In the Indian ocean, samples of surface water separated suspension have different compositions. There, the prevailing form is the sorbed one (46.2-89.7%), then

TABLE 17

Concentration of suspended manganese in oceanic and sea waters

Region	Layer	(µg/l)	References
Atlantic Ocean			
North Sea, shelf	Surface	2.4-20.3	Hunter (1980)
North-west, New Scotland shelf	Surface	0.14	Bewers et al. (1976)
	Bottom	0.62	
North-west, New England shelf	Surface	0.330±0.030	Wallace et al. (1977)
North-west, ~39°-42°N	"	0.019-0.029	
North of Bermuda Is.	"	0.001-0.006	
Northern tropical zone	Surface	0.0035	Buat-Menard and Chesselet (1979)
	Intermediate and deep water	0.0048±0.0021	
", Barbados Is. region	400	0.0002	Brewer et al. (1980)
(material of sedimentary	1000	0.0002	
traps)	3800	0.0017	
Ocean as a whole	Surface	0.0033-0.0155	Lambert et al. (1984)
	Intermediate	0.0045-0.0188	
	Deep	0.0033-0.0261	
	Bottom	0.007-0.234	
Pacific Ocean			
Sea of Okhotsk	0-100	0.13	Demina (1982)
	100-1500	0.05	
	below 1500	0.03	
Eastern part	Surface	0.09	Gordeev and Khandros (1971)
Central part	"	0.07	
Eastern tropical zone	0-300	0.008±0.004	Baker et al. (1979)
	450-1000	0.004±0.001	
	below 2000	0.003-0.006	
	nepheloid layer	0.010-0.016	
Northern part	0-40	0.019	Tsunogai et al. (1982)
	40-200	0.035	
	200-1000	0.035	
	1000-3000	0.035	
	3000-4600	0.037	
	5000	0.074	
Mexico shelf, 18°N, 108°W	0-1955	0.001-0.03	Martin and Knauer (1984)
Indian Ocean			
Tropical zone	Surface	0.1-0.3	Lisitsin (1961)
Western part	"	0.04	Morozov et al. (1979)
Central part	"	~0.041-0.100	Vizhensky and Shnykin (1984)
Basins with reducing environment			
Black Sea	Surface	0.2-0.5	Emelyanov et al. (1976)
	Surface	0.04-0.40	Morozov et al. (1976)
	Surface	0.03-0.57	Spencer et al. (1972)
	10-150	0.01-18.75	"
	175-2100	0.02-57.9	
Baltic Sea	0-40	0.2-1.0	Emelyanov and Struyk (1981)
	80-185	1.3-25.6	
Canada, Jervis Inlet	Bottom	3.6-39.4	Grill (1978)
", Saanich Inlet	Surface/bottom	0.27-687	Grill (1982)
", St.Lawrence estuary	1-32	0.33-7.3	Yeats et al. (1979)

TABLE 18

Manganese concentration in oceanic and marine suspension

Region	Depth (m)	Mn (%)	References
		Atlantic Ocean	
Entire ocean	Surface	ca 0.01	Emelyanov (1976)
North-western part, shelf	"	0.062	
Open ocean, profile from Narragansett to Bermuda Is.	"	0.0016-0.017	
Sargasso Sea	Surface	0.016	Honjo (1978)
	2786	0.044	
	Nepheloid layer	0.059	
"	Bottom layer	0.0855-0.108	Spencer et al. (1978)
"	976	0.0053	Brower et al. (1980)
	3694	0.0768	
Northern part, near Barbados Is.	389	0.0045	
	988	0.0071	
	5086	0.0462	
Northern tropical part	Surface	0.014	Buat-Menard and Chesselet (1979)
	Deep water	0.032	
		Pacific Ocean	
Eastern part (material of sedimentary traps)	Surface	0.012	Demina (1982)
	110	0.0033	Tsunogai et al. (1982)
	1040	0.0067	
	2160	0.0158	
	4380	0.0397	
	5250	0.1850	
Same (filtration material)	0-40	0.0055	
	40-200	0.0420	
	200-1000	0.062	
	1000-3000	0.073	
	3000-4600	0.079	
	5000	0.180	
California shelf (35°40'N, 123°50'W)	50	0.0023	Martin and Knauer (1983)
	100	0.0160	
	200-2000	0.0366-0.0500	
Mexico shelf (18°N, 108°W)	50-1955	0.002-0.012	
		Indian Ocean	
Northern part (filtration suspension)	Surface	0.003	Morozov et al. (1979)
Tropical zone	Surface	0.0085-0.1828	Demina et al. (1984)
		Seas and Bays	
Baltic Sea, Bornholm Deep	1-30	0.0229-0.4609	Kremling and Peterson (1978)
Baltic Sea	Surface	0.004-0.08	Emelyanov and Struyk (1981)
Riga Bay	"	0.29	
St. Lawrence estuary	1-323	0.075-1.832	Yeats et al. (1979)

TABLE 19

Forms of Mn in oceanic separational suspensions

NN St.	Region (coordinates)	Total content of Mn in suspension ($\mu g/l$)	Organic complexes extracted by pyrophosphate	Albumen and carbohydrate matter extracted by chlorinated spirit	Adsorbed and carbonate forms	Amorphous hydrooxides	Residue mineral form
			(% with respect to total content)				
			Pacific Ocean (Demina, 1982)				
5	33°36'- 33°49'N 140°59'-128°48'W	88	30.8	14.8		42	12.4
30	04°30'- 01°39'N 105°14'-119°04'W	49	32.4	16.4		40	11.2
24	07°59'- 13°00'S 110°06'-102°42'W	52	31.4	20.8		41.7	6.1
20	12°02'- 10°33'S 77°14'- 79°58'W	242.1	2.4	54.9		15	27.7
Average		108	24.2	26.7		34.7	14.4
			Indian Ocean (Demina et al., 1984)				
5	07°57'- 03°03'N 117°00'-108°21'E	1822	2.5	89.7		3.8	3.9
6	0°49'- 07°16'S 105°03'-104°11'E	470	8.4	60.6		1.3	29.7
11	26°19'- 29°19'S 104°43'-104°40'E	165	17.9	55.1		7.7	19.3
18	05°56'- 02°39'N 79°47'- 80°00'E	85	9.4	58.8		12.9	18.9
27	13°56'- 16°52'S 66°58'- 66°03'E	203	6.4	46.2		21.4	26.0
Average by 20 samples		280	7.7	60.9		13.6	17.8

follows the residual (4-29%) and amorphous hydrooxide ones (1.3-24%), whereas the organic form is a subordinate form (2.5-17.9%) (Demina et al., 1984).

As the suspension sinks down, its components are considerably transformed, which affects its aggregate state, mineral and chemical composition. The ratio of various elements to aluminium serves as a good index of chemical changes in the material, because aluminium is believed to be an inert element, a so-called "benchmark", that reveals terrigenous nature.

An average Mn/Al ratio in the crustal rocks is about 0.01 (Vinogradov, 1962; Taylor, 1964). This ratio is 5.8 in the suspension from the upper waters of the Atlantic, 4.8 in the suspension from the intermediate layers and 3.0 from the abyssal suspensions (Lambert et al., 1984) Thus, manganese tends to accumulate in oceanic suspension (with respect to aluminium) at general decrease in Mn/Al coefficient with depth. Earlier it was shown that some accumulation of manganese occurs in suspension with depth, however, aluminium being accumulated more intensively.

Chemical and biochemical factors that control manganese behaviour in the transformation of oceanic suspension will be considered in the chapter devoted to manganese deposition from sea water.

CHAPTER III

MANGANESE IN SEA ORGANISMS

Manganese content and distribution in marine organisms is of special interest in many aspects, taking into account its biogenic functions, problems of ocean environment protection and mechanisms of manganese extraction and deposition out of sea waters.

It was in the 18th century that the first pieces of information about the occurrence of manganese in living matter were obtained, when Karl Sheele, Swedish chemist, recognized it in plants (Sheele, 1774; quoted in Vernadsky, 1954). Since that time, the range of biogenic objects investigated for manganese content has been widening, embracing marine organisms as well.

The data on this problem collected for a century and a half (to the 1920's) were generalized by Vernadsky in his monograph Geochemistry published in 1924 in Paris and reedited with additions many times (Vernadsky, 1954). Later, the work was developed by Vinogradov (1935, 1937, 1944, 1953) and by many other Soviet scientists and, recently, some generalized papers appeared in the West (Goldberg, 1957; Bowen, 1966, 1979; Baseline, 1972; Bryan, 1976; Fortescue, 1980; Eisler, 1981).

The biogenic functions of manganese are numerous, and it is believed to be one of the major biogenic microelements. It activates oxydation processes and thus is a component of some ferments and vitamins, playing a significant role in the processes of photosynthesis, respiration, mass and energy exchange in general (Voinar, 1953; Saenko, 1968; Bumbu, 1976; Ehrlich, 1976; Nozdrukhina, 1977). Long ago, it was found experimentally that slight additions of manganese, as dissolved salts, stimulate evolution of algae and other aquatic organisms (Guseva, 1937).

When investigating the problems of water as a medium, one takes advantage of the ability of some organisms to selectively accumulate this metal extracting its radioactive isotopes from solutions and suspensions. However, considerable increase in manganese content over the natural one might be toxic to some organisms (Guseva, 1937; Slowey et al., 1965; Tikhomirov et al., 1970; Fowler and Oregioni, 1978; Noro, 1978; Phillips, 1978; Sanders, 1978).

The ability of organisms to bioassimilate manganese accounts for the major role organic material and organic components play in the transfer of manganese from water to the bottom - this problem will be considered in the next section. Here, we shall confine ourselves to the most general data on manganese content and distribution in marine organisms.

First, we shall note that the main difficulty in comparing data by various authors arises not only from the different techniques used by them but also from the various forms of data presentation. The latter can be calculated either by dry weight, or by natural weight, or by ash content, the concentration of the water and ash being ignored as a rule, despite its great scatter even within one genus of organisms (Kizevetter, 1960, 1973; Morozov, 1983). So, from various data sources, ash concentration in macrophytes is from 1 to 15% in natural and from 15 to 52% in dry matter, whereas water content in fish flesh varies from 52 to 86%. Taking this into account, when compiling summary tables for organism composition, we prefer to consider the predominating data type, with the exception of those cases where their total amount was scarce.

Phytoplankton. The most representative data on manganese content in phytoplankton are given by Martin and Knauer (1973) who studied plankton along the profile from Hawaii to the Monterey Bay. They subdivided the entire material into

TABLE 20

Mn content in oceanic and marine plankton, mg/kg of dry weight

mple	Mn content		References
	Extrema	Average	
Phytoplankton			
tal phytoplankton (Hawaii-Monterey Bay profile)			
a) poor in titanium	2.1-13	–	Martin and Knauer, 1973
b) rich in titanium	7.4-32	–	"
c) rich in strontium	3.2-30	–	"
d) siliceous tests of phyto-plankton	2-3.5		
tal phytoplankton	–	10	Bryan, 1976
Texas shelf	4.4-41.5	21.7	Horowitz and Presley, 1977
tal phytoplankton			
a) marine	–	22	Morozov, 1983
b) oceanic	–	33	"
ue-green algae cultivated on a growth medium	132-700	352	Udelnova et al., 1974
ash	860-5530	–	"
atomaceous plankton from the Black Sea (ash)	50-300	–	Vinogradova and Kovalsky, 1962
Zooplankton			
anktonic foraminifera			Emiliani, 1955
	9.3-63.3		Krinsley, 1960
	40-290	130	Belyaeva, 1973
adiolarians	4.6-12.4	6.4	Martin and Knauer, 1973
iphauzide	2.2-4.5	3.6	"
opepodeans	2.8-10.4	4.4	"
tal zooplankton (Hawaii-Monterey Bay profile)	2.9-7.1	4.3	"
icroplankton	3.4-32.7	7.4	"
tal plankton, Pacific Ocean	5.3-11.5	8	Collier and Edmond, 1983
ooplankton of Black Sea (ash)	15-800	–	Vinogradova and Kovalsky, 1962

four groups, namely, specimens with minimum titanium content, with relatively high titanium content, with relatively high concentration of strontium, and with siliceous tests (Table 20). A minimum concentration of manganese is typical of siliceous particles of phytoplankton (about 2-3 mg/kg by dry weight). Higher manganese concentration (about 6 mg/kg, on average) was registered in specimens of the total phytoplankton low in titanium concentration, and its highest one in the specimens rich in titanium (about 20 mg/kg, on average). Some lower content of manganese was recorded in the samples rich in strontium (3.2-30 mg/kg). The manganese content of the total phytoplankton along the Texas shelf was 21.7 mg/kg on average (Horowitz and Presley, 1977), whereas 33 mg/kg was estimated by Morozov (1983), based on the total phytoplankton of the ocean. For the total phytoplankton in the seas within the U.S.S.R. territory, the manganese concentration is 218 mg/kg on average (6.6 mg/kg by natural weight). The manganese concentration in the phytoplankton ash from the Black Sea is 50-300 mg/kg (Vinogradov and Kovalsky, 1962).

In phytoplankton (blue-green algae, Cyanophyta) cultivated on a growth medium rich in manganese, the Mn concentration was 10-15 times higher than that of the phytoplankton from natural sea water. It was up to 700 mg/kg in dry weight and up to 5530 mg/kg in ash (Udelnova et al., 1974).

TABLE 21

Forms of Mn in oceanic plankton (Demina, 1982)

N of sample	Coordinate	Predominant plankton forms	Content, mg/kg of dry weight	Mn forms, mg/kg (%)			
				Metal-organic compounds	Adsorbed	Albumen and carbohydrate matter	Inert clastic form
B-542	10°33'S 79°57'W	Calanus plumchrus, Pyrocystis pseudonoctila, Thalassiosira subtilis, Peunatac sp.	9.1	4.7 (51.6%)	1.2 (13.2%)	2.8 (30.8%)	0.4 (4.4%)
B-547	9°45'S 79°22'W	Calanus cristatus, Pyrocystis pseudonoctila, Ceratum triochoceras	22.2	16.6 (75%)	5.5 (25%)	0 (0%)	0 (0%)
B-558	7°30'S 102°46'W	Pyrocystis sp., Ceratum carrience, Thalassiosira sp., Planctoniella sol.	21	17.2 (82%)	1.3 (6.2%)	2.2 (10.7%)	0.3 (1.1%)
Average			17.4	12.1 (69.5%)	2.6 (14.8%)	2.4 (13.8%)	0.3 (1.9%)

Zooplankton. Manganese concentration was estimated in organisms with calcareous shells (Foraminifera), siliceous skeleton (Radiolarian) and chitinous carapace (crustacea, euphauzides and copepods), and also in the total zooplankton.

The data on manganese concentration in planctonic Forminifera are not uniform. Krinsley (1960) estimated that it varies within 9.3-63.3 mg/kg, according to Belyaeva (1973) it is from 40 to 290 mg/kg. However, special analysis of Foraminifera sampled (using binoculars) from the surface suspension of the eastern tropical zone of the Pacific, revealed extremely low Mn/Ca ratios in them. This ratio is 15×10^{-6} in the total suspension, whereas in Foraminifera it is only 0.4×10^{-6} (Collier and Edmond, 1984). When calculating the Mn concentration using calcium carbonate it will be 0.16 mg/kg. Consequently, accumulation of manganese in planktonic Foraminifera occurs as they die off and sink to the bottom. Boyle (1983) arrived at the same conclusion when analysing the manganese calcareous films coating Foraminifera shells in reduced sediments.

The manganese content in Radiolarians varies from 4.6 to 12.4 mg/kg; in plankton crustacea it is from 2.2 to 10.4 mg/kg (Martin and Knauer, 1973). However, in euphauzide carapaces it increases to 11.7, and in faecal pellets it reaches 243 mg/kg (Fowler, 1977).

In the Pacific, average manganese concentration in the summary zooplankton is 4.3 mg/kg (Martin and Knauer, 1973), in total plankton (both phyto- and zoo-) it is 8 mg/kg, in microplankton (by various estimates) it is from 7.4 to 17.9 mg/kg (Martin and Knauer, 1973; Fowler, 1977).

In the Black Sea, manganese concentration in zooplankton ash varies within the range 15 to 800 mg/kg (Vinogradova and Kovalsky, 1962) at the average water concentration of 89% in natural zooplankton (Morozov, 1983).

Forms of manganese occurrence in the total plankton were determined by Demina (1982) in the south-eastern tropical part of the Pacific. In all samples analysed zooplankton predominated (about 65%). It was represented mainly by copepodeans. In phytoplankton composition it was diatoms and peridinians that predominated.

At a general concentration of manganese from 9 to 22 mg/kg in dry plankton matter, the better part is taken up by metal-organic compounds extracted by pyrophosphate (about 70% of the total manganese, on the average).

Adsorbed forms were determined from the processing of samples by the mixture of acetic acid and (25%-solution) of hydroxylamine hydrochloride according to the Chester-Hughes method (1967). The average relative concentration of manganese in this form was about 15% in the total plankton. The average concentration of carbohydrate and albumen extracted by chlorinated absolute spirit was the same - about 14%. The inert form of manganese remained is probably associated with terrigenous material and reaches only about 2%, on average (Table 21).

Macrophytes. Manganese in oceanic and sea algae, i.e., macrophytes, was analysed in detail, since it is of value for the food industry (Kizevetter, 1960). Manganese content in a dry algal material varies over a wide range, from 3 to 3400 mg/kg. The average manganese concentration varies from one alga species to another: from 4 to 700 in green algae (Chlorophyta), from 20 to 288 in red algae (Rhodophyta) and from 10 to 455 mg/kg in brown algae (Phaeophyta) (Table 22). The manganese concentration in algal ashes also varies within a wide range - from 20 to 1400 mg/kg in laminaria (Laminaria) and from 200 to 1300 in red algae (Saenko et al., 1977).

The content of manganese and also of other metals in macrophytes is a function of a number of factors, namely, age of the plant, season and region of sampling, intensity of tidal motions, and occurrence of lithogenic material in a sample (Bryan and Hummerstone, 1973; Khristoforova et al., 1976; Saenko et al., 1976; Sanders, 1978; Khristoforova and Maslova, 1983). Coefficients of manganese accumulation in algae relative to sea water were defined by many scientists. Its minimum was detected in Fucus species (about 4000), its maximum in laminaria (200000) (Van As et al., 1973; Foster, 1976; Yoshimura et al., 1976). Experimental research showed that, on the whole, manganese is slow to accumulate in algae, in contrast to uranium and base metals (Disnar, 1981).

Coelenterata. The manganese content in organisms of the coelenterata type has not been studied by many researchers.

In two representatives of Anthozoa - Alcyonium digitatum and Tealia felina - manganese concentration is 3.7 and 9.3 mg/kg by dry weight, respectively (Riley and Segal, 1970).

Table 22

Mn content in oceanic and marine algae-macrophytes,
mg/kg of dry weight

Algae	Mn content		References
	Extrema	Average	
Green			
Ulva	23.7-316	–	Sivalingan, 1978
"	11-50	30	Gryzhankova et al., 1973a, b; 1975
Codium	300-316	310	"
Enteromorpha	600-800	700	"
"	–	4	Wong et al., 1979
Chaetomorpha	–	57	"
Caulerpa	4.6-9.5	7	Khristoforova and Bogdanova, 1980
Halimeda	6.4-7.4	7	"
Red			
Anfeltia	–	288	Gryzhankova et al., 1973a
Tichocarpus	–	51	", 1973b
Porphyra	–	29	Preston et al., 1972
Brown			
Agarum	331-386	350	Gryzhankova et al., 1973b
Ascophyllum	–	25	Young and Longille, 1958
"	16-21	18	Foster, 1976
Chorda	16-19	18	Gryzhankova et al., 1973b
Chordaria	–	455	"
Coccophora	–	20	"
Dyctiola	–	26	"
Fucus	94-692	363	Bryan and Hummerstone, 1973; 1977
"	71-103	90	Preston et al., 1972; Foster, 1976
"	–	10.4	Gryzhankova et al., 1973a
"	25-50	–	Young and Langille, 1958
Macrocystis	30-120	61	Black and Mitchell, 1952
"	–	9.5	Boothe and Knauer, 1972
Pelvetia	–	30	Gryzhankova et al., 1973b
Podina	–	99	Stevenson and Ufret, 1966
Sargassum	23-42	–	Gryzhankova et al., 1973b
"	4.5-181	–	Trefrey and Presley, 1976
"	12-89.5	31.4	Horowitz and Presley, 1977
"	3.4-12.7	–	Ishii et al., 1978
Flower			
Spartina	30-330	–	Williams and Murdock, 1969
Thalassia	–	49	Stevenson and Ufret, 1966
Zostera	–	47.9	Gryzhankova et al., 1973a
"	122-366	–	Harris et al., 1979
Macrophytes in general			
near England coast	8.9-164	–	Black and Mitchell, 1952
"	3.8-73	–	Riley and Roth, 1971
near Norway	4-164		Lunde, 1970
near Iceland	13-680		Munda, 1978
near India	25-3421		Agadi et al., 1978
"	37-586		Zingde et al., 1976
near Korea	15-191		Pak et al., 1977
near Japan	4.4-41		Ishii et al., 1978

Table 23

Mn content in corals, mg/kg of dry weight

Species	Forms	Region and depth	Mn content
1	2	3	4
Open ocean, deep-water zone			
Solenosmilia variabilis	Black pigmentation	Mid-Atlantic Ridge, 45°N, „ 1000-1500 m	1100
"	"	"	2100
"	Brown pigmentation	"	6
"	Black pigmentation	Reykjanes Ridge, 60°N, 1100-1300 m	80
"	Brown pigmentation	"	34
"	White pigmentation	"	2
Desmophyllum cristogalii	Black pigmentation	Mid-Atlantic Ridge	380
"	Brown pigmentation	Reykjanes Ridge	23
Caryophyllia clavus	Black pigmentation	Mid-Atlantic Ridge	180
C. communis	Brown pigmentation	Equatorial Atlantic, 1°N, 1400-2400 m	3
Trochocyathus sp.	Black pigmentation	Mid-Atlantic Ridge	300
Not defined	Brown pigmentation	Reykjanes Ridge	8
"	"	"	155
Open ocean, shallow zone			
Dendrophyllia sp.	Grey pigmentation	Equatorial Atlantic, 1°N, 150-300 m	155
Madracis asperula	"	"	465
Cladocera patriorca	"	"	60
Anomocora cf.A.fecunda	"	"	40
Bathycyathus maculatus	"	"	35
Not defined	"	"	8
"	"	"	22
"	White	"	2
Coastal shallow zone			
Meandrina areolata	Alive, white	Florida Bay	5
M.braziliensis	"	Brazil coast	2
"	Grey, with traces of borers	"	2
Madracis sp. cf. M.pharensis	White	"	4
Porites porites	Light brown pigmentation	Florida Strait	3
", inner part	White	"	4
Madracis mirabilis	Alive	Jamaica	2
"	"	"	9
Montastrea annularis	"	"	2
"	"	"	5
Scolmia cubensis	"	"	3
"	"	"	30
Phyllangia americana	Alive with admixture of clay	"	130

In ascidians, manganese concentrations turned out to be higher. It was determined, by natural weight, as being 6.3 mg/kg in the samples from the Barents Sea
(Morozov, 1983), 6-70 mg/kg in the samples from the Sea of Okhotsk, 8-100 mg/kg
in the samples from the Black Sea (Kovalsky et al., 1974). The minimum manganese
content was detected in medusa (physalian) - 1 mg/kg of natural weight (Morozov,
1983); however, it would increase by over one order if we calculate it for a dry
weight.

More detailed analyses were performed to detect manganese contents in corals.
In fossil corals collected in relatively deep parts of the open ocean (1000 to 2400 m)
manganese content varies from 2 to 1100 mg/kg, in shallow waters of the open ocean
it is from less than 2 to 465 mg/kg and in the near-shore zone from less than 2 to
130 mg/kg (Table 23). Maximum manganese concentration was registered in fossil
corals with ferromanganese pigmentation, a lesser but still high concentration was
determined in corals with admixture of lithogenic material, and its minimum was recorded in pure corals (Livingstone and Thompson, 1971).

Molluscs. Molluscs are the most numerous representatives of sea fauna (over
100000 species) and their composition has been described in numerous papers.

In gastropods and bivalved molluscs, the occurrence of manganese and other
elements was detected in their shells and flesh. In gastropods, manganese concentration is from less than 1 to 500 mg/kg, in bivalved molluscs from less than 1 to
410 mg/kg by dry weight (Table 24). Distribution of manganese can be uniform in
one case, whereas in others, manganese may be accumulated either in shell or in
flesh.

When manganese is accumulated in flesh, its favourite organs are kidneys,
gills and the gastric gland (Eisler, 1981), its maximum concentration - up to 10% -
was found in kidney stones of mulluscs formed by a complex calcium-manganese-zinc
phosphate (Table 25) (George, 1980).

In mollusc-filtrators, manganese content depends, to a certain degree, upon
the composition of bottom deposits (Graham, 1972). Within some sea areas where
contamination of the water is high, such as sea-ports, common mussels, for instance,
contain a considerably greater amount of manganese in contrast to the same species
from clean waters (Flower and Oregioni, 1976); however, in other cases, no similar
correlation is observed (Phillips, 1978). When systematic analyses of molluscs for
manganese content were made, seasonal fluctuations in manganese content were
detected which are associated with fluctuations in water temperature and salinity
(Rucker and Valentine, 1961; Bryan, 1973; Patel et al., 1973; Phillips, 1978).

In some species of bivalved molluscs manganese content increases with their
growth; however, in some species this dependence is missing (Eisler et al., 1978).

Manganese found in molluscs comes from the water which they filter and from
the food they consume, which contains manganese in various forms. One group of
molluscs better assimilates dissolved manganese, whereas the other deals better with
suspensions (Pentreath, 1873; Orlando and Mauri, 1978).

One should note that the occurrence of radioactive [54]Mn in the water stimulates
accumulation of general manganese in molluscs (Chipman and Thommeret, 1970). Shells
of the living molluscs in shallow sea basins have a film coating of manganese oxide
(Allen, 1960). In laboratory experiments researchers even managed to cultivate
manganese micronodules on the shells (Zavarzin, 1964).

Benthic crustacea are represented mainly by shrimps, spiny lobsters (Palinurus),
lobsters (Homarus) and crabs. Manganese concentration in various organs and flesh
of crustacea varies within the range from 0.2 to 267 mg/kg by natural weight and
from 1 to 3800 mg/kg by dry weight (Table 26). In most cases, manganese is concentrated in the carapaces of crustacea, reaching there, for instance in lobsters,
98% of its total content (Bryan and Ward, 1965).

The opinion has been voiced that high concentration of manganese in the
carapaces of crustacea is caused by its sorption and formation of oxides (Tennant
and Forster, 1969). However, it became clear from experiments that crustacea bodies
can sorb manganese from water only very slowly and it is mainly consumed with food.
If an excess of manganese is introduced in a body it is extracted with fecal deposits
(to 40%) and through the body surface (40-80%) (Bryan and Ward, 1965).

Echinodermata are represented by starfishes, sea-urchins, ophyurians and
holoturians. Attempts to define manganese content in these organisms are scarce.
In starfishes manganese content is 6.5 to 43 mg per one kilogram of dry weight,

TABLE 24

Mn content in oceanic and marine mollusks, mg/kg of dry weight

Species	Mn content		References
	shells	flesh	
Acmaea digitalis	19.3–30.7	24.5–25.1	Graham, 1972
Anodonta sp.	540	–	Segar et al., 1971
Aplicia benedicti	101–106	4–21	Patel et al., 1973
Buccinum undatum	1.2	1.7	Segar et al., 1971
Cardium edule	2.0	6.3	"
Cerastoderma edule	–	6.2–44.6	Bryan and Hummerstone, 1977
Chlamys opercularis	17–18	4–158	Segar et al., 1971; Bryan, 1973
Crassostrea virginica	33–121	24–51	Windom and Smith, 1972
"	–	3.6–60	Galtsoff, 1942, 1953
"	–	5.9–41.9	Goldberg et al., 1958
"	21–227	–	Rucher and Valentine, 1961
"	505	–	Frazier, 1975
Crepidula fornicata	2.1	17	Segar et al., 1971
Glicymeris glicymeris	1.5	34	"
Littorina littorea	5.5–8.3	18.6–104.6	Ireland and Watton, 1977
"	–	18–133	Bryan and Hummerstone, 1977
Mercenarya mercenarya	1.5	–	Segar et al., 1971
Meretrix lamarckii	–	5.4	Ishii et al., 1978
Modiolus modiolus	20	47–160	Segar et al., 1971
Mya arenaria	–	29–70	Eisler, 1977
Mytilus californianus	8.4–14.2	2.5–17	Graham, 1972; Goldberg et al., 1978
Mytilus edulis	3.3	3.5	Segar et al., 1971
"	9.3–45.8	6.1–28.4	Graham, 1972; Simpson, 1979
"	–	5.2–35.4	Bryan and Hummerstone, 1977
"	–	4.9–91.7	Phillips, 1978
M. edulis aoteanus	1	1–105	Brooks and Rumsby, 1965
Nucella lapillus	1.1	12.0	Segar et al., 1971
"	–	11.4–16.8	Bryan and Hummerstone, 1977
Ostrea angasi, total	7.8–16.7		Harris et al., 1979
Ostrea edulis	–	6	Watling and Watling, 1976
"	–	61–85	Fukai et al., 1978
Ostrea sinuata	2	2–11	Brooks and Rumsby, 1965
Patella vulgata	8.6	13.0	Segar et al., 1971
"	–	5.4–42	Preston et al., 1971; Bryan and Hammerstone, 1977
Pecten maximus	4.9–12	4–410	Segar et al., 1971
P. novae-zelendiae	1	2–353	Brooks and Rumsby, 1965
Protothaca stamnea	16.8	11.5	Graham, 1972
Tapes semidecussata	24.7	24.7	"
Tegula funebralis	5.3–8.9	5.0–10.1	"
Thais emarginata	5.3–6.0	8.8	"
Thais lapillus	2.8–4.2	5.9–17.3	Ireland and Wootton, 1977

in various organs of sea-urchins it is 0.5 to 88 mg/kg (Table 27). In one species of a starfish, the manganese content was a bit higher - 51.7 mg/kg by natural weight (Eustace, 1974). In ophyurians and sea-lily the manganese content was lower, 3.3–3.6 mg/kg by natural weight (Morozov, 1983). The same is true for holoturians in the ashes of which 8.9 mg/kg manganese was recognized (Chipman and Thommeret, 1970).

Laboratory tests with [54]Mn showed that holoturians can extract manganese out of water actively - it seems to stimulate their growth (Ichikawa, 1961). On the other hand, high concentration of manganese has a negative effect upon reproductivity of the echinodermata (Young and Nelson, 1974).

TABLE 25

Mn in kidneys of mollusks, mg/kg of dry weight

Species	Sample	Mn content	References
Agropecten gibbus	Kidney stones	23643-24000	Carmichael et al., 1971
A. irradians Chlamis opercularis	Kidney	17300	Bryan, 1973
Ostrea sinuata	"	2	Brooks and Rumsby, 1965
Pecten maximus	"	15300	Bryan, 1973
"	Kidney stones	100000	George et al., 1980
Pecten novae-zelandiae	Kidney	2660	Brooks and Rumsby, 1965
Pinna nobilis	"	37000	Ghiretti et al., 1972
"	Kidney stones	80000	Hignette, 1979
Tridacna maxima	"	20000	"

TABLE 26

Mn content in marine crustacea, mg/kg

Crustacea	Mn content	References
Shrimp		
as a whole, dry weight	6.1	Knauer, 1970
", natural weight	0.23-2.2 (cp.1.4)	Capelli et al., 1983
flesh, dry weight	14.9-21.3	Zingde et al., 1976
"	1.5-8	Horowitz and Presley, 1977
"	1.0	Ishii et al., 1978
", natural weight	0.3	Morozov, 1983
", ashes	24-100	Lowman et al., 1966; Eisler, 1981
carapace, dry weight	17.8-89.8	Horowitz and Presley, 1977
", natural weight	1.0	Morozov, 1983
inner parts, dry weight	14.2	Horowitz and Presley, 1977
Spiny lobster, natural weight		
flesh	0.6	Morozov, 1983
"	0.22-0.27	Van As et al., 1973, 1975
gills	1.3	Morozov, 1983
caviar	0.8	"
carapace	12.5	"
Lobster (Homarus), natural weight		
flesh	0.3-23.3	Brian and Ward, 1965
carapace	187-267	"
Crab, natural weight		
flesh	0.4	Morozov, 1983
"	0.8-28.7	Greig et al., 1977a, b
gastric gland	0.6-3.0	"
gills	5.7-22.1	"
carapace	6.3	Morozov, 1983

Fishes. Fishes, in particular prey-fishes, belong to the highest stages in the general ecosystem of the ocean and, therefore, it is interesting to know the content of microelements in them for two reasons: ecology and usage of fish resources.

Manganese content in lamellybranchean fishes, like sharks and jelly-fishes, is low in general and is 0.1-1.9 mg/kg by natural weight in various organs and flesh. Its higher concentration - 2.7 mg/kg - was recognized only in a jelly-fish skin. When calculating manganese concentration in a shark's flesh by dry weight, we obtain 6.6-10.6 mg/kg (Table 28).

In hard-skeleton fishes, the manganese concentration is 0.1 to 8.8 mg/kg by natural weight and to 170 mg/kg in ashes.

In some organs of hard-skeleton fishes, manganese content varies within the range from less than 0.1 to 27.4 mg/kg by natural weight. The higher concentration was detected in bones, gills, skin and liver.

Manganese penetrates to a fish organism mainly with food. In a cod-fish which was fed by ^{54}Mn-charged nereid worms, a halfperiod of manganese exchange in the medium was 35 days (Pentreath, 1973). In the Black Sea, plankton-eating fishes accumulate larger amounts of manganese as compared with benthic fishes (Petkevich, 1967). Liver of pelagic fishes contains more manganese with respect to the continental shelf fishes (Pearcy and Osterberg, 1968).

Manganese is one of the vital microelements for the growth of young fishes (Pentreath, 1976); however, some large mature specimens contain less manganese than small specimens of the same species (Chernoff and Dooley, 1978).

Sea mammals form a comparatively small group of sea inhabitants and, therefore, data on manganese content in them are scarce.

In some organs and flesh of a seal, manganese content varies from less than 0.04 to 8 mg/kg by natural weight, the highest content being registered in heart, spleen and brain (Duinker et al., 1979). In Californian sea-lions, manganese content was determined in dry flesh and varied from 2.7 to 19.2 mg/kg. Its maximum content was recognized in the liver of mature specimens, the minimum in the kidney of calves (Martin et al., 1976).

Sea birds. Determinations of manganese content in sea birds are also scarce. The bones of living mature sea-gulls contain from 0.1 to 4.2 mg of manganese per 1 kg of natural weight, the liver 0.5 to 5.4, kidney 0.5 to 4.1, flesh 0.1 to 1.3. In the nestlings, manganese content is the same in general, namely, 2 in bones, 2.0 to 2.3 in kidneys and liver, 1.2 to 2.4 mg/kg in flesh. Higher manganese concentration (3.4 to 11.3 mg/kg) was detected in the matter in a sea-gull's stomach that contains remnants of fish (Hulse et al., 1980).

TABLE 27

Mn content in echinodermata, mg/kg

Echinodermata	Mn content	References
Starfish, dry weight	6.5	Riley and Segar, 1970
Asterias rubens	6.5	Riley and Segar, 1970
Henricia sanguinolenta	34	"
Porania pulvillus	22.0	"
Solaster papposus	31–43	"
Starfish, natural weight	1.3–18.0	Morozov, 1983
Patiriella regularis	51.7	Eustace, 1974
Sea-urchin, dry weight		
Echinometra locunter, skeleton	20.0	Stevenson and Ufret, 1966
Echinus esculentus, gonads	0.5	Riley and Segar, 1970
", thorns	4.5	"
", other organs	11–88	"
Spatangus purpurens, gonads	82	"
", skeleton and thorns	27	"
Tripneustes esculentus, skeleton	14	Stevenson and Ufret, 1966
Ophyurians, natural weight	3.6	Morozov, 1983
Sea-Lily, natural weight	3.3	"
Holoturians, ashes	8.9	Chipman and Thommeret, 1970

TABLE 28

Mn content in fishes, mg/kg of natural weight

Parts of body	Mn content	References
Lamellybranchean		
Sharks		
flesh	0.5-[2].6	Eustace, 1974
"	0.4	Greig and Wenzloff, 1977
"	0.1-0.6	Glover, 1979
", dry weight	6.6-10.6	Zingde et al., 1976
liver	0.4	Greig and Wenzloff, 1977
Jelly-fishes		
flesh	0.29	Pentreath, 1973
blood	0.16	
heart	0.27	
spleen	0.32	
liver	1.4	
kidney	1.9	
gonads	0.37	
intestines	0.82	
skin	2.7	
Hard-skeleton fishes		
Fish as a whole	0.9-8.8	Wolfe et al., 1973
"	0.2-6.0	Hall et al., 1978
"	0.11-1.3 (cp. 0.55)	Capelli et al., 1983
", ashes of benthic fishes	170	Lowman et al., 1970 (cited from (Eisler, 1981)
Flesh	0.18-1.15	Brooks and Rumsey, 1974
"	0.07-1.1	Van As et al., 1973, 1975
"	0.5-1.3	Eustace, 1974
"	0.1-2.0	Hall et al., 1978
"	0.18-0.44	Plaskett and Potter, 1979
Heart	0.1-0.8	Brooks and Rumsey, 1974
Liver	0.9-6.7	"
"	0.1-2.5	Greig and Wenzloff, 1977
"	0.2-2	Hall et al., 1978
"	3.1	Morozov, 1983
Kidney	0.2-1.4	Brooks and Rumsey, 1974
Gonads	0.2-3.3	"
"	2.3	Morozov, 1983
Gills	5.1	"
"	0.6-11	Brooks and Rumsey, 1974
Bones (spines)	7.6	Morozov, 1983
Bones in general	1.0-27.4	Brooks and Rumsey, 1974
Average by 112 samples	7.5	Morozov, 1983

CHAPTER IV

MANGANESE DEPOSITION FROM OCEANIC WATERS

It was mentioned in the previous chapters that manganese enters the ocean from various sources and in various forms. Low contents of manganese in sea water demonstrates that manganese has a short residence time in the sea and sinks actively down to the bottom. The routes by which manganese arrives at the bottom have been considered many times with respect to chemical, biological and biochemical processes.

Chemical deposition of manganese. It is known from the chemistry of manganese that MnO_2 is its most stable form in oxygen rich natural waters (Stumm and Morgan, 1970, 1981); its solubility is extremely low - from one of the estimates, the SP (solubility product) is 10^{-56} (Charlot and Bezier, 1958). Most obviously, Mn^{2+} oxidation should be accompanied by its deposition from the water. In this regard R. Horn wrote in his book Marine chemistry that he - as a specialist in chemistry - wonders why the origin of manganese nodules at the oceanic bottom remains mysterious, since formation of MnO_2 films in laboratory retorts with manganese solutions provides for a direct answer to this question (Horn, 1969). However, geologists and geochemists involved in this problem find it much more complicated.

Thermodynamic modelling and simulations were tried to find out the manner of chemical deposition of manganese from oceanic waters.

Thermodynamic simulations performed by various authors (Morgan, 1964; Crerar and Barnes, 1974; Gramm-Osipov and Shulga, 1980; Savenko, 1981; Savenko and Baturin, 1981) give controversial results. From Morgan's data (1964) presented in the review by Murray and Brewer (1977), Mn^{2+} oxidation in sea water occurs in accordance with the pattern.

Mn (II) + O_2 = MnO_2 solid.

Mn (II) + MnO_2 solid = (Mn (II) x MnO_2) solid.

(Mn (II) x MnO_2) solid + O_2 = 2 MnO_2 solid.

The first and third reaction are slow, the second goes rapidly. On the whole, this reaction is an autocatalytical one of the first order with respect to O_2 and Mn^{2+} concentrations and of the second order with respect to OH^-. From equilibrium calculations in pH-Eh coordinates, Crerar and Barnes (1974) arrived at the conclusion that sea water is undersaturated by several orders of magnitude by MnO_2. However, if this were true, manganese nodules at the sea bottom should have dissolved. The inferences these authors made are based on Eh measurements in sea water by platinum electrode which, as was later discovered during special tests, fails to detect the actual oxidation-reduction potential (Peschevitsky et al., 1970; Vershinin et al., 1981).

Hydrochemical observations show that the content of dissolved oxygen in sea water is the most sensitive indicator of oxidation-reduction processes. Therefore, the oxygen partial pressure can serve as a more objective indicator of the oxidation-reduction state of a sea medium. From thermodynamic constants of the materials involved in the formations of manganese oxides (Briker, 1965; Naumov et al., 1971), diagrams were plotted for stability and solubility potential of these materials in sea water at various partial pressures of oxygen (Figures 4, 5).

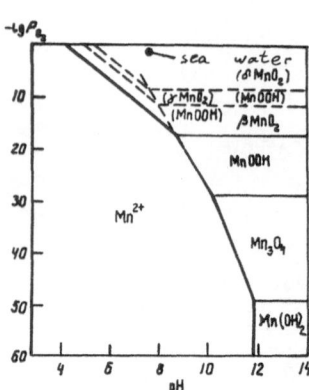

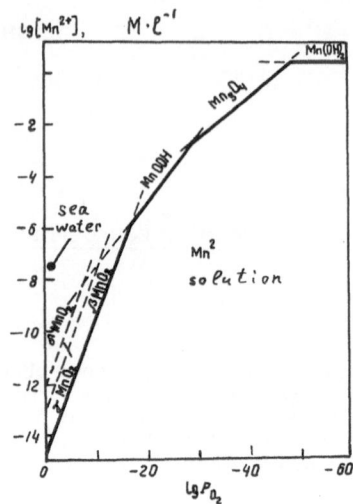

Fig. 4. Stability fields of manganese species in sea water in relation to partial pressure of oxygen and pH at a total manganese concentration of 2 µg/l. Dashed line shows metastable equilibrium (Savenko, 1981)

Fig. 5. Solubility of manganese species in sea water at pH=8 and various partial oxygen pressure. Dashed line shows metastable forms (Savenko, 1981).

One can see from these diagrams that in sea waters containing oxygen, tetravalent manganese compounds only can be considered as being stable mineral phases, whereas Mn^{2+} in a solution is not stable and tends to oxidation and deposition (extraction out of a solution).

This controversy between computed results and observed actual forms of manganese occurrence in a sea-water solution (Mn^{2+}, see Chapter II) can be explained by the effect of biogenic processes upon the sea medium - manganese is stabilized in sea water due to the occurrence of a dissolved organic material produced by phytoplankton photosynthesis. In the deep waters the effect of this activity seems to decrease sharply (Savenko and Baturin, 1981). A supposition has been also voiced that Mn^{2+} concentration in sea water is controlled by solubility of $MnCO_3$, i.e. ultimately, by total CO_2 concentration and pH. The estimated solubility of $MnCO_2$ - manganese in sea water was 1 to 5 m mol/l (Bischoff and Sayles, 1977), that is, one order higher than its real concentration. Thus, chemical deposition of manganese as $MnCO_3$ from natural sea water is less probable and can occur only in pore waters.

Experimental tests of manganese deposition from sea water are based on the attempts to obtain MnO_2 from a solution and to define solubility of manganese compounds in sea water.

Hem (1963) performed a series of tests to determine Mn^{2+} oxidation rate as a function of Cl^-, HCO_3^- and SO_4^{2-} concentrations. The results of his tests did not give any answer to the question of how manganese is deposited from sea water, as the latter differs considerably from experimental solutions in their salt composition.

Morgan (1964) performed special experimental work to verify the theoretical model of Mn^{2+} oxidation in sea water that he proposed (see above) and concluded that the composition of a solid phase appearing in the process of oxidation, namely the state of Mn (IV) and Mn (II) in it, are controlled by pH of the solution and by some other variables that govern the kinetics of the reaction. Extrapolating the results obtained to the sea water he arrived at the conclusion that in the ocean, at pH about 8, approximately 1000 years is necessary to oxidize most of the bivalent manganese. The attempts to model manganese deposition from sea water and formation of manganese nodules were made by Varentsov and others (1972, 1978) who used for their tests high-concentration metal solutions (to 1-2 mg/l) and excess of citric acid to prevent hydrolysis in them. In the solid phase formed manganese compounds exhibited a wide range of valencies - from 2 to 7. The results obtained are difficult to interpret from a thermodynamical standpoint; natural analogues of these compounds do not exist - all these are good reasons for criticism (Bazilevskaya et al., 1979).

A low degree of general oxidation of manganese (below $Mn_{1.33}$) obtained by Varentsov et al. can probably be caused by large amounts of Mn^{2+} introduced in the solutions that leads to its active sorption by MnO_2 sediment and formation of a compound with a great admixture of MnO (Savenko and Baturin, 1981).

Results of these tests, performed under conditions that differ sharply from the natural sea water, do not help us to find out the mechanism of manganese deposition in the ocean.

Another experimental approach to this problem is based on the analysis of manganese compound solubility by various solvents (Listova, 1961; Miller and Fisher, 1973; Afanasiev et al., 1979, 1982; Savenko, 1985). Some of the experimental results obtained aimed to study the solubility of a synthetic MnO_2 in acid solutions, solutions of Na_2SO_4 and sea water are given in Tables 29 and 30.

One can see the minimum concentration of a synthetic MnO_2 in a sodium sulphate solution of pH 8 - it is two orders higher than its concentration in the oceanic water. A similar concentration, 40 $\mu g/l$, was determined by Listova (1961) in sea water with pH = 7.90 (after a 260-day exposure).

To define solubility of a natural MnO_2 it would seem expedient to use natural manganese nodules; however, these contain not only MnO_2 but also a bivalent manganese in various forms which has been detected by numerous chemical and mineralogical investigations. In the experiments concerning manganese solubility, this bivalent manganese is obviously the first to be desorbed and dissolved and that prevents any reliable interpretation of the results.

Synthetic MnO_2 was believed to contain only MnO_2. However, when testing the composition of this standard chemical reagent some admixture of MnO was also revealed (Savenko, 1985); this makes one doubt the results of many earlier experiments.

Of prime concern is the problem whether neutral compounds of tetravalent manganese like $Mn(OH)_4^0$ may occur in sea water similarly to the known hydrooxides of other multi-charged cations (Ti, Zr, Hf, V, etc.). Sillen (1961) was the first to raise this question; nevertheless, no experimental work has been done yet.

Assuming that manganese oxidation follows the pattern

$$Mn^{2+} + 1/2\ O_2 + H_2O = MnO_2 + 2H^+,$$

manganese solubility is a direct function of pH: as pH changes by one unit manganese solubility changes tenfold. Experimental results (Table 30, Figure 6) show that

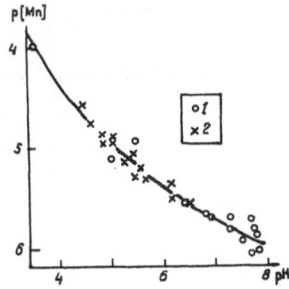

Fig. 6. Dissolved Mn equilibrium concentration as a function of pH in solution.
(1) solution of 0.2 M Na_2SO_4; (2) sea water with 35‰ salinity (Savenko, 1985).

TABLE 29

MnO_2 solubility in various acids and at various acidity

Conditions	Acid concentration M/l	Mn concentration mg/l
H_2SO_4, 25°C*	10^0	142
	10^{-1}	142
	10^{-2}	139
	10^{-3}	56
	10^{-4}	5.5
	10^{-5}	0.88
	10^{-6}	0.44
HCl, 25°C*	10^0	242
	10^{-1}	140
	10^{-2}	135
	10^{-3}	56
	10^{-4}	8
	10^{-5}	0.65
	10^{-6}	0.60
H_2SO_4, 25°C $P_{2} = 1$ atm**	5×10^{-2}	600
	5×10^{-3}	276
	5×10^{-4}	37
	5×10^{-5}	2.8
H_2SO_4, 0°C $P_{2} = 1$ atm**	5×10^{-2}	97
	5×10^{-3}	49
	5×10^{-4}	6.1
	5×10^{-5}	1.7
H_2SO_4, 25°C $P_{2} = 500$ atm*	5×10^{-2}	307
	5×10^{-3}	150
	5×10^{-4}	21
	5×10^{-5}	1.7
H_2SO_4, 8°C $P_{2} = 500$ atm**	5×10^{-2}	130
	5×10^{-3}	80
	5×10^{-4}	13
	5×10^{-5}	0.6
$0.2M$** Na_2SO_4		
0°C, P = 1 atm	not defined	1.7
25°C, P = 1 atm	"	2.8
8°C, P = 500 atm	"	1.1
25°C, P = 500 atm	"	1.7

* After Miller and Fisher, 1973.
** After Afanasiev et al., 1979.

TABLE 30

MnO$_2$ solubility in Na$_2$SO$_4$ solutions and in sea water

(Savenko, 1985)

0.2 M Na$_2$SO$_4$			Sea water, 35°/oo		
pH	Mn, 10^{-6}M/l	p(Mn)	pH	Mn, 10^{-6}M/l	p(Mn)
3.60	104.6	3.98	4.55	25.1	4.60
5.08	7.4	5.13	4.65	16.2	4.79
5.55	11.5	4.94	4.87	13.5	4.87
6.50	2.7	5.57	4.88	10.7	4.97
6.88	2.1	5.68	5.14	10.7	4.97
6.94	1.9	5.72	5.15	12.3	4.91
7.35	1.4	5.85	5.35	6.8	5.17
7.37	1.9	5.72	5.45	8.5	5.03
7.57	1.1	5.96	5.50	4.8	5.32
7.67	0.8	6.10	5.63	6.0	5.22
7.76	1.8	5.74	5.66	4.6	5.34
7.80	1.4	5.85	6.22	3.0	5.52
7.82	1.3	5.89	6.23	4.1	5.39
7.86	0.9	6.05	6.54	2.7	5.57

TABLE 31

Constants of apparent sorption equilibrium of Mn in sea water
at various composition of sorbent, % (Balistriery and Murray, 1984)

Sorbent components	Sorbent		
	Terrigenous sediments*	Bottom suspension**	Surficial suspension**
Alumosilicates	Predominate	55	10-25
CaCO$_3$	–	4	60-70
C$_{org}$	2	1.3	4-6
Mn	0.09	4.5	0.012
log K	4	5.3	7

* After Nyffeler et al., 1984.
** After Balistriery and Murray, 1984.
*** After Brewer et al., 1980; Honjo et al., 1982.

manganese solubility does depend on pH of a solution; however, it changes considerably less than might follow from the above reaction. One cannot exclude the possibility of formation of neutral manganese compounds.

On the whole, the data presented show that the question of solubility of tetravalent manganese compounds in sea water is far from any definite answer, since the
solubility of manganese oxides determined experimentally is much higher than the
actual content of manganese in sea water.

Coagulation. When investigating the mechanism of manganese deposition from
sea water, Goldberg (1954, 1965) put forward a supposition that manganese in the
ocean occurs as negative MnO_2 compounds that may coagulate and coprecipitate
alongside with positive colloidal iron hydrooxides, for instance.

Some researchers (Carnoll, 1958; Strakhov, 1960; Turekian and Imbrie, 1966;
Bender, 1972) have developed the concept that manganese arrives at the pelagic
part of the ocean as a component of finely dispersed particles of terrigenous suspensions that precipitate by coagulation.

A series of theoretical and experimental works were devoted to a coagulation of
suspended particles in the ocean. A vertical flow of the finest particles in the open
ocean is believed to be controlled by Brownian motion, whereas migration of larger
particles, in particular in the near-bottom oceanic waters and in the vicinity of
estuaries, is governed by orthokinetic laws (Stumm and Morgan, 1970, 1981; Lal and
Lerman, 1973; Murray and Brewer, 1977).

From experimental data on effective collision of clay particles (Edzwald et al.,
1974) and estimating their average concentration in sea water as 500 units per
milliliter (Eittreim, 1970), Murray and Brewer (1977) calculated the average size of
these particles as about 1.2 microns, the average rate of their deposition as 40 m/y
and their average residence time in oceanic water as 200 to 400 years.

Though a considerable role for this mechanism is recognized, many researchers
do not consider it as a leading one in manganese deposition.

Sorption. When Goldschmidt (1937) was engaged in the analysis of the mass
balance during continental weathering and oceanic sedimentation, he proposed the
process of sorption as the main mechanism of metal extraction from sea water.
Strakhov (1960) has developed this concept and pinpointed sorption as an important
mechanism of metal supply, in particular of manganese, in the pelagic zone of the
ocean.

Lately, this concept was supported by Li (1981) who believes that sea water
composition is determined by sorption balances.

Sorption of metals out of solutions, in particular out of sea water, depends
upon many factors; namely, upon number and composition of sorbing particles, their
surface dimensions, duration of their contact with a solution, pH of a solution, concentration of dissolved metals (Stumm and Morgan, 1970, 1981; Stumm et al., 1976;
Murray and Brewer, 1977). Theoretical and experimental aspects of manganese
sorption from sea water have been considered by many authors (Bender, 1972;
Balistriery et al., 1981; Li, 1981a, b; Balistriery and Murray, 1983, 1984).

So, Balistriery and Murray (1984) performed a number of experiments on metal
sorption from sea water by a near-bottom suspension and by suspended sediments
at various pH and various concentration of suspended particles. A near-bottom suspension was sampled specially for this purpose by the submersible Alvin in the zone
of hemipelagic sediments of the Guyana Basin (6°33'N, 92°48'W, depth 3600 m).

The majority of metals reached their sorption equilibrium during less than 20
days; however, manganese did not achieve it even after 40 days, which can probably
be explained by the low kinetics of manganese oxidation. On the whole, manganese
was sorbed on a suspension more actively than one could expect on the basis of its
hydraulic potential - this may result from the effect of its oxidation, which was also
pointed out by other researchers (Murray and Dillard, 1979; Sung and Morgan, 1981).

As the concentration of metals (manganese included) in a solution increased,
their sorption intensity decreased; whereas the latter increased at pH growth.

The result of experiments (at pH = 7.82) were given as constants of an apparent
sorption equilibrium

$$K_{sw} = \frac{Me_{ads.}/kg \text{ of suspension}}{Me_{diss.}/l} \times \frac{1}{N} ,$$

where N is the number of accessible centres of sorption (in mol/kg of suspension) determined from the tritium exchange method or by the cation exchange capacity.

Logarithms of these constants fall within the range from 2.008 for cesium to 6.082 for lead (for manganese it is 5.367). These results can be arranged as a sequence of sorption activity for the metals analysed:

Pb   Fe   Sn   Co   Mn   Cu   Be   Sc   Zn   Ni   Cd   Ba   Cs,

According to the Schindler model (1975), sorption residence time of metals in the ocean depends upon the residence time of sorbing particles and upon the distribution coefficients:

$$\frac{t_{met.}}{t_{susp.}} = \frac{Me_{tot}}{Me_{ads.}} = 1 + \frac{Me_{diss.}}{Me_{ads.}}$$

Combining this equation with the previous one we have

$$K_{sw} = \frac{1}{aN \left(\frac{t_{met.}}{t_{susp.}} - 1 \right)} \quad ,$$

where, a is the amount of suspended matter ($\mu g/l$), N is the number of accessible centres of sorption (M/kg of suspension).

This equation permits one to compare the results of the experiments on adsorption with independent results obtained by other authors, which seems to be of special interest, as far manganese is concerned.

Balistriery and Murray (1984) proposed use of the following estimates for this purpose:
- average concentration of suspended matter in the ocean as 15 $\mu g/l$ (Lal, 1977);
- residence time of the fine particles prevalent in the ocean having a low rate of deposition as 7.5 years (Bacon and Anderson, 1982);
- average sorption residence time for manganese as 51 years (Weiss, 1977);
- number of accessible centres of sorption on suspended matter as 0.5 to 3 M/kg.

According to one of the determinations, this value is 1 equ./kg for oceanic suspended particles with sizes from 1 to 53 microns (Bishop et al., 1977).

The use of these data results in a logarithmic value about 7 for the constant in question. It presupposes a more active sorption of manganese by suspended matter in the open ocean, rich in organic material. Following that, according to sorption capacity, is the hemipelagic near-bottom suspension, rich in manganese. The sorption capacity of terrigenous sediments in the San Clemente region turned out to be remarkably lower (Table 31).

Recently, a long series of experiments on manganese sorption by bottom sediments was performed by Soviet researchers (Gromov, 1975; Tikhomirov et al., 1979; Tikhomirov, 1982, 1984).

In one of the experiments, 2.5 $\mu g/l$ of [54]Mn was introduced to natural sea water until the general concentration of dissolved manganese reached 4.5 $\mu g/l$. The process of sorption was followed under static conditions in 50 millilitres of water to which from 10 to 500 mg of solid matter of natural oceanic sediment was added.

From the results of three parallel experiments, diatomaceous ooze sorbs 32% of manganese from water, calcareous sediments 78%, and red clay 95%. In the process of desorption, a variable amount of manganese returns back to solution: 45% from diatomaceous ooze, 20% from calcareous sediments and 0% from red clay (Table 32).

In another series of experiments it was found that if samples are processed by weak acid their sorption potential goes down sharply; for instance, red clay absorbs only 2%, not 88% as before, terrigenous sediments 3-6%, not 25-28% as a result of the loss of manganese hydrooxides as the main sorbent of the sediments (Tikhomirov, 1984). These data testify in favour of the important role the sorption mechanism plays in the extraction of manganese from oceanic waters; however, its intensity depends upon the sorbent composition and upon the variable parameters of different water layers. Arising from this supposition, and also from data on manganese geochemistry in sea water, in suspension, and atmospheric aerosols, the hypothesis was

TABLE 32

Mn sorption and desorption in the "bottom sediment-oceanic water" system (Tikhomirov, 1982)

Type of sediments	Sorption K_1	Sorption K_2	Share of sorbed Mn, %	Desorption K_1	Desorption K_2	Share of sorbed Mn, %
Diatomaceous	$(1.4\pm0.1)\times10^2$	$(0.8\pm0.1)\times10^2$	32 ± 3	$(5.5\pm0.5)\times10^3$	$(2.5\pm0.2)\times10^3$	55 ± 5
Calcareous	–	$(3.3\pm0.3)\times10^2$	78 ± 9	$(1.2\pm0.1)\times10^4$	$(8.5\pm1.0)\times10^3$	80 ± 5
Red clay	–	$(1.1\pm0.1)\times10^4$	95 ± 15	–	–	100

TABLE 33

Mn and other elements in material of sedimentary traps from the Atlantic (Brewer et al., 1980)

Layer, m	Concentration, % Mn*	Fe*	Al*	Alumo-silicates	CaCO₃**	Opal	Organic matter	Mn/Al	Mn flux, µg/cm²/y
	Site E, Sargasso Sea, 13°05'N, 54°00'W, 98 days								
389	0.0056 / 0.0034	0.34 / 0.18	0.76 / 1.03	12.1	53.5(62.7)	11.1	23.3(19.7)	0.0074 / 0.0033	0.1057 / 0.0642
988	0.0078 / 0.0065	0.81 / 0.60	1.31 / 1.76	21.0	51.1(55.1)	10.0	17.9(17.5)	0.0060 / 0.0037	0.1255 / 0.1046
3755	0.0330 / 0.0315	1.58 / 1.30	2.89 / 3.11	37.5	48.6(56.3)	6.5	7.5(10.3)	0.0114 / 0.0101	0.6455 / 0.6161
5086	0.0460 / 0.0464	1.91 / 1.84	3.64 / 3.79	45.9	41.2(49.2)	4.4	8.5(10.4)	0.0127 / 0.0123	0.8312 / 0.8384
	Site S west of Barbados Is., 31°34'N, 55°03'W, 110 days								
976	0.0046 / 0.0061	0.50 / 0.35	0.80 / 0.82	9.4	65.6(43.7)	6.5	18.2(17.6)	0.0057 / 0.0074	0.0128 / 0.0170
3694	0.0780 / 0.0756	0.75 / 0.75	2.18 / 2.14	25.9	62.6(68.0)	0	11.5(10.0)	0.0358 / 0.0353	0.5795 / 0.5617

* Numerator indicates atomic-adsorption definitions, denominator shows neutron-activation definitions.

** In brackets we give data by Honjo, 1980.

formed that manganese sorption mainly occurs not in a water body but at the bottom-sea water interface (Li, 1981a, b).

In its general form this idea was first promoted by Volkov et al., (1974) to explain the high Mn content of pelagic sediments.

Biogenic deposition. The occurrence of manganese in organisms and the role it plays in metabolic processes are connected with its bioassimilation by living matter, which is also accompanied by biodifferentiation, since relations between elements in living matter differ from those in sea water (Vinogradov, 1953). Since the average manganese content in oceanic water is 0.027 µg/l (Bruland, 1983) and in marine organisms it varies within the range from tenths of a miligram to hundreds milligrams per kilogram (Chapter III), the coefficient of manganese concentration by oceanic living matter is 10^4 - 10^7.

Considering this, the biosedimentation of manganese was proposed as one of the possible mechanisms of its deposition from oceanic water.

According to Lisitsin (1964), the C_{org}: Mn ratio in plankton and in suspension in the World ocean is 2000, on average. On this basis a yearly consumption of manganese by phytoplankton was calculated (Lisitsin et al., 1985). If the annual primary production of phytoplankton is 20 billion tonnes, it assimilates yearly 10 million tonnes of manganese - which exceeds by 20 times the annual supply of dissolved manganese by river discharge. However, in the zones of maximum biological productivity, assimilation of manganese probably occurs with less intensity due to its deficiency. At a primary productivity of 10 g of $C_{org}/m^2/day$, a theoretical consumption of manganese would be 5 mg/m^2/day, which requires an extraction of all manganese from a 200 m layer, but the water upwelling rate in the coastal areas does not exceed 1 to 2 meters per day (Hart and Currie, 1960; Wooster and Reid, 1963; Coastal Upwelling, 1983).

In sedimentation of organic matter that is recycled in the process of biological activity, the majority of organic material dissolves and only about 1% of the primary product reaches the bottom (Romankevich, 1977). The same result was obtained in the investigation of oceanic sedimentation by using sedimentary traps: after E. Suess (1980), a sedimentary flux of organic material is connected with the primary production and oceanic depth by the expression

$$C_{fl} = \frac{C_{prod}}{0.0238 \times depth + 0.212} \ g/m^2/yr$$

Thus, about 200 million tonnes of organic carbon is supplied yearly to the deep oceanic bottom surface. If the C_{org} : Mn ratio in sinking organic matter remains the same as in the surface suspension, about 100000 tonnes of manganese sink to the bottom by this mechanism.

However, as sedimentation of suspension and of pelletal material goes on, basic changes occur in their composition. From the data of sedimentary traps placed in various parts of the World ocean for long time, as the sedimentary material sinks to the bottom it becomes depleted in all organic components: organic matter, biogenic calcium carbonate and opaline silica. At the same time it becomes richer in lithogenic components: aluminium, silica, and then, iron and manganese - especially the latter - sometimes its content reaching 0.1-0.2% (Tables 33-35). In samples from near-bottom sedimentary traps in the Atlantic, west off Barbados, and in the North Pacific, the Mn/Fe ratio increases to 0.07-0.10, and the Mn/Al ratio to 0.035-0.086. In the Earth's crust this ratio is 0.02 and 0.01, respectively. However, in the samples from the Sargasso Sea this ratio, on the whole, is the same as in the Earth's crust (Spencer et al., 1978; Brewer et al., 1980; Tsunogai et al., 1982; Martin and Knauer, 1983).

However, these results do not give any evidence of an exact association of manganese with organic matter, as was believed earlier (Wangersky and Gordon, 1965). Some heavy metals that are not accumulating in living matter, meanwhile, are known to be concentrated in components of decomposing organic matter (Manskaya and Drozdova, 1964). So far, for manganese, this tendency is not so distinct, although in laboratory experiments it was extracted from sea water by a flocculating organic material derived from diatom decomposition (Hunt, 1983). Meantime, manganese was found to have a definite association with biogenic calcium carbonate in sediments.

When analysing the sedimentary matter trapped in the Pacific west of California, both suspension composition and manganese forms were determined; namely, leachable

TABLE 34

Mn forms and fluxes and composition of material in sedimentary traps
from the North Pacific, 47°51'N, 176°20'6"W (Tsunogai et al., 1982)

Depth	Forms	Mn, mg/kg (I)	(II)	(III)	SiO$_2$, %	Ca, %	Al, %	Fe, %	Ign.loss, %	Mn/Al 10^{-2}	Observed fluxes mg/m^2/yr Mn	Fe	Al	Normalized fluxes mg/m^2/yr Mn	Fe
110	Dissolved	3	8	5	0.58	26.0	0.04	0.02	–	–	–	–	–	–	–
	Non-solved	45	8	28	10.3	1.01	0.22	0.73	–	–	–	–	–	–	–
	Total	48	16	33	10.9	27.1	0.26	0.75	19.5	1.3	0.7	151	52	1.3	187
1040	Dissolved	24	24	24	0.59	9.6	0.2	0.01	–	–	–	–	–	–	–
	Non-solved	25	50	43	50.4	0.6	0.15	0.27	–	–	–	–	–	–	–
	Total	50	74	67	51.0	10.2	0.17	0.28	18.5	3.9	5.1	213	129	3.9	163
2160	Dissolved	49	43	45	0.22	2.2	0.1	0.02	–	–	–	–	–	–	–
	Non-solved	55	122	113	70.3	0.2	0.22	0.41	–	–	–	–	–	–	–
	Total	144	165	158	70.5	2.4	0.23	0.43	16.2	6.9	1.3	306	164	6.8	185
4380	Dissolved	50	64	62	0.25	8.2	0.001	0.002	–	–	–	–	–	–	–
	Non-solved	36	392	335	56.1	0.6	0.56	0.53	–	–	–	–	–	–	–
	Total	86	456	397	56.3	8.7	0.57	0.55	12.8	7.0	21.5	298	308	6.9	96
5250	Dissolved	270	1280	920	0.76	5.7	0.22	0.23	–	–	–	–	–	–	–
	Non-solved	1220	770	930	49.2	0.7	1.92	2.11	–	–	–	–	–	–	–
	Total	1490	2050	1850	50.0	6.4	2.14	2.34	16.6	8.6	23.1	292	267	8.6	108
Sedim. 0-3 cm	Dissolved	–	–	2260	1.3	0.42	0.87	0.36	–	–	–	–	–	–	–
	Non-solved	–	–	3510	50.2	0.79	5.76	3.68	–	–	–	–	–	–	–
	Total	–	–	5770	51.5	1.21	6.63	4.04	5.0	8.7	61	99	–	8.7	61

Note: (I) particles over 58µm, (II) particles less than 58 µm, (III) average.

TABLE 35

Mn forms and fluxes and composition of material in sedimentary traps
from the Pacific, 35°40'N, 123°50'W (Martin and Knauer, 1983)

Depth, m	mg/m²/day				Mn, μg/m²/day (% ratio)				Mn μg/g			
	Sus-pension	Alumo-silicates	CaCO₃	Organic matter	Inert	Dis-solved	Exchange-able	Total	Inert alumo-silicates	Dissolved CaCO₃	Exchangeable Org. matter	Total suspension
50	720	3.3	89	440	1.5(9.2)	2.9(18)	12(73)	16	450	33	27	23
100	210	2.8	53	130	1.6(4.3)	8.1(23)	25(72)	35	570	150	150	160
200	170	8.5	53	130	6.2(9.0)	22(32)	22(32)	69	730	770	170	410
300	120	14	39	63	7.4(14)	32(59)	15(28)	54	530	820	240	450
500	100	22	41	39	13(23)	39(70)	3.9(7.0)	56	590	950	100	560
600	110	29	43	45	12(23)	37(70)	3.5(6.7)	52	410	860	78	480
700	120	33	24	33	15(28)	31(58)	7.4(14)	53	450	1300	220	440
900	150	43	26	61	21(36)	33(57)	4.3(7.4)	58	490	1300	70	390
1100	170	47	30	45	21(35)	37(61)	2.5(4.1)	60	450	1200	156	360
1700	120	39	29	30	16(30)	36(67)	1.4(2.6)	53	410	1200	47	440
2000	100	37	18	31	16(32)	27(53)	7.5(15)	60	430	1500	240	500

(in 25% acetic acid); exchangeable (dissolved in trap solution); and refractory (insoluble in week acid) (Table 35). It turned out that when calcareous material is partly dissolved, manganese is the first element to be transferred into solution and is subsequently trapped again by suspended particles: this process recurs many times until a complete dissolution of biogenic carbonate happens. Then the manganese concentration in solution goes up sharply. This presents unambiguous evidence for a surface sorption effect, i.e. manganese sorption at an active surface of biogenic calcareous fragments (Martin and Knauer, 1983). A similar inference was adduced from other experimental data (Hager, 1980; Franclin and Morse, 1983). This mechanism of manganese sedimentation can be considered as a biogenic one since such extraction of manganese by sorption is connected with biogenic material (though it is completely independent of its absolute amount).

Another mechanism of manganese extraction out of sea water is a microbiological one. It has been determined that some marine bacteria species and their spores can oxidize bivalent manganese or act as a catalyst in its oxidation and fix the manganese oxide formed in bacterial capsules (Emerson et al., 1982; Rosson and Nielson, 1982; Cowen and Silver, 1984). Bacterial spores with a manganese coating, sampled from sea water, are shown on the photograph (Figure 7).

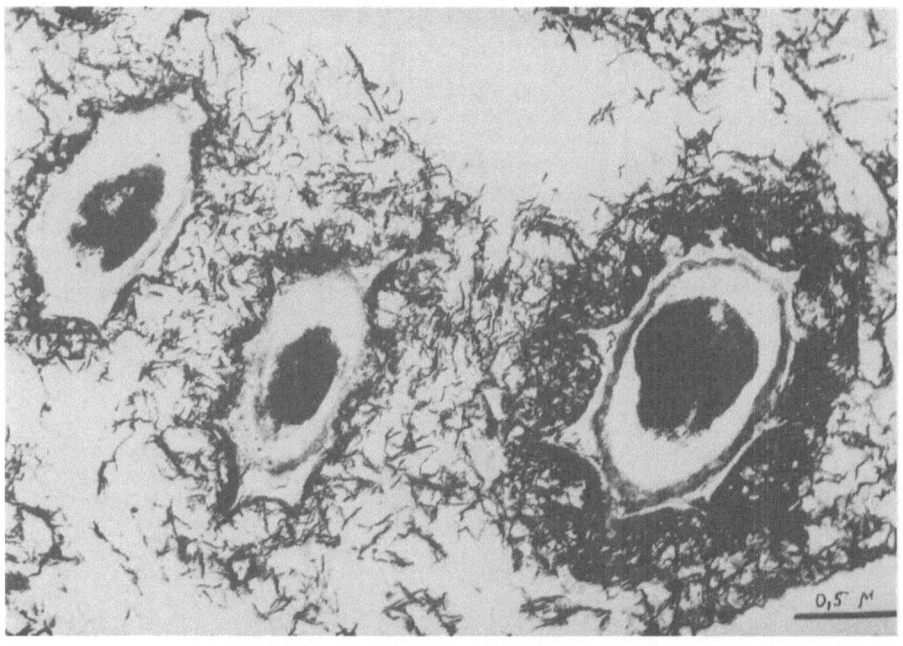

Fig. 7. Spores of marine bacteria in manganese dioxide capsules. Horizontal bar corresponds to 0.5 μm (Rosson and Nielson, 1982).

In a microbiological analysis of suspended and trapped material from the northern tropical Pacific the ratio of bacteria with those without capsules was up to 0.5% (Table 36). It is important to note that no capsules occur in the euphotic layer and in the oxygen minimum layer (to 140 m), whereas they occur permanently in the intermediate and deep water layers contributing to the biogenic manganese deposition in the ocean, though the quantitative input has not yet been estimated yet. When analysing samples from sedimentary traps this supply is added to that from other sources. It has also been determined that some bacteria in a water layer play a role as a catalyst in the oxidation of manganese dioxide which may stimulate its accumulation in suspension (Diem and Stumm, 1984).

When analysing samples collected by sedimentary traps, another set of problems arises from regularities in deposition and re-deposition of a matter in the ocean. In particular, it turned out that in some oceanic regions a flux of material, especially lithogenic, in the lower water layers considerably exceeds the sedimentation rate as determined by independent radiometric techniques. One such case is demonstrated in Table 34. This made the authors of the cited research normalize the observed fluxes of the material and of individual components to aluminium so as to reconcile fluxes with sedimentation rates. The high fluxes in the lower water layers seem to result from suspension of sediments by bottom currents, advective water mass transportation, and from turbid waters sloping down the continental rise. It should be kept in mind that a sedimentation rate is a summary parameter of a sedimentary process over a prolonged period, whereas sedimentary traps give momentary observational data on the process with its considerable seasonal and secular variations.

Manganese deposition from oceanic waters to the bottom of deep ocean as determined from the data of sedimentary traps varies from 0.06 (Spenser et al., 1978) to 0.8 $\mu g/cm^2/y$ (Brewer et al., 1980; Tsunogai et al., 1982). At oceanic depths to 2000 m the deposition rate can be considerably lower, about 0.006 $\mu g/m^2/day$ (Martin and Knauer, 1983). Manganese flux to the bottom of the oceanic pelagic area of 280 million square kilometers, calculated from these data, is 0.17 to 2.24 million tonnes per year, which does not contradict the estimates obtained by different techniques.

TABLE 36

Capsuled and non-capsuled bacteria in suspension samples
from the Pacific 18°00'N, 107°30'W (Cowen and Bruland, 1985)

Depth, m	Capsules with cells		Capsules with no cells		Non-capsuled bacteria	
	10 μm	0.2-10 μm	10 μm	0.2-10 μm	10 μm	0.2-10 μm
80	0	0	0	0	271	53
100	0	0	0	0	242	521
120	0	–	0	–	446	–
140	3	–	0	–	349	–
200	1	1	0	1	287	871
300	33	–	0	–	918	–
400	8	1	0	1	497	64
580	6	–	1	–	435	–
700	7	–	1	–	610	–
770	10	1	3	1	407	623
900	4	–	0	–	738	–
1293	10	–	5	–	202	–
1816	10	1	5	1	248	23
2000	5	–	3	–	134	–
2282	1	–	1	–	51	–
2500	22	4	12	4	294	261
2758	22	2	11	6	373	82
2948	11	–	13	–	416	–
3048	7	–	9	–	193	–
3250	13	–	15	–	184	–

MANGANESE IN MARINE AND OCEANIC SEDIMENTS

The study of manganese in oceanic bottom sediments, with which manganese nodules
are genetically associated in one way or another, was started after the Challenger
(1872-1876) and Gauss (1901-1903) complex oceanological expeditions, when sedimentary material was first sampled. The first analyses showed that manganese concentration in pelagic sediments varies from 0.03 to 3.11% (Gebbing, 1909; Murray and
Lee, 1909; Philippi, 1910).

For the following 50 years manganese in sea and oceanic sediments was analysed
as far as the general composition and origin of the latter was concerned (Clarke,
1924; Correns, 1937; Revelle, 1944; Pettersson, 1945; Strakhov, 1954). However,
several papers appeared with a special accent on this element (Klenova and Pakhomova,
1940; Pakhomova, 1948; Ostroumov, 1954, 1955).

Later, since the 1960's, these researches have acquired a systematic character
and the volume of information collected has increased sharply.

This chapter is devoted to the data on manganese concentration, distribution
and forms of occurrence in marine and oceanic sediments.

CONCENTRATION AND DISTRIBUTION OF MANGANESE IN THE UPPER SEDIMENTARY LAYER

Manganese concentration in the upper sedimentary layer of the seas and oceans varies
from 0.003 to 9%.

A certain dependence can be observed in manganese content in marine sediments
upon the granulometric composition of the latter: coarse-grained sediments are mainly
poor in manganese, whereas fine-grained sediments are enriched by it (Table 37).
This dependence is most distinct when determining manganese content on a lithogenic
matter basis. It happens because biogenic calcium carbonate and opal serve as a
diluent of the bulk portion of the metal associated with the abiogenic sedimentary
fraction.

The average manganese content in the main sedimentary types of various seas
is 0.017 to 0.032% in shelly sediments, 0.02-0.04% in calcareous sands, 0.01-0.08% in
terrigenous sands, 0.03-0.07% in coarse silt, 0.04-0.17% in fine silt, 0.03-0.21% in
silt-pelitic muds, 0.05-0.38% in pelitic muds (Table 37). In sand-silt volcanic sediments the manganese content is noticeably higher in contrast to that in terrigenous
sediments and especially in biogenic calcareous sand. Thus, in the Santorin Inlet
(the Mediterranean Sea) the manganese content in sediments is 0.10 to 0.62% as determined by various authors (Butuzova, 1969; Emelyanov et al., 1979).

Manganese distribution in the surface layer of marine sediments is considerably
irregular - this is especially true for the northern and Far East seas.

In the White Sea (Figure 8) a minimum of manganese (less than 0.01 to 0.025%)
was observed in the sands of the northern area and in the Onega Inlet. A belt of
higher manganese concentration (over 0.025%) can be traced along the Kola peninsula
coastline which results from the occurrence of coarse-grained polymictic sediments
accumulated there in the course of the weathering of basic rocks enriched with heavy
minerals. Zones under the influence of river discharge (for instance, of the Onega
and Northern Dvine Rivers) also have slightly higher concentrations of manganese.
In the White Sea basin, the manganese content in sediments increases in a concentric-

zonal way from the shore to the centre. The zone of the highest manganese concentration is bounded by a 100 m isobath; within the zone (towards the centre) the manganese concentration is over 1% and in some samples even 7% (Nevessky et al., 1977).

In the Barents Sea, manganese concentration varies from 0.01 to 0.5% (Figure 9). In the southern and northern parts of the Sea, sediments of an approximately similar granulometric spectrum are distributed. However, sediments from the southern region contain below 0.02% of manganese, whereas in the northern region, in the area of oxidized brown clay, manganese concentration increases to 0.5% which probably results from its remobilization from the underlying sediments. In the oxidized brown clay of the Kara Sea this contrast is even more distinct: manganese concentration reaches 1.53% (Klenova and Pakhomova, 1940; Pakhomova, 1948; Klenova, 1948; Kulikov, 1961; Gorshkova, 1966, 1967).

In the Arctic basin sediments, the maximum manganese content (0.5-1%) is confined to the northern sea slope of the Asia continent and adjacent parts of the Nansen, Makarov and Beaufort depressions (Figure 10). The belt of high manganese concentration extends further along the western slope of the Lomonosov Ridge, reaching the central part of the Makarov depression. Within the major part of the Beaufort basin and in the regions adjacent to North America, sediments are relatively poor in manganese which can probably be explained by peculiarities in the supply and distribution of terrigenous material and its rate of accumulation (Belov and Lapina, 1961; Belov et al., 1968; Strakhov et al., 1968).

In the Baltic Sea sediments, manganese is convined to deep sea deposits of depressions (Figure 11). High concentrations of manganese (over 0.1%, and maximum 1.7%) were recorded in highly reduced deep-sea sediments, on one hand, and, on the other, in oxidized clay and silt mud of the Riga and Bothnia Gulfs. In reduced shallow sediments, the manganese concentration is usually below 0.05% (Blazhchishin, 1976; Blazhchishin et al., 1982).

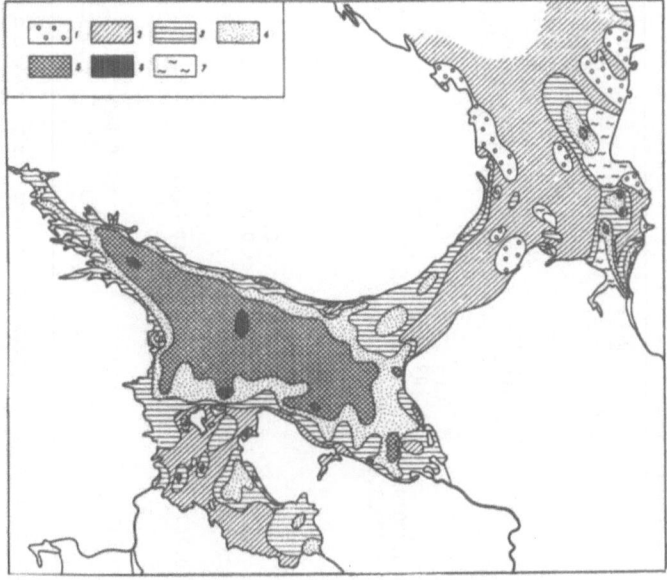

Fig. 8. Mn distribution (%) in the White Sea sediments (Nevesskiy et al., 1977):
(1) less 0.01; (2) 0.01 to 0.025; (3) 0.025 to 0.05; (4) 0.05 to 0.1;
(5) 0.1 to 1.0; (6) over 1; (7) bench.

TABLE 37

Mn average concentration (%) in surficial marine sediments*

Sea	Shelly sediments	Sands	Types of sediments				References
			Coarse silts	Fine silts	Silt-pelitic mud	Pelitic mud	
White	–	0.02–0.04	0.05	0.04–0.08	0.13	0.38	Nevessky et al., 1977
Barents	–	0.014	0.017	0.027(south) 0–083 (north)	0.217(north)	–	Klenova, 1948
Kara	–	–	0.102	0.284	0.733	0.750	Gorshkova, 1957
"	–	–	0.056	0.227	0.429	1.04	"
Baltic	–	0.04	0.03	0.08	–	0.11	Blazhchishin, 1976
Okhotsk	–	0.065	0.066(0.073)	0.116(0.137)	0.119(0.168)	0.433(0.677)	Strakhov and Nesterova, 1968
Bering	–	0.04	0.04	0.05	0.06	0.34	Lisitsin, 1966
Japan	–	0.08	0.03	0.09	0.21	0.42	Gramm-Osipov, 1973
Black	0.032(0.144)	0.060(0.070)	–	0.066(0.070)	–	0.045(0.075)	Glagoleva, 1961
Azov	0.032(0.138)	0.030(0.062)	–	0.049(0.137)	–	0.060(0.069)	Khrustalev and Scherbakov, 1974
Mediterranean:							
terrigenous	–	0.08	0.07	–	0.09	0.11	Emelyanov et al., 1979
low calcareous	–	0.04	–	0.17	0.09	0.11	
calcareous	–	0.04	0.04	0.08	0.09	0.09	
high calcareous	–	0.02	0.03	0.05	0.08	0.07	
volcano-clastic (Santorini Inlet)	–	0.12	0.15	0.35	–	–	
Northern Caspian	(0.157)	(0.049)	(0.049)	(0.051)	–	(0.093)	Khrustalev, 1968
Central Caspian	–	–	–	0.073		0.118	Glagoleva and Turovsky, 1975
Entire Caspian	0.017	0.022	0.052	0.13	0.152	–	
Aral	–	(0.08)	(0.11)	–	–	(0.07–0.18)	Khrustalev et al., 1977

* In brackets: Mn concentration on an abiogenic matter basis (biogenic carbonate and opal excluded).

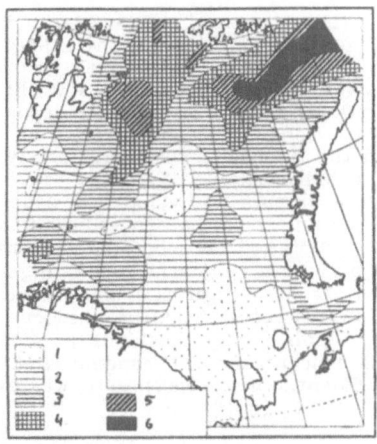

Fig. 9. Mn distribution (%) in the Barentz Sea sediments (Pakhomova, 1948):
(1) less 0.02; (2) 0.02 to 0.03; (3) 0.03 to 0.05; (4) 0.05 to 0.1;
(5) 0.1 to 0.3; (6) over 0.3.

In the sediments over most of the Sea of Okhotsk, the manganese concentration is less than 0.1% (Figure 12). The minimum content (0.005%) was recorded in sands near Sakhalin. In sand-silt sediments near the Kuril Islands, the manganese content reaches 0.1-0.5% and over the major area of the southern deep-sea basin it reaches 1-2%. The maximum manganese content (2.87-3.07%) was registered by Ostroumov in the muds of the Deryugin basin (Ostroumov, 1954, 1955; Bezrukov, 1960; Petelin and Ostroumov, 1961; Strakhov and Nesterova, 1968).

In the Bering Sea sediments, the manganese concentration varies from 0.02 to 2.02%. In the platform part of the sea, in the Bay of Anadyr, its concentration increases off-shore to the centre, i.e. from sand-silt deposits towards silt-clay muds, where it reaches 0.05% (Figure 13). In the southern geosynclinal part of the sea, the manganese concentration in sands and silt is higher than that in the Bay of Anadyr; as these sediments give way to finely dispersed ones, the manganese content first decreases slightly and then sharply rises up to 1.2% in oxidized deep-sea sediments. In the northern parts of the basin, where reduction processes prevail in the upper sedimentary layer, the manganese concentration in clayey deposits does not exceed 0.04-0.06%. As one approaches the Kuril Range, the manganese concentration increases and reaches 0.1% and over in sand-silt sediments (Lisitsin, 1959, 1966; Gershanovich, 1962).

Before we consider the southern seas, we should first note that manganese distribution in their sediments is considerably less irregular than in the northern and Far East seas.

On the chart of manganese distribution in the Black Sea sediments (Figure 14a), zones of minimum concentration (below 0.04%) are recognized in its north-western section, and also near the Kerch Strait and in the western and eastern basins. The belt of high manganese concentration (over 0.04%) is confined to the Asia Minor coast and closely approaches the Caucasus coastline. Over the rest of the Black Sea area, the manganese distribution in sediments is relatively regular. When determining manganese concentration on a carbonate-free basis, the pattern of this distribution changes fundamentally (Figure 14b). The minimum manganese concentration is localized in the coastal areas, whereas its maximum is in the deep-sea parts. Small areas of high manganese content (over 0.1%) are also found near the Caucasus shore and in the estuarine areas of the rivers that drain manganese-rich deposits. The occurrence of a belt of high manganese content in the north-western part of the sea can probably be explained by the presence of shells with a 90% $CaCO_3$ content. Since a part of the manganese is associated with carbonates, as was shown in Chapter II, a carbonate-free basis gives an overestimated concentration in the lithogenic portion of the sediment In the sediments of the Azov Sea, the minimum manganese concentration (below 0.025%) is confined to the coastal zones where sands and silt are spread (Figure 15). The high concentration (over 0.1%) is confined to clay deposits of the deep-sea part of the basin. The maximum manganese concentration on a carbonate-free basis appears to be associated with the lithogenic portion of shelly sediments such as in the Black Sea (Khrustalev and Scherbakov, 1974).

In the Mediterranean Sea sediments the manganese concentration varies from below 0.01 to 0.46%. Over the major part of the sea sediments with manganese concentration from 0.05 to 0.1% are distributed (Figure 16). Low manganese concentrations (below 0.5%) are confined to peripheral and arid zones, high concentrations (over 0.05%) to the deepest parts of the Sea. High manganese concentrations in the deep part of the Tyrrhenian Sea seem to result from the chemical weathering of volcanic rocks, products of which are supplied from the land.

In the Santorin Inlet the concentration of manganese in volcanic-hydrothermal sediments increases to 0.2-0.6% and, in one sample, it was even 9%; the latter, how-

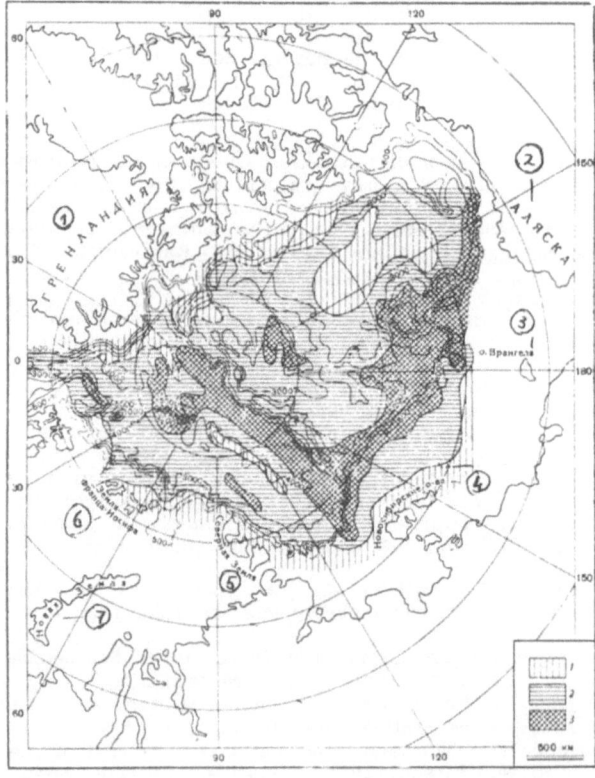

1 Greenland   5 Northern Islands
2 Alaska   6 Franz-Joseph Land Islands
3 Vrangel Island   7 Novaya Zemlya Islands
4 Novosibirsk Islands

Fig. 10. Mn distribution (%) in the Arctic basin sediments (Belov and Lapina, 1961): (1) less 0.2; (2) 0.2 to 0.5; (3) 0.5 to 1.0.

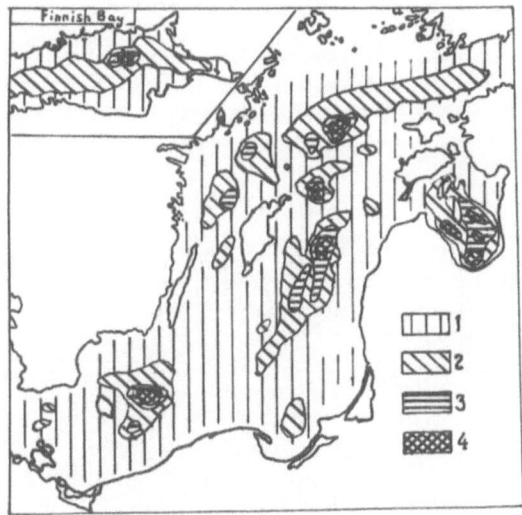

Fig. 11. Mn distribution (%) in the Baltic Sea sediments (Blashchishin, 1976):
(1) less 0.05; (2) 0.05 to 0.10; (3) 0.1 to 0.2; (4) over 0.2.

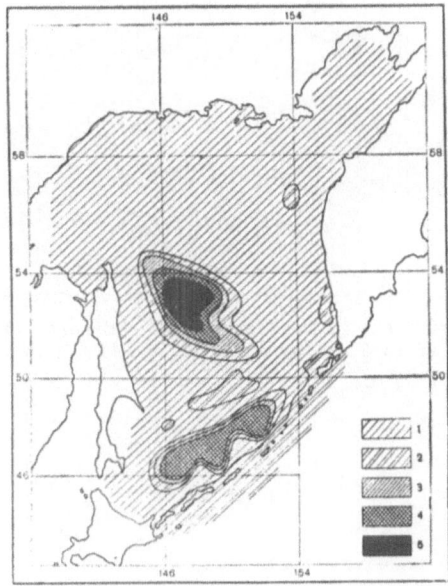

Fig. 12. Mn distribution (%) in the Sea of Okhotsk sediments (Ostroumov, 1954):
(1) less 0.1; (2) 0.1 to 0.5; (3) 0.5 to 1.0; (4) 1.0 to 2.0; (5) over 2.0.

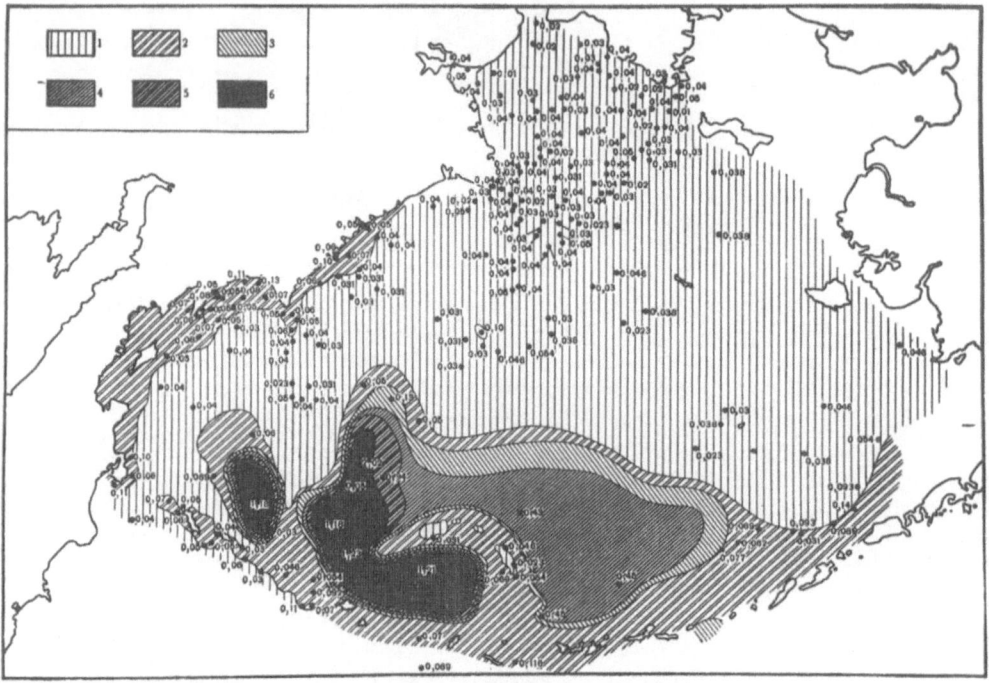

Fig. 13. Mn distribution (%) in the Bering Sea sediments (Lisitsin, 1966): (1) 0.2 to 0.05; (2) 0.05 to 0.1; (3) 0.1 to 0.2; (4) 0.2 to 0.5; (5) 0.5 to 1.0; (6) 1 to 2.

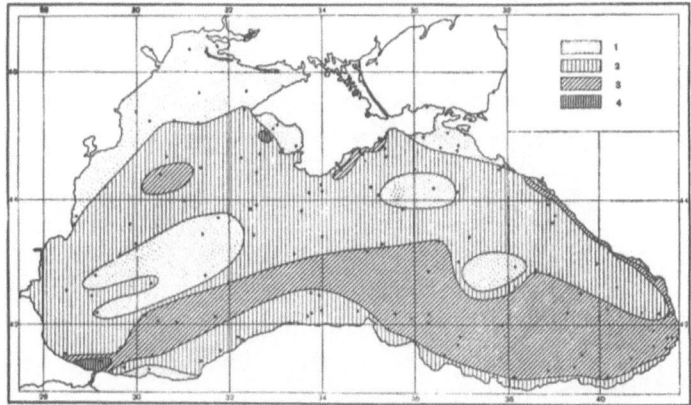

Fig. 14a. Mn distribution (%) in the Black Sea sediments (Glagoleva, 1961):
(1) less 0.04; (2) 0.04 to 0.06; (3) 0.06 to 0.10; (4) over 0.10.

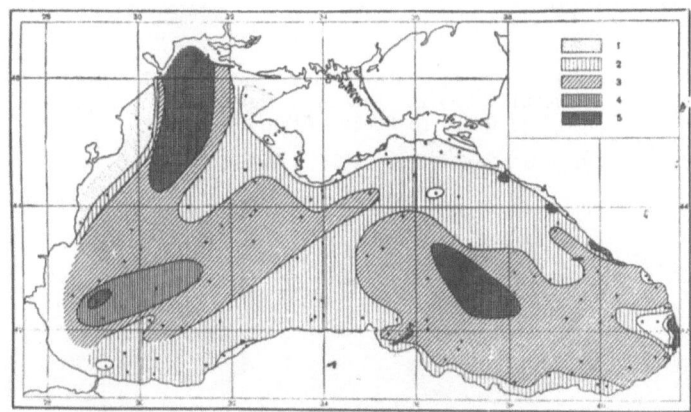

Fig. 14b. Mn distribution (%) in carbonate-free matter of the black
Sea sediments (Glagoleva, 1961): (1) less 0.05; (2) 0.05 to
0.075; (3) 0.075 to 0.10; (4) about 0.1; (5) over 0.10.

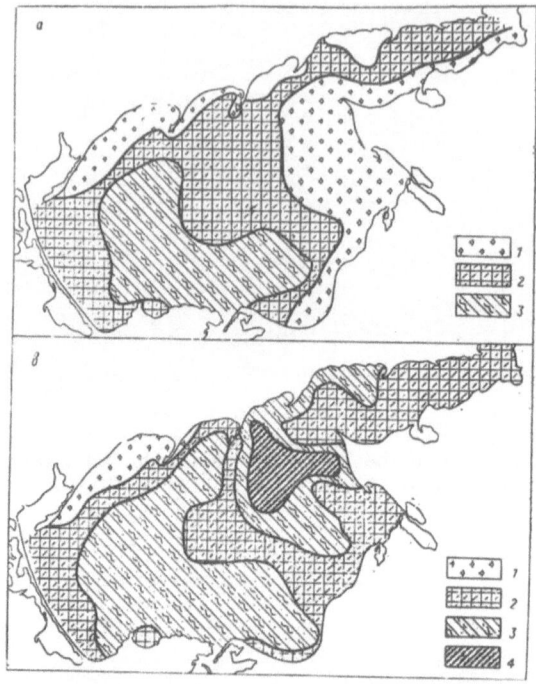

Fig. 15. Mn distribution (%) in the Azov Sea (Khrustalev and Sherbakov, 1974):
(a) in natural sediments; (b) in carbonate-free matter. (1) less 0.025;
(2) 0.025 to 0.035; (3) 0.035 to 0.050; (4) over 0.050.

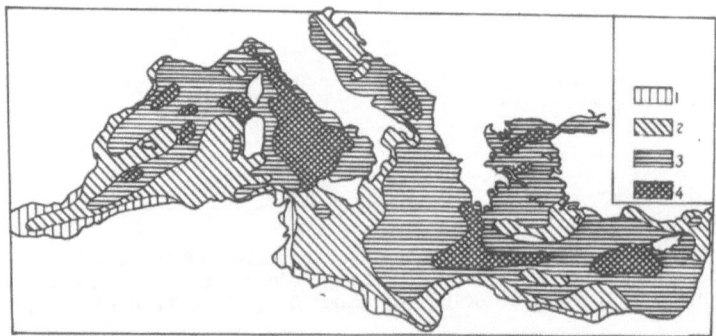

Fig. 16. Mn distribution (%) in carbonate-free matter of the Mediterranean sediments
(Emelyanov et al., 1979): (1) less 0.05; (2) 0.05 to 1.0; (3) 0.1 to 0.2;
(4) over 0.2.

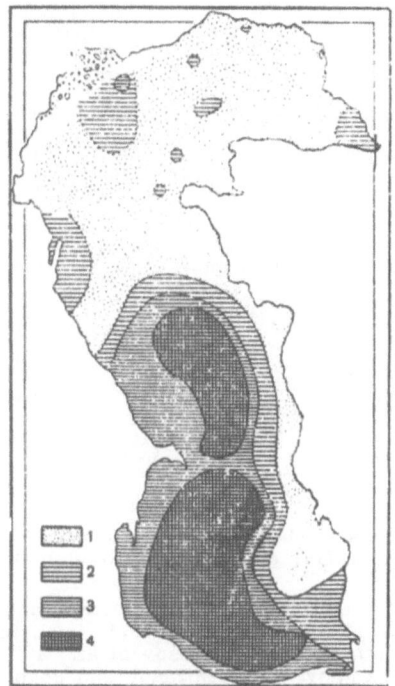

Fig. 17. Mn distribution (%) in the Caspian Sea sediments (Pakhomova, 1948):
(1) less 0.03; (3) 0.03 to 0.05; (3) 0.05 to 0.10; (4) over 0.10.

ever, has a narrow, local significance (Emelyanov, 1975, 1982; Emelyanov et al.,
1979).

In the Caspian Sea, the manganese content varies from 0.017% in shelly sediments
to 0.15% in clayey and clayey calcareous deposits. The general picture of manganese
distribution over the entire sea bottom (Figure 17) shows that its concentration is
low in shallow coarse-grained sediments and higher in deep-sea clayey deposits
(Pakhomova, 1948; Bruevich and Vinogradova, 1949; Strakhov, 1954; Maev and
Lebedev, 1970; Lebedev et al., 1973; Glagoleva and Turovskyi, 1975; Kholodov and
Turovskyi, 1985).

In the Aral Sea, deposits depleted of manganese (less than 0.1% on a carbonate-
free basis) are localized in silty coastal sediments of the northern bays and in sands
of the eastern shallow areas (Figure 18). As they change to deeper clayey-calcareous
muds, the manganese concentration increases to 0.20-0.34%. An increase in manganese
concentration is observed in the estuarine zones of the Syr Darya and Amu Darya Rivers
(Khrustalev et al., 1977).

On the whole, as one can see from this brief review, conditions for manganese
differentiation and relative concentration in sediments are less favourable for recent
arid marine lithogenesis than for the humid one. In the northern and Far East seas
the average manganese concentration in sediments is 0.3 to 1%, and it is only
0.06-0.18% in the sediments of the Black Sea, Azov Sea, Mediterranean, Caspian and
Aral Seas, even on a carbonate-free basis.

The data on manganese concentration and distribution in bottom sediments of
various parts in the open ocean are given in: Correns, 1937; Revelle, 1944; Goldberg and Arrhenius, 1958; Wedepohl, 1960; El Wakeel and Riley, 1961; Lisitsin, 1961,
1978; Skornyakova, 1961, 1964, 1970; Landergren, 1964; Turekian and Imbrie, 1966;

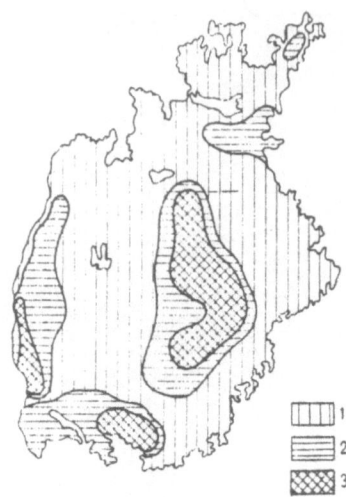

Fig. 18. Mn distribution (%) in carbonate-free matter of the Aral Sea sediments
      (Khrustalev et al., 1977): (1) less 0.1; (2) 0.1 to 0.2; (3) over 0.2.

Swanson et al., 1967; Bender and Schulz, 1969; Bostrom and Pettersson, 1969;
Chester and Messiha-Hanna, 1970;Dvoretzkaya, 1971; Pushkina, 1971; Piper, 1971;
Emelyanov, 1975, 1982; Strakhov, 1976; Migdisov et al., 1979; Bischoff et al.,
1979; Sevastyanova, 1982; Svalnov, 1983; Rozanov and Sokolov, 1984; Lisitsin et
al., 1985; and others.
      Some of these data on manganese contents in the main types of sediments in
the Pacific, Atlantic and Indian oceans are given in Tables 38-40.
      When considering these data, it turned out that in some similar types of bottom
sediments in all oceans, the manganese concentration is almost the same, whereas
in other similar types it is different.
      In the terrigenous coarse-grained sediments of the Atlantic, the average man-
ganese content is 0.03, in the Indian ocean it is 0.06, in the Pacific 0.08 to 1.0%;
in shallow silts it is 0.15, 0.08 and 0.19% respectively; in fine-grained sediments
0.2 to 0.4, 0.3 to 0.4, 0.3 to 0.5% respectively. Thus, the manganese content in fine
grained sediments in all three oceans is almost the same.
      In the biogenic calcareous sediments of the Pacific the average manganese con-
tent increases from coarse- to fine-grained varieties as follows: 0.04-0.06-0.09-0.16-
0.31% In the biogenic calcareous sediments of the Indian ocean, in general, the manganese concentration is lower: 0.06% in foraminiferal and 0.16% in coccolithic ooze.
The data on the Atlantic, calculated for carbonate-free matter, show the following
increases of manganese content from coarse- to fine-grained sediments: 0.11-0.23-
0.36%; for the Pacific, this sequence is 0.28-0.50-0.72%, i.e. twice as great, on
average.
      In the biogenic siliceous sediments of the Pacific the average manganese content
is 0.19% in diatomaceous ooze and 0.49% in radiolarian ooze. In the radiolarian ooze
of the Indian ocean, the average manganese content is 0.57% which is close to that
of the Pacific. In the Atlantic, diatomaceous sediments proper are mainly spread over
Namibian shelf and manganese concentration in them is extremely low, 0.01%, on average.
Red clay of the Pacific contains 0.76% of manganese, on average, that of the Atlantic 0.71%, and that of the Indian Ocean 1.28%. The latter seems to be unexpected,
since in all other sedimentary types the maximum manganese concentration is observed in the Pacific. In the miopelagic clays of the Indian ocean the average
manganese concentration is 0.64%; combining mio- and eupelagic clays in one group
gives a decrease to 0.96%, which does not seem to change the matter in principle.

TABLE 38

Mn concentration (%) in the Pacific surficial sediments (Skornyakova, 1970)*

Types of sediments	Mn in natural sediments		Mn on the abiogenic matter basis		Number of samples
	from/to	average	from/to	average	
Terrigenous:					
Sands	0.075-0.23	0.102	–	–	30
Coarse silts	0.052-0.19	0.085	0.06-0.19	0.091	17
Fine silts	0.01-0.81	0.193	0.04-1.01	0.225	27
Silt-pelitic	0.02-2.04	0.30	0.02-2.16	0.33	77
Pelitic	0.05-1.89	0.50	0.07-2.32	0.53	52
Low siliceous diatomaceous:					
Coarse silts	0.04-0.065	0.057	0.05-0.08	0.066	3
Fine silts	0.06-0.95	0.23	0.058-1.04	0.29	10
Silt-pelitic	0.032-1.49	0.41	0.04-1.74	0.43	37
Pelitic	0.05-1.27	0.44	0.06-1.60	0.50	21
Diatomaceous ooze	0.04-0.44	0.19	0.06-0.75	0.40	11
Radiolarian ooze	0.11-1.40	0.49	0.12-1.43	0.52	24
Calcareous:					
Sands	0.0048-0.075	0.037	0.07-0.63	0.27	11
Coarse silts	0.01-0.20	0.065	0.07-0.65	0.29	12
Fine silts	0.01-0.43	0.091	0.08-2.66	0.53	18
Silt-pelitic	0.03-0.43	0.16	0.14-2.55	0.68	38
Pelitic	0.04-1.15	0.31	0.097-1.85	0.77	29
Red clay:					
Silt-pelitic	0.16-1.55	0.49	0.24-1.57	0.52	20
Pelitic	0.18-1.68	0.60	0.18-1.68	0.62	109
Red clay in total	0.16-3.00	0.76	0.18-3.49	0.78	165
Volcanic sand-silt sediments	0.06-0.52	0.25	0.07-0.63	0.29	18
Metalliferous sediments from the SE Pacific	0.21-3.76	1.50	2.38-12.9	5.89	24
" (Bostrom and Peterson, 1969)	0.1-4.5	1.3	0.6-8.8	3.9	21
" (Migdisov et al., 1979)<10% $CaCO_3$	–	3.93	–	–	24
" 10-50% $CaCO_3$	–	3.89	–	6.38	11
">50% $CaCO_3$	–	1.21	–	5.96	86

* Using data from Revelle, 1944; Goldberg and Arrhenius, 1958.

TABLE 39

Mn concentration in the Atlantic surficial sediments on the abiogenic matter basis (Emelyanov, 1975)

Types of sediments	Mn concentration, %		Number of samples
	from/to	average	
Terrigenous (< 10% CaCO₃):			
Sands	< 0.01–2.0	0.10	128
Coarse silts	< 0.01–0.07	0.03	38
Fine silts	< 0.01–0.10	0.03	42
Silt–pelitic	0.02–0.82	0.15	13
Pelitic	0.02–0.94	0.38	7
	0.03–2.00	0.20	28
Terrigenous low calcareous (10–30% CaCO₃):			
Sands	0.01–3.25	0.16	54
Coarse silts	< 0.01–0.10	0.04	10
Fine silts	< 0.01–0.16	0.05	11
Silt–pelitic	0.01–3.25	0.31	13
Pelitic	0.02–0.20	0.10	7
	< 0.01–1.63	0.24	13
Biogenic calcareous (30–50% CaCO₃):			
Sands	< 0.01–0.74	0.20	50
Coarse silts	< 0.01–0.12	0.06	5
Fine silts	< 0.01–0.33	0.16	7
Silt–pelitic	< 0.01–0.74	0.23	10
Pelitic	0.06–0.55	0.19	14
	0.08–0.59	0.23	14
Biogenic high calcareous (> 50% CaCO₃):			
Sands	< 0.01–2.98	0.29	110
Coarse silts	< 0.01–0.96	0.14	22
Fine silts	0.13–0.28	0.17	5
Silt–pelitic	< 0.01–0.66	0.23	28
Pelitic	< 0.01–2.98	0.36	44
	< 0.01–1.56	0.54	11
Volcanic sediments:			
Sands	0.06–0.36	0.17	30
Coarse silts	0.06–0.18	0.15	6
Fine silts	0.11–0.26	0.16	5
Silt–pelitic	0.11–0.36	0.19	12
	0.09–0.26	0.15	7
Low siliceous terrigenous:			
Namibia shelf	< 0.01–0.23	0.06	9
Antarctic zone	< 0.01–0.04	0.02	5
	0.01–0.23	0.11	4

	Mn concentration, %		Number of samples
	from/to	average	
Diatomaceous sediments (> 30% SiO_2 am.):			
Namibia shelf	0.01–1.67	0.15	13
Antarctic zone	0.01–0.04	0.02	11
	0.14–1.67	0.90	2
Edaphogenic sediments of rift zones	0.13–0.16	0.15	8
Deep oceanic red clays	0.23–3.62	0.71	14

TABLE 40

Manganese concentration in the surficial sedimentary layer
in the eastern part of the Indian Ocean (Svalnov, 1983)

Types of sediments	Mn concentration		Number of samples
	from - to	average	
Volcanic and clayey sediments			
Coarse silt	0.06-0.07	0.065	2
Fine silt	0.07-0.10	0.08	3
Tuffite tephra	0.09-0.21	0.14	2
Silt-clayey	0.06-0.51	0.4	6
Hemipelagic	< 0.01-2.23	0.31	38
Miopelagic	0.13-1.34	0.64	36
Eupelagic	0.66-2.57	1.28	34
Biogenic oozes			
Foraminiferal	< 0.01-0.29	0.06	15
Coccolithic	< 0.01-0.60	0.16	21
Radiolarian	0.25-0.77	0.57	7
Ethmodiscoidal	0.06-0.15	0.10	6
Radiolarian-clayey	0.32-1.02	-	34
Ethmodiscoidal-clayey	0.14-0.40	0.25	6
Clayey-calcareous	0.09-0.91	0.34	9
Calc-clayey	0.06-1.51	-	6
Clayey-radiolarian	0.18-0.78	0.38	21
Clayey-ethmodiscoidal	0.16-0.34	0.20	4
Variegated detrital	< 0.01-0.01	< 0.01	7

The previous estimates of manganese averages in the pelagic clays of the Pacific varies within the range from 0.37 to 1.02% (Revelle, 1944; Goldberg and Arrhenius, 1958; El Wakeel and Riley, 1961; Swanson et al., 1967; Cronan, 1969). From these data, Cronan (1969) suggested that manganese averages of 0.478% should be assumed as the general average.

The data presented show that this value seems to be underestimated by a factor of 1.5. It is interesting that the first estimate of the manganese average in pelagic clays of the ocean - 0.76% - (Clarke, 1924) coincides with its recent estimate in the pelagic clays of the Pacific.

Metalliferous sediments of the World Ocean are of special concern as they have an extremely high manganese content (over 1%) and, as has recently been recognized, are widespread. These sediments were described in many publications (Böstrom and Peterson, 1966, 1969; Böstrom, 1973; Lisitsin et al., 1976; Heath and Dymond, 1977; Lisitsin, 1978; Migdisov et al., 1979; Cronan, 1980; and many others).

The map of manganese contents in the surficial sediments of the World Ocean compiled by Lisitsin (Figure 19) provides a good illustration on the role of these sediments in oceanic sedimentation. As can be seen from the map, most of the oceanic bottom, especially in the Pacific, is covered by sediments that contain over 0.5% of manganese (calculated on a mineralgenic basis). This probably can explain the high average concentration of manganese in the eupelagic clays of the Indian ocean (Table 40), since sediments with an admixture of hydrothermal material rich in manganese could occur among the samples analysed.

The highest manganese concentrations in oceanic sediments are confined to active mid-oceanic ridges. It is a belief that the higher the rate of spreading in the ridges, the more active is the supply of hydrothermal manganese which, in its turn, results in the increase of manganese content in sediments and an expansion of the area of their occurrence. Irregularity in the manganese distribution in oceanic sediments has not hindered the estimation of its average concentration in the surface layer; the result of 1500 samples analysed is 0.30% (Lisitsin, 1978).

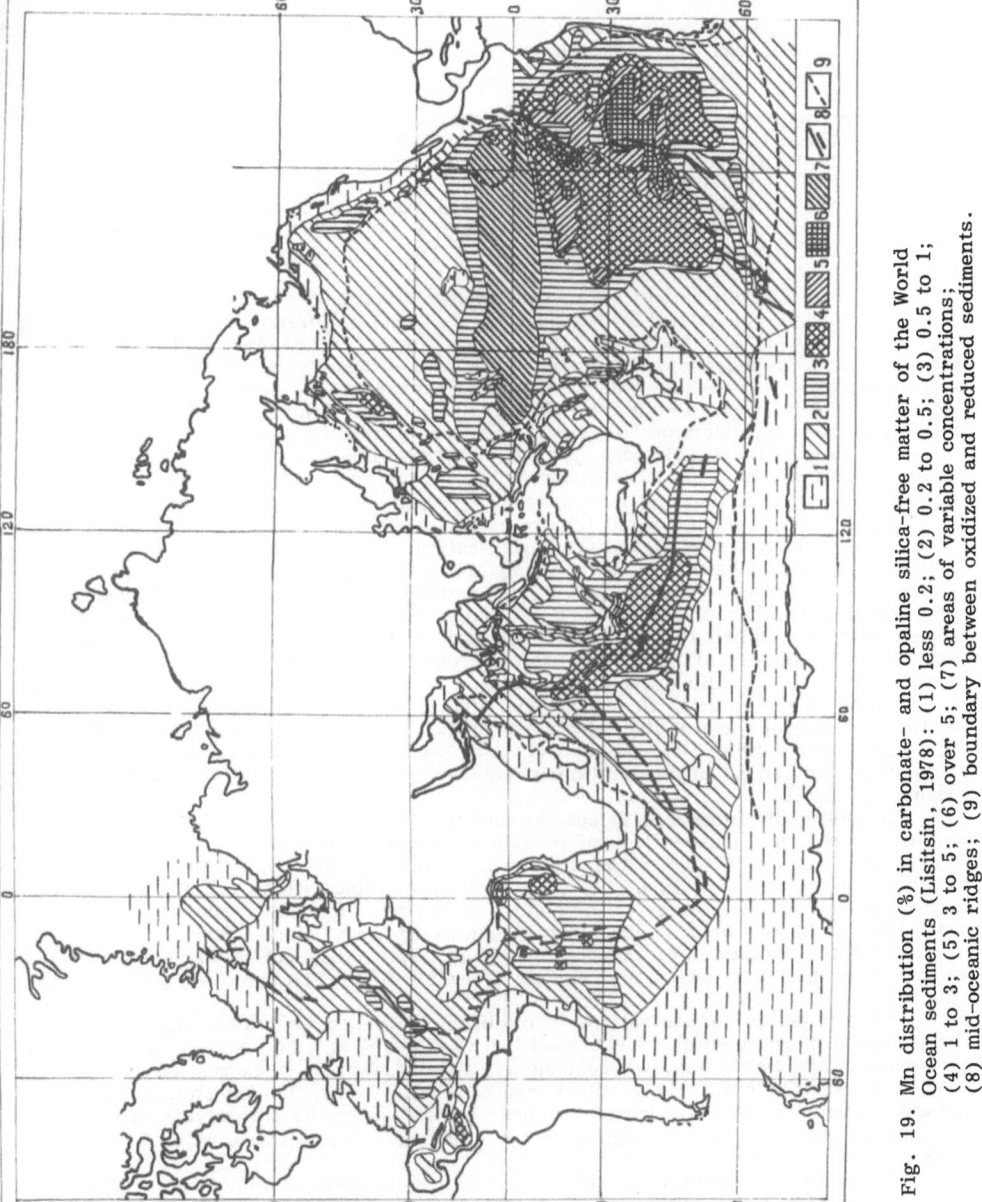

Fig. 19. Mn distribution (%) in carbonate- and opaline silica-free matter of the World Ocean sediments (Lisitsin, 1978): (1) less 0.2; (2) 0.2 to 0.5; (3) 0.5 to 1; (4) 1 to 3; (5) 3 to 5; (6) over 5; (7) areas of variable concentrations; (8) mid-oceanic ridges; (9) boundary between oxidized and reduced sediments.

MANGANESE CONTENT AND DISTRIBUTION IN SUBSURFACE SEDIMENTS

The study of manganese occurrence in subsurface marine and oceanic sediments is more extensive but less systematic than of that in the surface layer. This results from the abundance and great variety of material from ordinary and deep-sea drilling cores.

Data on the vertical distribution of manganese in marine sediments are available for the White Sea (Nevessky et al., 1977), the Baltic Sea (Hartmann, 1964; Emelyanov, 1981; Blazhchishin et al., 1982); the Black Sea (Volkov, 1973; Emelyanov et al., 1978, 1980, 1982; Mitropolsky et al., 1982); the Mediterranean (Shimkus, 1981); the Caspian Sea (Lebedev and Maev, 1973; Glagoleva and Turovsky, 1975); the Red Sea (Miller et al., 1966; Baturin et al., 1969; Bischoff, 1969; Butuzova and Lisitsina, 1983; Butuzova, 1984, 1985); the Sea of Japan (Kato et al., 1983).

Manganese distribution in oceanic sedimentary cores is given in Landergren (1964); Lynn and Bonatti (1965); Bonatti et al. (1971), Rozanov et al., (1972, 1976); Glagoleva et al. (1975); Glagoleva (1979); Migdisov et al., (1979), Volkov (1980); Isaeva (1982); Tsunogai and Kusakabe (1982) and in many other references. Data on manganese content in deeper layers of oceanic sediments are given in the Initial Reports of the Deep-sea Drilling Project, results of the "Glomar Challenger" cruises, and in numerous publications. A review of them is given by Cronan (1980).

According to the lithological type of the sediments, the physical and chemical conditions and the history of their accumulation, one can distinguish several patterns of vertical manganese distribution; namely, a relatively monotonic one with low concentration of manganese in completely reduced terrigenous and biogenous sediments; a relatively monotonic one with high concentration of manganese in completely oxidized deep-sea sediments; a contrasting one with high concentration in the upper oxidized layer and low concentration in the underlying layer; and an irregular one with alternating oxidized and reduced sediments downward the section. A specific pattern of sedimentation is associated with basal metalliferous deposits that represent ancient analogues of the recent metalliferous sediments in spreading zones (Böstrom et al., 1972; Horowitz and Cronan, 1976; Meylan et al., 1981).

Manganese distribution in marine and oceanic sedimentary cores gives evidence for the high diagenetic mobility which determines its dispersion or concentration in accordance with definite physical and chemical conditions. Tables 41 and 42 give examples of manganese differentiation in various types of oxidized and reduced oceanic sediments.

The ability of manganese to concentrate in the upper oxidized layer, or film, coating the reduced sediments is of particular relevance to manganese geochemistry in seas and oceans. It allows the manganese mobilized out of a sedimentary sequence to accumulate in the vicinity of the water/bottom interface which creates conditions for marine and hemipelagic types of nodule formation. Figures 20 and 21 illustrate manganese accumulation in sediments of the White Sea and the Pacific.

Under the conditions of the oceanic pelagic zone, the limit of diagenetic manganese accumulation in other than nodule formations is 4-5% (Table 42) or even 11% (Rozanov et al., 1972).

High manganese concentrations in oxidized sediments are often connected with the presence of micronodules. The maximum concentration of dispersed manganese in sediments of the Baltic Sea, where diagenetic redistribution is extremely active, reaches 8-14% (Hartmann, 1964; Blazhchishin et al., 1982). The hydrothermal manganese content can be considerably higher. In the manganite interlayers of ore deposits in the Red Sea, it reaches 25-39% (Bischoff, 1969; Butuzova and Lisitsina, 1983).

The manganese average in the oceanic sediments from ordinary cores (1200 samples) calculated by Lisitsin (1978) is 1.09%, in the DSDP cores (669 samples) it is 0.84%, which exceeds threefold the manganese average calculated for the surficial sediments (0.30%).

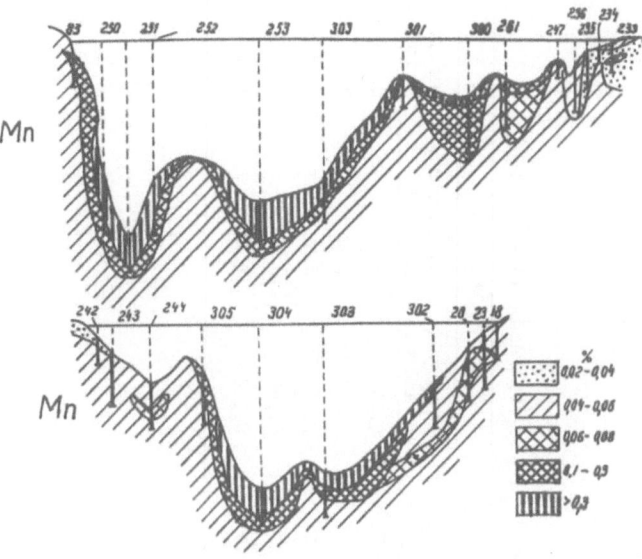

Fig. 20. Mn distribution in vertical cross-section of the White Sea sediments
(Nevesskiy et al., 1977).

FORMS OF MANGANESE OCCURRENCE IN SEDIMENTS

The problem of the forms of manganese occurrence in deep-sea sediments was first
posed by Murray and Irvine (1895) who supposed that at the oceanic bottom manganese is highly mobile.

    At present many techniques, such as granulometric, mineralogical, chemical,
etc., are used to attack this problem; due to these techniques it is possible to determine to a certain degree how manganese is associated with various components of
sediments.

    Terrigenous coarse-grained components are usually poor in manganese. This is
true both for marine and oceanic basins (Tables 37-40). What role is played by a
fine-dispersed terrigenous matter in the general pattern of manganese transfer? This
question was debated for a long time (Turekian, 1965, 1967; Strakhov, 1960, 1976;
Lisitsin, 1977, 1978; Lisitsin et al., 1985) since Revelle had already detected, in 1944,
that a colloidal fraction of deep-sea sediments is rich in manganese. This was also
verified by recent data, though the colloidal sedimentary fraction in the central
Pacific has lower content of manganese with respect to that of a more coarse fraction
(Table 43). However, manganese accumulation in this material results from a specific
condition of oceanic lithogenesis and has nothing to do with terrigenous sources
(Lisitsin et al., 1985).

    The volcanic sand-silt components are richer in manganese with respect to
terrigenic ones (Tables 37-39); however, the manganese concentration in them is
below 0.3 to 0.6% and spatial distribution over the bottom is local.

    The main biogenic components of sediments are calcium carbonate, amorphous
silica and organic matter.

TABLE 41

Mn distribution in cores of transitional sediments from the North-western Pacific (Volkov, 1980)

N of site	Type of sediments	Thickness of oxidized layer	Surficial layer, cm	Mn concentration, %				
				Surficial layer	Average in the core	Average in reduced l.	Average in oxidized l.	Maximum in oxidized l.
6163	Hemipelagic	1.5	0-0.5	0.134	0.054	0.054	0.232	0.273
6164	"	7.0	0-3	0.31	0.055	0.054	0.21	0.31
6168	Miopelagic	14.0	0-2	0.33	0.112	0.085	0.53	0.68
6169	"	13.0	0-5	0.72	0.17	0.106	0.80	0.85
6171	"	65.0	0-2	0.51	0.45	0.128	1.63	11.75

TABLE 42

Mn distribution in sedimentary cores from the Pacific
(Tsunogai and Kusakabe, 1982)

Site	Coordi-nates	Depth, m	Type of sediments	Layer*, cm		Mn concentration, %	
						total	mobile
KH-74-4-1	35°23'N 149°34'E	6052	Red clay	I	0-5	0.424	0.379
				II	30-35	4.150	4.120
				III	150-960	0.093	0.032
KH-74-4-12	34°51'N 175°03'E	4397	Calcareous	I+II	0-5	0.752	0.72
				III	265-417	0.039	0.026
KH-74-4-21	26°46'N 139°42'E	3850	Turbidite	I	0-5	0.229	0.144
				II	44-49	0.654	0.585
				III	74-195	0.151	0.038
KH-74-4-24	37°26"N 137°00'E	5050	Red clay	I	0	0.396	0.353
				II	58-61	5.64	5.53
KH-75-4-5	52°00'N 161°58'E	5805	Siliceous	I+II	0-5	0.267	0.225
				III	512-1007	0.0669	0.0254
KH-71-2-9	22°00'N 125°00'E	5700	Red clay	I	0-2	0.276	0.247
				II	24-26	0.447	0.409
7010-11	31°22'S 156°58'E	4090	Calcareous	I	0-2	0.1010	0.0836
				II	22-24	0.989	0.901
FB-7-1	42°24'N 140°32'E	102	Silt mud	I	0-5	0.0401	0.0076
				II	40-45	0.0771	0.048
				III	50-96	0.0347	0.0075

* (I) surficial layer, (II) layer of Mn maximum, (III) layer with even Mn distribution.

In general, living matter contains considerably less manganese than do sediments, as was described in Chapter III; however, after marine organisms die and sink to the bottom they, as biogenic components, are able to accumulate many of the metals, including manganese.

The nature of the association of manganese with biogenic calcium carbonate in sediments may be two-fold. On one hand, manganese may be sorbed on the surface of carbonate particles, in a process similar to that which occurs in the water layer (Martin and Knauer, 1983). On the other hand, calcareous limestone fragments may serve as nuclei of manganese accretion, thus forming films or micronodules which may be accompanied by a complete replacement of carbonate by manganese hydrooxides (Margolis, 1973).

A supposition has been also expressed that bivalent manganese may replace calcium in calcite during the process of diagenesis (Wangersky and Joensuu, 1964).

Biogenic opal seems to play a similar role; however, it differs from carbonate by its greater geochemical stability in pelagic conditions. When analysing magnetic and non-magnetic sub-fractions of a silt fraction in the sediments of the central Pacific, it was observed that a non-magnetic sub-fraction, which is composed almost entirely of radiolarians, contains 0.08% of manganese, on the average (Table 44), whereas siliceous tests of phytoplankton from sea suspension contain only 0.0002-0.0004% of manganese (Martin and Knauer, 1973). Initially, manganese was probably sorbed on biogenic opal in sediments, it may then have formed an independent mineral phase on the latter (Margolis, 1973), or a specific stable compound that undergoes no further diagenetic redistribution (Marchig and Gundlach, 1977).

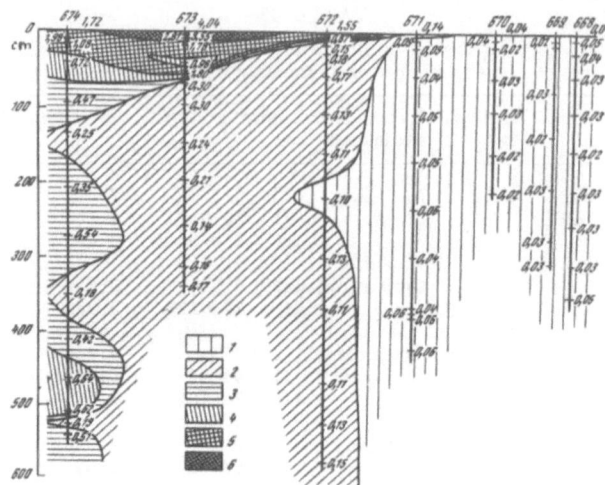

Fig. 21. Mn concentration (%) in carbonate- and opaline silica-free matter of
         Pacific sediments; vertical cross-section adjacent to the California Bay
         (Rozanov et al., 1976): (1) less 0.1; (2) 0.1 to 0.3; (3) 0.3 to 0.5;
         (4) 0.5 to 1.0; (5) 1.0 to 2.0; (6) over 2.0.

The probable association of manganese with organic sedimentary material was
supposed because some sedimentary layers are rich in both these components. In
particular, a higher concentration of manganese was discovered in the samples of
ancient black shales from deep-sea drilling cores in the North Atlantic (Murdmaa et
al., 1979).

Fractional analysis of sediments from the Tokyo Bay revealed that the peroxide
fraction which extracts organic matter and related elements contains 0.0068-0.369%
of manganese; the correlation coefficient between manganese and organic carbon in
this fraction is 0.81 (Yasushi et al., 1980).

However, the share of manganese associated with organic matter of shallow and
deep-sea sediments is not considerable. The manganese concentration in humic acids
extracted from sediments is 0.0008 to 0.01% (Nissenbaum and Swaine, 1976) and the
total share of manganese associated with humic and fulvic acids of coastal sediments
does not exceed 0.55% of its total amount in the samples analysed (Langsten, 1982).

From numerous studies it follows that manganese in sediments also forms independent mineral phases represented by films, flakes, thin coatings and by microconcretions. Granulometric analysis of sediments sampled from the central Pacific revealed that the maximum number of microconcretions is concentrated in fractions of
10 to 20 microns (Table 43). In microconcretions from a magnetic subfraction over
63 microns, the manganese content is 26.5% (Table 44). Among minerals of manganese and iron that occur in sediments in dispersed form, vernadite, feroxyhyte,
lepidocrocite and ferrihydrite were identified (Chukhrov et al., 1981). In reduced
sediments, manganese carbonate was detected, in particular in the North Pacific
(Logvinenko et al., 1972) and in the Baltic Sea (Suess, 1979; Emelyanov et al.,
1982).

To determine the mobility of elements and to recognize forms of their occurrence
in marine and oceanic sediments several methods of selective leaching have been proposed based on various combinations of reactive leaching solutions (Goldberg and
Arrhenius, 1958; Arrhenius, 1963; Chester and Hughes, 1967; Butuzova et al., 1967;
Gurvich and Shurygina, 1982). The last one (Gurvich and Shurygina, 1982) was
applied to samples from the central Pacific. It revealed that, when samples are processed by acetic buffered solution which dissolves carbonate, the share of extracted
manganese is from 0.1% (zeolitic clays) to 5-7% (radiolarian ooze and foraminiferal
sediments). Upon successive processing of samples by 0.5 M hydroxylamine, which
dissolves most of the hydrooxides, 78 to 83% of total manganese is transferred to the

TABLE 43

Distribution of metals in granulometric fractions of the Central Pacific sediments*
(Halbach et al., 1979)

Element	Fractions, μm							
	< 1	1-2	2-4	4-10	10-20	20-40	40-63	> 63
Mn	$\frac{0.16-0.40}{0.30}$	$\frac{0.18-0.68}{0.50}$	$\frac{0.21-1.03}{0.68}$	$\frac{0.29-1.14}{0.85}$	$\frac{0.29-2.78}{1.53}$	$\frac{0.23-1.97}{0.88}$	$\frac{0.20-1.28}{0.64}$	$\frac{0.22-0.69}{0.53}$
Fe	$\frac{5.56-6.28}{5.90}$	$\frac{4.40-4.96}{4.68}$	$\frac{3.63-4.17}{3.80}$	$\frac{2.90-3.47}{3.15}$	$\frac{2.19-2.73}{2.41}$	$\frac{1.13-2.49}{1.73}$	$\frac{0.67-1.70}{1.15}$	$\frac{0.54-1.19}{0.89}$
Ni	$\frac{137-403}{252}$	$\frac{151-310}{252}$	$\frac{200-390}{316}$	$\frac{210-592}{446}$	$\frac{300-1162}{751}$	$\frac{255-848}{453}$	$\frac{251-817}{390}$	$\frac{164-464}{298}$
Cu	$\frac{487-655}{596}$	$\frac{299-522}{421}$	$\frac{320-479}{400}$	$\frac{320-712}{524}$	$\frac{324-1312}{756}$	$\frac{323-951}{522}$	$\frac{306-651}{437}$	$\frac{149-537}{313}$
Co	$\frac{132-263}{191}$	$\frac{156-260}{208}$	$\frac{178-285}{238}$	$\frac{195-318}{237}$	$\frac{160-538}{305}$	$\frac{159-406}{231}$	$\frac{143-380}{209}$	$\frac{128-255}{188}$

* $\frac{\text{from-to}}{\text{average}}$; Mn and Fe concentration is in %; other elements are in 10^{-4}%.

TABLE 44

Concentration of metals in magnetic and non-magnetic fractions of the Central Pacific sediments*
(Marchig and Gundlach, 1977)

Element	Fraction < 63 μm (89.7%)		Fraction > 63 μm Non-magnetic (9.2%)		Slightly magnetic (0.5%)		Strong magnetic (0.1%)	
	1	2	1	2	1	2	1	2
Mn	0.71	90	0.08	1.0	7.43	5.3	26.50	3.7
Fe	3.63	98.5	0.25	0.7	4.36	0.7	3.34	0.1
Ni	0.0325	89.8	0.0074	2.2	0.2830	4.3	1.23	3.7
Cu	0.0475	93.4	0.0104	2.0	0.2460	2.6	0.8860	2.0
Co	0.0151	92.5	0.0060	4.1	0.0713	2.7	0.1380	0.7

* (1) element concentration in fraction, %; (2) relative element concentration (% of total) in fractions.

TABLE 45

Mn forms in the Central Pacific sediments (Gordeev, 1986)

Type of sediments	Mn, %	Relative Mn concentration in fractions*				Number of samples
		I	II	III	IV	
Clayey-radiolarian	0.48	5.5	80.5	9.4	4.5	8
Radiolarian-clayey	0.42	7.0	78.0	10.8	4.2	11
Diatomaceous-radiolarian	0.45	1.8	92.2	3.9	2.1	1
Radiolarian	0.75	0.6	90.5	8.0	0.9	2
Foraminiferal-coccolithic	0.145	6.6	78.6	12.5	2.3	2
Miopelagic	0.80	3.3	86.1	7.4	3.2	17
Eupelagic	1.10	1.1	91.0	7.0	0.9	8
Zeolitic clay	0.95	0.1	93.0	6.2	0.7	1

* Fractions after processing by: (I) acetic buffered solution at pH=4.7; (II) 0.5M hydroxilamine; (III) 5M hydroxilamine and 3M HCl; (IV) residue.

solution from all types of sediments. At the third stage, samples are processed by a mixture of 5 M hydroxylamine and 3 M HCl to destroy oxides and hydrooxides of iron; 6.2 to 12.5% of the total manganese is transferred to this fraction. The share of manganese in the residual alumosilicate fraction is 1 to 11% (Table 45).

When analysing forms of metals in sediments less detailed fractioning is used in most cases to extract a mobile (reactive) fraction out of sediments (Chester and Hughes, 1967). This method reveals that the share of mobile manganese in sediments increases in the ocean from 10-70 in shallow to 77-99% in pelagic regions (Lisitsin et al., 1976, 1985; Migdisov et al., 1979; Volkov et al., 1980; Förstner and Stoffers, 1981; Pfeiffer et al., 1982). The content of manganese in the reactive fraction of pelagic sediments from the central Pacific is about 0.6%, and only 0.04-0.05% in the residual fraction (Piper et al., 1979).

The degree of manganese oxidation (by 0:Mn ratio) in the upper sedimentary layer of the Pacific is 1.80-1.95 (Glagoleva, 1972; Rozanov et al., 1972); in the near-surface layers of sediments from the north-eastern tropical Pacific, where manganese concentration in pore water increases, the degree of manganese oxidation in a solid phase goes down to 1.35-1.60 (Kalhorn and Emerson, 1984).

So, arising from these and many other similar data, most researchers believe that mobile manganese resources in oceanic sediments exceed by many times the amount of manganese required to provide for the process of nodule ore formation.

MANGANESE IN PORE WATER OF MARINE AND OCEANIC SEDIMENTS

The first information on the peculiar composition of pore water, obtained by filtering through a dense canvas, was published by Murray and Irvine in 1893.

At present, the problem of manganese occurrence in pore water of marine and oceanic sediments is very acute, since it concerns the problem of manganese nodule genesis and the source of metals in them.

As was shown in the previous chapter, manganese behaviour in sea sediments depends upon the type of the basin, the lithological composition of sediments, and oxidation-reduction conditions; therefore we shall consider manganese behaviour in marine and oceanic oxidized and reduced sediments, respectively.

INLAND BASINS

Manganese occurrence in pore water was studied in the Baltic and Black Seas, Gulf of California, Saanich Inlet (Vancouver, Canada) and Loch Fyne (Scotland).

In the Baltic Sea depressions, where H_2S toxification of lower water horizons often happens, no oxidized film occurs at the surface of sediments. The manganese content in the pore waters of these depressions varies from 820 to 31800 μg/l (Table 46) according to various authors; whereas, in the surface sea water, it is no more than several micrograms per litre.

The Gulf of California is another case where no oxidized film at the surface of sediments was discovered during the period of sampling. In nine samples of the two sediment cores analysed there, the manganese concentration was 310 to 3850 μg/l, and in one sample from 5-20 cm horizon it was 33000 μg/l (Pushkina, 1980).

According to another series of determinations, manganese concentrations in the cores of fine-grained sediments decrease from 33-106μg/l in the upper layers to 12-17 μg/l in the 50 cm horizon (Brumsack and Gieskes, 1983).

In the pore water of Saanich Inlet the manganese concentration is considerably lower - from 40 to 100 μg/l (Presley et al., 1972).

In the reduced Baltic Sea sediments coated by an oxidized layer or film, manganese concentration in interstitial water varies from less than 40 to 80 μg/l in sands and silts near Klipeda and from 50 to 30500 μg/l in the sediments of the Riga Inlet. These variations are from 120 to 17000 μg/l in the pore water of the sediments taken from the main sea area.

The same range of manganese concentration is typical of the pore water in Loch Fyne, where it is 260 to 15220 μg/l (Calvert and Price, 1972).

Within the oxygenated zone of the Black Sea, the manganese concentration in pore water was determined both in the upper oxidized film and in the lower reduced sediments. In the samples of the first type, manganese content was 20 to 90 micrograms per litre, and it was 240 to 1760 micrograms per litre in the second type of samples, but for one: a concentration of 20 μg/l was determined in the 140-160 cm horizon (Volkov and Sevastyanov, 1968).

Manganese occurrence in pore water was also determined in sediments of many gulfs, bays and inlets (Chesapeake Bay, Long Island, U.S.A.; Funca Bay, Japan), river mouths and other shallow oceanic localities. Similarly to the previous cases, manganese concentrations in them vary from 30 to 2200 μg/l (Holdren et al., 1975;

TABLE 46

Mn content (μg/l) in pore water of sediments from inland basins

Basins	Types of sediments	Depth (m) from - to	Layer (cm) from - to	pH from - to	Eh from - to	Mn content from - to	Number of samples	References
Baltic Sea			Reduced sediments					
Gotland depression	Clayey mud	184-240	0-136	7.4-8.2	-230--70	820-20000	19	1, 2, 3
Landsort depression	"	465	12-168	6.8-6.9	-	11300-31800	4	3
Gulf of California	"	3260	0-310	7.3-7.7	-270--100	310-33000	10	4
Saanich Inlet	Clayey-diatomaceous ooze	200	0-250	7.6-8.0	-140--60	40-1000	21	5
		Reduced sediments overlayed by oxidized sediments						
Baltic Sea								
Klipeda region	Sands, silts	23-36	0-20	7.2-8.1	-160-+400	40-80	7	2
Riga Bay	Sand-silt-clayey sediments	30-56	0-418	7.3-8.3	-260-+150	50-30500	48	1, 2
Major sea area	"	70-194	0-820	7.2-8.3	-250-+320	120-17000	53	1, 2, 3
Loch Fyne	Sand-clayey	40-200	0-123	-	-	260-15220	30	6
Black Sea	Oxidized clayey	70-100	0-1.5	-	-	20-90	3	7
	Reduced clayey	70-100	2-290	-	-	240-1760	16	7

From (1) Hartmann, 1964; (2) Gorshkova, 1970; (3) Shishkina et al., 1981; (4) Pushkina, 1980; (5) Presley et al., 1972; (6) Calvert and Price, 1972; (7) Volkov and Sevastyanov, 1968.

Sanders, 1978; El Ghobary, 1982; Lyons and Fitzgerald, 1983; Uematsu and Tsunogai, 1983; Sundby and Silverberg, 1985).

PERIPHERAL ZONES OF THE OCEAN

Sediments of oceanic peripheral zones can be subdivided into three groups: namely, reduced sediments in zones of coastal upwelling; reduced sediments with an oxidized upper layer within shelves and continental slopes; and transitional sediments which are intermediate between reduced and completely oxidized sediments according to their physical and chemical parameters (Volkov et al., 1980; Rozanov et al., 1980). In this case, sediments with variable but mainly positive values of the oxidation-reduction potential were confined to the third type.

The reduced biogenous and biogenous-terrigenous sediments from coastal upwelling zones, which often have no oxidized film on them, are characterized by a low manganese concentration in the pore water (less than 1 to 93 µg/l); this pattern is generally regular and typical of all the three zones of large oceanic upwellings considered, namely, California, Peru and Benguela (Table 47).

The reduced sediments covered by an oxidized coating are mainly represented by terrigenous and terrigenous-calcareous varieties. The manganese concentration in the pore water of these sediments varies over a wide range - from less than 50 to 15000 µg/l.

Sediments of a transitional type, moderately to intensively oxidized, are also represented by mainly terrigenous and terrigenous-calcareous varieties. The manganese distributions in the pore water of these sediments are as irregular as in the reduced sediments, the range being almost similar - from less than 1 to 14500 µg/l.

Sediments from the northern seas such as the Barents Sea, Kara Sea, or Norwegian Sea belong to this category as well. Manganese concentrations in pore water of their sediments vary from 170-190 µg/l to 1-29 mg/l (Pavlova, 1982).

PELAGIC ZONES OF THE OCEAN

Manganese behaviour in the pore water of pelagic sediments is of special theoretical and practical concern since the area of their occurrence embraces the largest fields of manganese nodules. Pelagic calcareous, siliceous and polygenic sediments (red clay), though with a different set of rock-forming components, have some common features: low rates of sedimentations, low concentration of organic matter, high degree of oxidation.

The manganese concentration in pore water of pelagic calcareous clay varies from 0.35 to 192 µg/l (Table 48), according to various authors. The lowest manganese concentration (less than 1 µg/l) was determined in the pore water of the calcareous muds in the eastern equatorial Pacific (Callender and Bowser, 1980), the highest one (over 100 µg/l) being in pore water of the lower horizons of the same muds in the northern parts of the Atlantic and Indian oceans.

In pore water of the siliceous (mainly radiolarian and clayey radiolarian) sediments of the Pacific, the manganese concentration varies mainly within the range 0.4 to 6.2 µg/l and sometimes it reaches 150 µg/l in siliceous-clayey-calcareous mud of the south-eastern zone (Pogrebnyak and Krendelev, 1976), sometimes it goes down to 1-10 µg/l within the radiolarian belt (Schnier et al., 1978; Klinkhammer et al., 1982).

In the radiolarian ooze of the northern Indian ocean, from data by Zvolskiy et al., (1979), the manganese concentration in pore water turned out to be many times higher - from 160 to 12000 µg/l - which probably results from an insufficient accuracy of the technique applied.

Manganese concentration was also determined in pore waters of deep-sea red clays in the Pacific; it varies from 0.7 to 140 µg/l. The minimum concentration (0.7-3.3 µg/l) was recorded in the northeastern equatorial zone, its maximum (20-140 µg/l) being in the central part of the northern oceanic zone (Table 48). A higher manganese concentration in pore water of red clays (about 10 mg/l in the lower horizon of one column) was reported by Japanese researchers (Tsunogai and Kusakabe, 1982).

TABLE 47

Mn concentration (µg/l) in pore water of sediments from the peripheral oceanic zones

Region	Types of sediments	Depth (m) from-to	Layer (cm) from-to	pH from-to	Eh from-to	Mn content from-to	Number of samples	References
Reduced sediments of coastal upwelling zones								
The Pacific:								
Californian region	Black clay	567	0-45	7.8	- 90	< 1-3	8	1
Peru region	Biogenous-terrigenous sediments	70-1060	0-345	7.1-8.2	-	<25-75	14	2
The Atlantic, Namibia shelf	Diatomaceous ooze	75-120	0-218	7.3-7.9	-230--45	2-93	23	3
Reduced sediments of peripheral zones (coated by oxidized layer)								
The Pacific:								
Peru slope and trench; Californian region	Clayey and clay-calcareous-siliceous ooze	140-6250	0-555	7.3-7.9	-350-+110	140-13600	20	2, 4, 5
Eastern equatorial zone	Foraminiferal ooze	-	0-490	-	-	50-3690	19	6
The Atlantic, northern and southeastern part	Calcareous and terrigenous sediments	115-3328	0-682	7.2-7.9	-330--20	55-15000	63	3, 7
Indian ocean, northern part	Terrigenous sediments	2180-2680	0-745	7.18-7.59	-275-+350	240-11700	13	8
Arctic ocean	Clayey ooze	3800	0-288	7.95-8.53	-	50-9200	19	9
Sediments of the transitional type								
The Pacific, Californian and eastern equatorial regions	Terrigenous and biogenous ooze	3100-5160	0-1600	7.2-8.1	+130-+600	< 1-13200	181	1, 2, 4, 10
The Atlantic, south-eastern part and Puerto Rico trench	Terrigenous and calcareous sediments	3880-6510	0-810	7.3-7.7	+380-+530	< 1-14500	53	3, 10, 11
Indian ocean, northern part		3954-4855	0-830	7.4-8.9	- 35-+375	240-11700	25	8

From: (1)Presley et al., 1967; (2) Pogrebnyak and Krendelev, 1976; (3) Emelyanov et al., 1973; (4) Pushkina, 1980; (5) Rozanov et al., 1980; (6) Michard et al., 1974; (7) Bischoff and Ku, 1971; (8) Zvolsky et al., 1979; (9) Li et al., 1969; (10)Klinkhammer, 1980; (11) Addy et al., 1976.

TABLE 48

Concentration of manganese ($\mu g/l$) in pore water of oceanic pelagic sediments

Region	Type of sediments	Depth (m) from-to	Layer (cm) from-to	pH from-to	Eh from-to	Mn content from-to	Number of samples	References
			Calcareous sediments					
The Pacific: Eastern part	Calcareous and calc-clayey	3070-4430	0-650	6.6-7.9	+400-+550	25-100	49	1
Northeastern equatorial zone	Calcareous ooze	4380	0-23	7.18-7.75	-	< 0.7-3.3	10	2
"	Foraminiferal-cocco-lithic ooze	4788	0-20	-	-	6-25	6	3
", eastern part	Calcareous ooze	4430-4460	0-30	7.54-8.04	-	0.35-192	10	4
The Atlantic, northern part	"	4625-5320	0-10	-	-	27-143	4	5
Indian ocean, northern part	"	3980-4600	0-444	7.2-7.5	+200-+250	110-170	8	6
			Siliceous sediments					
The Pacific: Northeastern equatorial zone	Radiolarian and clay-ey-radiolarian ooze	4100-5214	0-977	7.13-8.05	+395-+640	0.4-62.4	218	2, 7, 8
"	"	4910	0-25	7.72-7.97	-	0.04-0.09	9	4
"	"	4675-5466	0-185	7.05-7.60	-	5-42	52	3
Southeastern part	Siliceous and silice-ous clayey-calca-reous ooze	3350-5030	0-150	6.9-7.8	+375-+490	25-150	12	1
Indian ocean, northern part	Radiolarian ooze	4250-5415	0-450	7.3-7.6	+250-+375	160-1200	10	6

Table 48 (continued)

Region	Type of sediments	Depth (m) from-to	Layer (cm) from-to	pH from-to	Eh from-to	Mn-content from-to	Number of samples	References
			Deep-oceanic red clays					
The Pacific: Northeastern equatorial zone	Clayey ooze	5000	0-62	-	-	8-100**	17	9
"	"	"	0-60	-	-	70-12100***	47	9
"	"	"	0-490	-	-	< 5-82	74	10
"	Clayey ooze with radiolarians	5150	0-20	7.50-7.71	-	0.7-3.3	10	2
"	Clayey ooze	4788-5830	10-425	7.20-7.68	-	10-61	26	3
Profile from Japan to California	"	3200-3550	0-430	6.7-7.6	+540-+610	20-140	30	11
Southeastern part	"	4200-5160	0-137	7.5-7.8	+560	25	5	1
The Atlantic, south-eastern part	"	5470	28-192	7.7	+420-+460	65-148	3	5

From: (1) Pogrebnyak and Krendelev, 1976; (2) Callender and Bowser, 1980; (3) Baturin et al., 1986; (4) Klinkhamer et al., 1982; (5) Emelyanov et al., 1973; (6) Zvolsky et al., 1979; (7) Gundlach et al., 1977; (8) Hartmann and Muller, 1982; (9) Raab, 1972; (10) Michard et al., 1974; (11) Pushkina, 1980.

** Samples were kept and processed at T=4°C.
*** Samples were kept and processed at T=22°C.

METALLIFEROUS SEDIMENTS

One can find numerous works devoted to metalliferous sediments; however, a few papers can be found describing manganese behaviour in the pore water of these sediments.

Manganese in the pore water of metalliferous sediments of the Pacific was investigated in the Bauer Depression and in the East Pacific Rise zone. In pore water within the Bauer depression, manganese concentration is 100 to 11000 µg/l according to one series of determinations (Bischoff and Sayles, 1972) and 25 to 84 µg/l from the other (Pogrebnyak and Krendelev, 1976). The second series of determinations was performed by a more precise method than the first.

In the pore water of metalliferous layers and inter-layers that occur in the cores of biogenic sediments, manganese concentration varies from 30-200 to 50-3200 µg/l according to various authors (Table 49, Figure 22). A higher manganese concentration was also recognized in the pore water of hydrothermal sediments in the Galapagos zone (Bender, 1983).

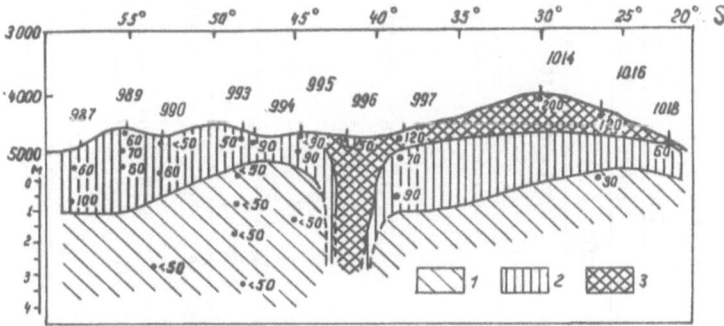

Fig. 22. Mn distribution in pore water of metalliferous sediments from the south-eastern Pacific along the 100°W profile (Shishkina et al., 1979). Mn concentration, µg/l: (1) less 50; (2) 50 to 100; (3) over 100.

In the pore water of ore sediments in the Red Sea, the manganese concentration turned to be a maximum when compared to all other types of marine and oceanic sediments; it is 50-191 mg/l in the Atlantis II Deep and 22-264 mg/l in the Discovery Deep.

GENERAL REVIEW OF THE DATA PRESENTED

At least three difficult points appear when summarizing the data collected over a 30-year period (Tables 46-49); they concern nomenclature of sediments, grading of their physical and chemical parameters and determination of manganese content in pore water.

Some works that contain data on pore water avoid mentioning the composition of the sediments analysed. In such cases, sediments were identified by the schemes of oceanic sedimentation given by Lisitsin (1974, 1978).

Physical and chemical parameters of sediments, primarily their oxidation-reduction potential (Eh), were used as the main criterion for the classification of transitional sedimentary types. However, in some works devoted to pore water, these data are absent, so the sediments were classified by general schemes of sediment distribution in the ocean, and by the available data on their C_{org} content or biogenous component occurrence in pore waters.

The Eh of oceanic sediments measured by platinum electrodes does not seem to give the real oxidation-reduction conditions of the medium (Vershinin et al., 1981)

TABLE 49

Mn content (μ g/l) in pore water of metalliferous and ore sediments

Region	Types of sediments	Depth (m) from-to	Layer (cm) from-to	pH from-to	Eh from-to	Mn content from-to	Number of samples	References
The Pacific								
Bauer Depression	Clayey and calcareous muds	–	0-900	7.48	+100-+200	100-11000	110	1
"	"	3090-4750	0-1020	6.7-8.0	+306-+610	25-84	37	2
East Pacific Rise, northern part	Iron clays	4200-4490	0-700	7.0-7.5	+580	50-3200	6	2
Zone of EPR	Metalliferous sedimentary layers in biogenic ooze	3070-5030	50-1220	6.7-7.7	+ 18-+472	600-2830	14	2
"	Metalliferous sediments	2830-4900	0-200	7.1-7.7	+550-+650	30-200	45	3
The Red Sea								
Atlantis II Deep	Ore sediments	2050	15-858	5.90-7.32	+290-+481	$50 \times 10^3 - 191 \times 10^3$	52	4
"	"	2050-2071	0-555	5.6-5.9	+ 50-+290	$80 \times 10^3 - 137, 5 \times 10^3$	21	5
Discovery Deep	"	2180	0-400	6.08-6.66	+230-+570	$22 \times 10^3 - 97 \times 10^3$	36	4
"	"	2205	0-1010	6.0-7.2	+ 70-+530	$29, 7 \times 10^3 - 264 \times 10^3$	14	5

From: (1) Bischoff and Sayles, 1972; (2) Pogrebnyak and Krendelev, 1976; (3) Shishkina et al., 1979; (4) Brooks et al., 1969; (5) Pushkina et al., 1981.

and does not always give consistent results. So, in the column for Site 670, near the coast of Mexico, the interval of Eh-variations is from -100 to +250 following Pushkina (1980) and from -350 to +100 according to Rozanov et al., (1980). In this case, the second result was used, since the sediments were coated by an oxidized film.

Clayey muds in the Puerto Rico trench with positive Eh do not seem to be red clays (Addy et al., 1976) but sediments of the transitional type.

The reliability of the analytical methods used to determine manganese content in pore water cannot always be considered with confidence; however, in general, they improve their sensitivity and precision. The sensitivity of the first determinations of manganese in pore waters was usually 300-500 µg/l (Gorshkova, 1970; Bischoff and Ku, 1971). In the other works it was 100 (Bischoff and Sayles, 1972; Zvolskiy et al., 1979); 20-40 (Pogrebnyak and Krendelev, 1976; Shishkina et al., 1979; Pushkina, 1980); 1-5 (Presley et al., 1967; Li et al., 1969; Emelyanov et al., 1973; Baturin et al., 1985); 0.4-1 (Callender and Bowser, 1930; Klinkhamer, 1980; Hartmann and Müller, 1982); and 0.02 µg/l (Klinkhammer et al., 1982).

Such factors as temperature, duration of compression and gas regime during the process of compression of sediments greatly affect the result of manganese concentration in pore water samples. It may increase by factors of ten at a temperature increase from 4 to 22°C, or when samples are kept for a long time before processing (Gorshkova, 1970; Raab, 1972); this makes it necessary to squeeze samples as soon as they are lifted to the surface and to process them in an inert-gas atmosphere using a cooling apparatus.

Despite the ambiguity caused by disadvantages of the method applied, the data given in the Tables show that the manganese content in pore water of various marine and oceanic sediments can fluctuate from 0.04 µg/l to 264 mg/l; that is, over seven orders of magnitude, and, in general, exceeds considerably the manganese content in sea and oceanic water (see Chapter II).

Considerable variations of manganese content in pore water are typical of each of the basins in question, of various parts of one and the same basin, and of various horizons in an individual sediment core, which present several distributional patterns for dissolved manganese: maximum in a surface layer, maximum in a sub-surface layer or in one of the deeper layers, irregular peak-like, and relatively monotonic distribution.

Pore waters of deep oceanic sedimentary horizons recovered by deep-sea drilling have been analysed less thoroughly, though a wide range of dissolved manganese concentrations was also recognized there (Manheim, 1976; Gieskes and Reese, 1980).

To interpret these data one should consider a complex set of interrelated factors, including lithological composition of sediments, concentration of manganese and forms of its occurrence in the solid phase, content and composition of organic matter, oxidation-reduction regime, diffusion and processes of mineral formation.

The lithological type of sediments affects manganese concentration in pore water only indirectly, since the range within which it varies is considerably wide in each type. Nevertheless, according to manganese behaviour in pore water, sediments can be subdivided into three groups, namely: sediments of inner basins and marginal oceanic zones; pelagic sediments of the ocean; and metalliferous sediments.

Pore waters of sediments of the first group have a wide range of manganese concentration, with numerous high values; pore waters of sediments of the second group have a relatively narrow range with predominantly low concentrations; and that of the third group have a wide range of concentration with extremely high values, this obviously being the result of the activity of hydrothermal springs located in the vicinity.

Manganese concentration in the solid phase of marine and oceanic sediments varies generally within the same wide range as in the liquid phase, from some thousandths of one percent to tens percent (see the previous Chapter). In other cases, these fluctuations in the solid and liquid phases follow the same pattern: on Namibia shelf, a low manganese content in diatomaceous ooze (0.01-0.02%) corresponds to its low content in pore waters; in some samples of metalliferous sediments, high manganese content in an ore phase corresponds to its high concentration in a solution. However, a dissonance between high concentration of manganese in sediments and its low concentration in pore water is distinctive in pelagic red clays. In some samples of metalliferous sediments, manganese concentration in pore water can also be lower than in terrigenous and biogenous deposits poor in manganese (Tables 47, 49)

In the cores of marine and oceanic hemipelagic sediments with contrasting physical and chemical parameters, an inverse relation between manganese content in the solid and liquid phases can be often observed: a manganese minimum in pore waters is confined to the oxidized interlayers rich in manganese, whereas its maximum is limited to the reduced interlayers of the solid phase with low manganese content.

An inverse correlation can also be recognized between manganese concentration in pore water and the relative content of reactive manganese in the host sediments, the latter increases in the ocean from the peripheral towards pelagic zones (along with a general increase of total manganese content), whereas the former decreases (as organic matter decreases and oxidation degree of sediments goes up). Thus, the higher the content of labile manganese in pelagic sediments (in a chemical aspect), the less mobile it is in fact and the more fixed it is in the solid phase, mainly in a tetravalent form. Therefore, the content of reactive manganese in sediments is to a great degree the indicator of its hydrogenic origin, rather than evidence of its actual mobility in highly oxidizing conditions of a pelagial.

The physical and chemical environment of a diagenesis system, controlled by the organic matter content in sediments, and by conditions of its transformation, seems to be the determining factor for manganese behaviour in pore water of marine and oceanic sediments.

The increase of manganese concentration in pore waters of sediments of inland basins and peripheral oceanic zones is controlled by the reaction

$$2 \, MnO_2 + C_{org} + 2H_2O = 2 \, Mn^{2+} + CO_2 + 4OH^-.$$

If an oxidized film is missing over reduced sediments, manganese can diffuse freely into bottom water which becomes rich in this element. Pore waters of completely reduced sediments from coastal upwelling zones have thus extremely low concentrations of manganese when compared with other sediments from peripheral oceanic regions.

Migration of dissolved manganese from sediments was observed, in particular, in the Baltic Sea, in the bottom waters of which manganese concentration increases sharply to several hundredths of micrograms per litre as stagnation occurs and the water is poisoned by hydrogen sulphide (Hartmann, 1964; Shishkina et al., 1981). Meanwhile, the manganese concentration in pore water of sediments from the Baltic Sea depressions remains high in contrast to that in sediments from the upwelling zones; this can be explained by the sporadic character of a short-period intoxication by hydrogen sulphide, high rate of sedimentation with respect to diffusion rate, low gradient of changes in physical and chemical parameters (pH and Eh) in near-bottom and pore waters, and high concentration of dissolved manganese in the bottommost stagnant water, which lowers or hinders manganese diffusion out of sediments.

If the reduced sediments are coated by an oxidized layer, the highest concentration of dissolved manganese is detected below the upper boundary of the reduced sediments and, thus, the neighbouring oxidized layer is enriched by manganese due to its falling out of solution (Hartmann, 1964; Manheim, 1976; Volkov et al., 1980; Shishkina et al., 1981).

A set of similar oxidation-reduction reactions occurs in sediments of a transitional type which contain considerable amounts of reactive organic matter.

In oceanic pelagic zones, the flux of organic matter from the water to the bottom is considerable (Suess, 1980). However, due to low sedimentation rates, oxidation-reduction reactions occur mainly within a surface film which affects considerably the degree of manganese oxidation (Kalhorn and Emerson, 1984; Murray et al., 1984).

Another factor that stimulates solubility of manganese in oceanic pelagic sediments is the lower pH (Savenko and Baturin, 1981), this is mainly related to abyssal red clays.

A mathematical model of manganese behaviour in diagenesis was simulated from the data on manganese concentration in solid and liquid phases of sediments from the eastern equatorial Atlantic (Burdige and Gieshes, 1983) - this model can probably be extrapolated to the other regions of the ocean.

The forms of occurrence of manganese in pore waters of marine and oceanic sediments have scarcely been analysed. However, in two cases, a correlation was observed between the content of dissolved manganese and organic carbon in pore water of oxidized oceanic sediments that provoked a supposition on the occurrence of manganese-organic complexes (Hartmann and Müller, 1974; Pogrebnyak and Krendelev, 1976).

Special investigations of manganese-rich and organic-rich pore water were carried out with reduced sediments of fjords and water reservoirs by ultra-filtration and high precision techniques; the bulk of manganese in solution is in an ionic form, whereas only 1-7% of the total manganese is associated with organic matter (Krom and Sholkovitz, 1978; Krasintseva et al., 1982). In oxidized oceanic sediments, this ratio has not been determined specifically.

The behaviour of manganese in pore water of the sediments that underlay and incorporate manganese nodules is of particular concern.

In the reduced sediments containing buried nodules, the content of dissolved manganese increases towards the location of nodules (Calvert and Price, 1972) which probably depends on their dissolution. In the oxidized oceanic pelagic sediments, manganese concentration is noticeably lower in places where nodules occur (Pogrebnyak and Krendelev, 1976; Gunglach et al., 1977; Schnier et al., 1978; Pushkina, 1980; Hartmann and Müller, 1982) which probably gives evidence of manganese scavenging by nodules. However, some cases were reported where manganese concentration was higher near nodules (Michard et al., 1974) which is difficult to interpret since no information on the physical and chemical parameters of the sediments was reported.

A diffusive flux of elements out of sediments is defined by diffusion coefficients and concentration gradients according to the formula:

$$Q = D \, \frac{\Delta c}{\Delta x} \, \Delta t S,$$

where Q is the flux of elements, D is the diffusion coefficient, Δc is the concentration gradient in pore and bottom water, Δx is the thickness of sediments considered, t is time, and S is the area (Manheim, 1976).

According to estimates by various authors, a diagenetic flux of manganese out of sediments to bottom water in a pelagial varies from 0.8 to 180 $\mu g/cm^2/1000$ yr (Table 50).

In oceanic peripheral zones, a flux of manganese out of sediments may be at least two orders higher which, in total, exceeds by many times the amount of dissolved manganese in river discharge.

From the data given in Table 50, a diagenetic flux of manganese out of pelagic sediments can be assessed as an average of 50-70 $\mu g/cm^2/1000$ yr. For oceanic peripheral zones this flow can be 5-10 and over, up to 100-1000 $mg/cm^2/1000$ yr (Trefrey and Presley, 1982; Sawlan and Murray, 1983; Shulkin and Bogdanova, 1984; Sandby and Silverberg, 1985).

If we assume that the pelagic zone of the ocean is 272.6 million km^2, and the peripheral zone is 87.6 million km^2 (Lisitsin, 1974), the yearly maximum potential fluxes of manganese out of these zones are 0.13 and 4.35 million tonnes, respectively. However, in fact, a diagenetic flux of manganese seems to be many times lower; in particular because the reduced coastal and hemipelagic sediments are often separated from bottom water by a surface oxidized film or layer that blocks manganese diffusion. However, the notably lower concentration of manganese that is often observed in the reduced sedimentary layer shows that terrigenous material can in fact lose up to 30% of its initial content of manganese in the process of diagenesis. This seems to be of especial interest, since the bulk of manganese that diffuses to bottom water out of shallow sediments can be supplied to the open ocean (Sundby et al., 1981).

TABLE 50

Mn diffusive fluxes from pelagic sediments to bottom water

Region	Types of sediments	Sedimentary layer (m)	Ratio of Mn concentrations in pore water versus bottom water		Diffusive flux µg/cm²/1000 yr	References
			Gradient, µg/l	Concentration coefficient		
Entire ocean	–	100	1000	1000	44	Manheim, 1976
"	–	20	–	–	70	Elderfield, 1976
Central Pacific	Siliceous-clayey	2	3.5	19	180	Hartmann and Muller, 1982
"	"	10	2.6	40	11-32	Callender and Bowser, 1980
"	Siliceous ooze	12	1.8	30	9-40	"
"	"	8	0.02	2	0.8	Klinkhamer et al., 1982
"	Calcareous ooze	1	1.4	23	124	Callender and Bowser, 1980

PART II

GEOCHEMISTRY OF MANGANESE NODULES

The geochemistry of manganese nodules is associated directly with the geochemistry of underlying and parent sediments, pore water, oceanic water, i.e. with the geochemistry of the ocean as a whole. Part 1 was devoted to the behaviour of a "key" constituent of nodules - manganese - in oceanic sedimentation. Part 2 describes the geochemistry of a nodule itself, which is the result of oceanic sedimentation and lithogenesis. The material considered in this part is given in the following order: (1) general information on distribution, morphology, structure and mineral composition of manganese nodules; (2) description of the main chemical composition of nodules - major rock-forming and ore elements; (3) problems of content, distribution and behaviour of rare and trace elements, subdivided according to the groups of Mendeleev's periodic system; (4) problems of dating, accretion rate and origin of manganese nodules.

CHAPTER VII

MANGANESE NODULES: DISTRIBUTION AND STRUCTURE

DISTRIBUTION OF NODULES

Manganese nodules are spread over all oceanic basins and in some sea basins, the greater part of them being in the Pacific, then in the Indian and Atlantic oceans, with total reserves in the Pacific of about 1.7×10^{12} tonnes (Mero, 1965).

Accumulation of nodules is most frequent in the northern near-equatorial zones and within three latitudinal belts in the southern hemisphere, namely, 15-20°, 30-40° and 50-60° (Table 51).

TABLE 51

Latitudinal distribution of manganese nodule abundances in the ocean (Andreev et al., 1984)

Ocean	Northern hemisphere	Southern hemisphere
Pacific	4-30°	20-46° 55-63°
Atlantic	18-35°	30-40° 50-60°
Indian	0-11°	15-20° 32-40°

To describe locations of nodules Andreev et al. (1984) proposed recognition of individual regions, fields, belts and megabelts of nodule occurrence, together with spatial density (percent of bottom area covered by nodules), and weight density (weight of nodules per square unit, kg/m^2).

Nodules of the first and smallest subdivision, i.e. of the region, have an individual pattern of geochemical composition. Fields of nodules are delineated by their relation to large morphostructures of the oceanic floor, the tectonic setting of the area, and by the geochemistry they share in common, together with the density of occurrence of nodules and their composition.

Fields, regions and individual areas of occurrence of nodules can be combined in the belts within one ocean, whereas belts of similar type in two or three oceans can form a megabelt, such as the Northern global megabelt in the latitudinal range of 4-30°N.

A general pattern of distribution of manganese nodules over the floor of the World Ocean is given in Figure 23. There are some other versions of maps and schemes for nodule distribution over the World Ocean or over its separate zones; they are becoming more specific as research, reconnaissance and survay improve (Mero, 1965; McKelvey and Wang, 1969; Skornyakova and Andruschenko, 1970; Fraser and Fisk, 1977; Rowson and Ryan, 1978; McKelvey et al., 1983; Manganese nodules..., 1984).

In the Pacific eight fields of manganese nodule occurrence can be delineated, five of which are parts of the Northern megabelt (Andreev et al., 1984).

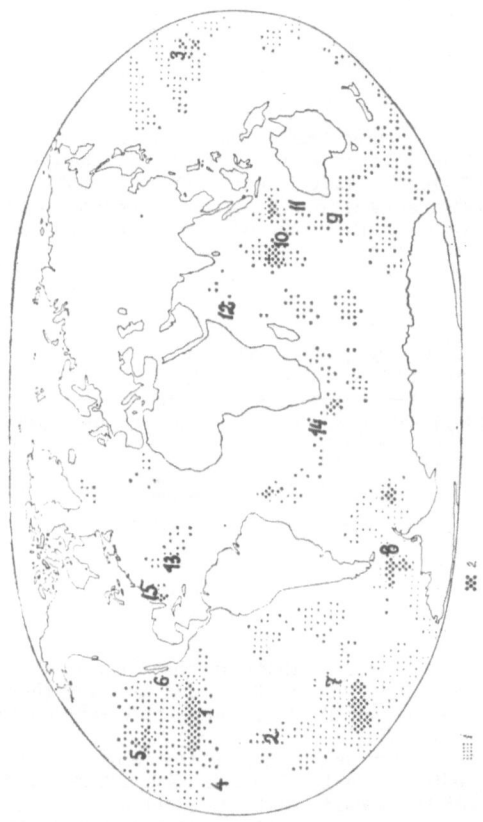

Fig. 23. Distribution of manganese nodules over the World Ocean (Cronan, 1980). Nodule fields (by Andreev et al., 1984); (1) Clarion-Clipperton; (2) South Pacific; (3) Wake-Necker; (4) Central Pacific; (5) Hawaii; (6) California; (7) Menard; (8) Drake Passage-Scotia Sea; (9) Diamantina; (10) Central Indian ocean; (11) West Australia; (12) Somali; (13) North America-Guyana; (14) Cape-Agulhas; (15) Blake Plateau.

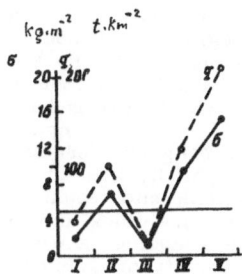

Fig. 24. Variations of nodule coverage, kg/m² (σ) and total metal resources, t/km²
(q) with respect to the bottom topography (Andreev et al., 1984).
(I) abyssal plain; (II) deep between hills; (III) rise of abyssal hills;
(IV) slopes of abyssal hills; (V) near summit part of abyssal hills.

(1) The sublatitudinal Clarion-Clipperton field extends for some 3500 km from
116°W in the east to the Line Ridge in the west and is bounded by the Clarion and
Clipperton fracture zones. The field is located in the centre of the Northeast Pacific
Basin at a depth of 5000-5200 m; the floor of the basin is an abyssal plain, compli-
cated by near-fault depressions and also by ranges of sea mountains and abyssal
hills. From data by various researchers, manganese nodules within the Clarion-
Clipperton field are distributed extremely irregularly (Horn et al., 1972; Frazer and
Fisk, 1977; Skornyakova et al., 1981).

In some zones of the western, central and eastern parts of the field, 145-157°,
130-144°, 120-127°W, respectively, nodules cover over 50% of the sea floor surface.

The weight density of nodules varies as a function of the sea-floor topography
within the range of 0.4 to 22.2 kg/m² (Figure 24). The density minimum is recorded
within the abyssal plain, at the foot of abyssal hills and sea mountains. The maximum

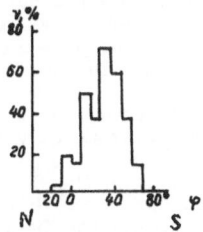

Fig. 25. Frequency of nodule occurrence (ν) in the Indian ocean as a function of
geographical latitude (φ) (Andreev et al., 1984).

density of nodules is confined to depressions between hills, to slopes and near-summit parts of hills. As a rule, increasing weight density of nodules causes a decrease
of the content of manganese, nickel and copper in them. Nodule distribution over
this field is thoroughly described in papers by Skornyakova (1984-1986) and
Skornyakova et al. (1984, 1985).

(2) The Central Pacific field is bounded by sea ridges Marcus-Necker in the
north, Line in the east, and Phoenix-Tokelau in the southwest. The ocean in the
eastern part of the field is as deep as 4800-5200 m, and 5600 m in its western part.
Hills and mountains are typical of most of the sea-floor topography within the field
area. The weight density and morphology of nodules within the field are extremely
variable; the former can be from 1 to 30 kg/m² (Mizuno and Chujo, 1975; Mizuno
and Moritani, 1977; Mizuno, 1981).

(3) The Wake-Necker field lies within the sub sea volcanic Marcus-Necker Ridge characterized by rugged sea-floor topography. No nodules occur at flat plains coated by turbidite sediments. In the regions of hilly topography, nodules are spread over the near-summit part of the hills, on their slopes and in between the hills. The distribution of nodules over the bottom is irregular: from 1-2 to 26 kg/m². The greatest perspective is the zone of southern ranges of the Marcus-Necker Ridge which has a slightly hilly topography. Manganese crusts also occur there.

(4) The Hawaiian field coincides spatially with the Hawaiian volcanic sea ridge which has a rugged topography of volcanic and tectonic origin. Bottom sediments in the Hawaii Archipelago surroundings are not thick and are composed of volcanic, calcareous and clay materials, manganese nodules and crusts that reach several centimeters in thickness. The weight density of nodules is 5-8 kg/m².

(5) The Californian field is bounded by the Clarion fracture in the north and northwest, and by the eastern flank of the East Pacific Rise. The depth is 3700-4800 m, the topography is rugged. In the western part of the field, abyssal red clays with zeolith and admixture of volcanic clastic material predominate; in the eastern abyssal part, the dominant species are hemipelagic slightly calcareous and slightly siliceous sediments; in the eastern coastal region, volcanic-terrigenous and biogenous-terrigenous sediments are widespread (Lisitsina and Butuzova, 1979). The weight density of nodules often exceeds 10 kg/m² (Rowson and Ryan, 1978).

(6) The South Pacific field is the greatest nodule field in the southern part of the ocean; the Tuamotu, Society, Line and Cook ridges, crowned by the islands, mark the centre of the field. Within the field the sea-floor topography is variable though bottom sediments are homogenous in general and are represented mainly by abyssal red clays. Nodules of 3-4 cm in diameter predominate in the central part of the field; in the marginal regions of the field they are larger - up to 6-8 cm. The maximum weight density of nodules reaches 40-70 kg/m². No definite correlation has been recognized yet between bottom topography and weight density of nodules within the field of interest.

In the northern part of the field, a depression with abundant nodules was outlined - Penrhyn Basin, located between the Manihiki and Line Rises. Nodules from Penrhyn, Line-Tuamotu and Rarotonga regions have certain differences in their composition (Landmesser et al., 1976; Pautot and Melguen, 1979; Glasby, 1981).

(7) Within the south-eastern part of the South Pacific Basin, the Menard nodule ore field extends northwestward for over 2000 km. In its northern part, nodules are rich in iron; in the southern part they are rich in manganese. The spatial density of nodules reaches 60% in some places (Glasby and Lawrence, 1974; Rawson and Ryan, 1978). East of that field, in the Chile Basin, an individual nodule area can be delineated.

(8) The field of the Drake Passage-Scotia Sea is confined to the Pacific-Atlantic boundary. Its northern border approximately coincides with the boundary of the area of iceberg occurrence.

Nodules are spread in abyssal regions in the Scotia Sea, on sea hills in the Drake Passage and over a continental slope. However, no correlation has been inferred yet between nodule distribution within the field and its bottom morphology, depths and composition of bottom sediments. Pebble and other coarse-grained material of glacial sedimentation form the core of the nodules.

The Bellinsgausen depression is one more area of nodule occurrence in the sub-Antarctic Pacific belt. There, nodules are mainly spread along its northern flank covering the belt of about 350 km along 60°S. Pebble of glacial origin and partly edaphogenic material form the core of the nodules (Watkins and Kennett, 1976).

Manganese nodules are also known in the Peru basin.

The Indian ocean. Manganese nodules are mainly spread over its southern part (15-50°) where a W-E sequence of fields is distinguished: Agulhas, Mozambique, Madagascar, Crozet, Masquaren, Central Indian, Western Australian, Cocos, Diamantina, Southwestern Australian, South Australian. Besides, the Australian-Antarctic area is recognized in the southern part of the ocean, and Arabian and Somali regions are distinguished in its northern part.

The general pattern of nodule distribution in the Indian ocean is controlled by the latitudinal and vertical zoning; it is illustrated by the graphs of nodule frequency occurrence as a function of latitude and depth to the bottom (Figure 25, 26), the latter varies from less than 2000 to 6780 m (Andreev et al., 1984).

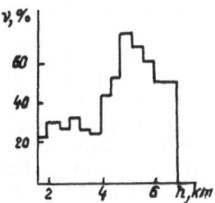

Fig. 26. Frequency of nodule occurrence (ν) in the Indian ocean as a function of depth (h) (Andreev et al., 1984).

In the Indian ocean four major nodule fields are delineated.

(1) The Somali nodule field is located in the central and eastern parts of the Somali basin. In the eastern part of the basin near the Masquaren Ridge the topography is regular; near the Arabian-Indian Ridge it is rugged. The relative altitude of individual sea mountains in the region reaches 3000-4000 m.

Within the field, a set of manganese formation types was determined. Nodules of a variety of shape, size and composition occur in the zones of calcareous foraminiferal ooze, pelagic red clays and siliceous radiolarian ooze. Manganese crusts and botryoidal aggregates of micronodules with a high content of cobalt are spread over the slopes of sea mountains and the Seychelles Ridge (Frazer and Wilson, 1980; Kachanov, 1980; Shnukov and Orlovskiy, 1980; Cronen and Moorby, 1981).

(2) The Central Indian field is located within the southern part of the Central Basin, south of 5°S. The bottom topography is considerably rugged in the greater part of the area. In the northern part of the field (15-18°S) nodules rest on siliceous radiolarian ooze, in the southern part they rest on over pelagic red clays, and in some places on calcareous coccolith and foraminiferal ooze. The weight density of nodules varies within a wide range, from 0.1 to 14 kg/m². In the northern part of the field where depths are 5100-5500 m, nodule composition is mainly manganese with a high concentration of base metals. Nodules from the zones of red clay and calcareous ooze are poor in these metals and have a monotonic composition (Frazer and Wilson, 1980; Levitan and Gordeev, 1981; Andreev et al., 1984).

(3) The West Australian-Cocos field bounds the submarine margin of Australia in the east, the submarine West Australian Ridge in the south, and the Ninetyeast Ridge in the west. In the north, it embraces a part of the Cocos Rise and Cocos Basin. Submeridional ridges of 99° and 100° are the major features of the field topography; west of these ridges a hilled plane is located, its topography being complicated by gradual swells and ranges of hills with relative altitude of 400-500 m; east of it an abyssal plain is situated. The zone between ridges has a complex rugged topography.

Within the field, biogenous siliceous sediments (in the northern part), pelagic red clays (in the southern part) and biogenous calcareous sediments (at depth less than 5000 m) are defined.

The spatial density of nodules is 20-70%; the weight density varies from 0.2 to over 50 kg/m². The maximum depth from which nodules were dredged is 6780 m. Manganese ore crusts up to 16 cm thick cover basic rock outcrops and blocks of basalt breccia on sea mountains (Skornyakova et al., 1979; Frazer and Wilson, 1979; Levitan and Gordeev, 1981).

(4) The Diamantina field is outlined in the eastern part of the Amsterdam Basin, southwest of Australia. From the north, it is framed by the Diamantina fault. Within the greater part of the area, bottom topography is relatively flat, the depth is mainly 3800-5000 m. Three zones are recognized within the field confined to its western, central and north-eastern regions. In the western zone nodules occur in the depressions at a depth of 3780-4240 m and their weight density varies from 0 to 24 kg/m²; in the central zone nodules are confined to depth of 4100-4300 m and their weight density varies from 0.8 to 5 kg/m²; the northeastern zone has a gradual hilly topography at a depth of 4020 to 5000 m. Nodules form a continuous coating with weight density of 5.2 to 54.5 kg/m² (Kennett and Watkins, 1975; Frakes, 1982; Andreev et al., 1984).

In the Atlantic three major nodule fields are recognized: the Blake Plateau, North American-Guyana, Cape-Agulhas.

(1) The Blake Plateau is located at a relatively gradual part of a continental slope northeast of Florida peninsula at a depth of 400-800 m. The area is some 5000 km². Loose sediments are almost absent over the plateau which results from a strong longshore current with velocity of to 30 cm/s. In the northern and north-western parts of the field, phosphorite nodules are spread; in its central part manganese crusts (often coating phosphorite) occur, they are about 7 cm thick on the average; in the southern and southeastern parts of the field, manganese nodules occur. Nodules were also discovered over the continental slope at a depth to 2700 m. Shell and limestone fragments often form the cores of the nodules (Pratt and McFarlin, 1966; Pratt and Manheim, 1967; Pratt, 1971).

(2) The North American-Guyana field embraces a number of geomorphologically different areas, namely, the Nereis abyssal plain, the region of abyssal hills west of the Mid-Atlantic Ridge and flanks of the Ridge itself, and also the southeastern slope of the Bermuda Rise. The field is bounded by the Atlantis and Muir fractures in the north and by the fracture of 17°40'N in the south. Sediments of three types are distinguished within the field: pelagic red clays, calcareous clays and calcareous oozes with over 50% of CaCO3. Some 52% of all nodules are confined to red clays, 43% to calcareous clays and only 5% to calcareous ooze. Meanwhile, nodules in red clays occur only in the regions with rugged topography and are virtually absent on an abyssal plain.

The weight density of nodules within the field is not great, as a rule, and does not exceed 3-5 kg/m² (Andreev et al., 1984).

(3) The Cape Agulhas field is located near the southwestern termination of Africa and is bound by the Walvis Ridge in the west and north-west; in the east, it is framed by the Agulhas uplift and Ridge. The bottom topography of the field is complicated and irregular due to the great number of sea mountains and ridges. Sediments are mainly calcareous with an admixture of terrigenous and volcanic material. Nodules and occasional manganese crusts are distributed irregularly over the field (Summerhayes and Willis, 1975).

Minor zones of nodule occurrence in the ocean are described by many authors (Mero, 1965; Cronan, 1980; Manganese nodules..., 1984).

A lot of works have also been devoted to manganese nodules and crusts in marine and lake basins. These works are of scientific value only since the content of valuable components in the nodules is low and their amount is relatively small (Samoilov and Titov, 1922; Samoilov and Gorshkova, 1924; Gorshkova, 1931, 1957, 1967; Manheim, 1965; Shterenberg et al., 1966, 1975, 1985; Volkov and Sevastyanov, 1968; Strakhov et al., 1968; Calvert and Price, 1970, 1977; Shterenberg, 1971; Blazhchishin et al., 1976; Kulesza-Owsikowska, 1979; Morten et al., 1980; Bostrom et al., 1982; Ingri, 1985; and others).

Nodules and nodule interlayers within oceanic sedimentary layers recovered by cores and deep-sea drilling are considerably less widely spread with respect to those in the surface layer and they have been studied less thoroughly. Information about them can be found in Menard (1964), Goodell (1965), Cronan and Tooms (1967a), Meylan (1968), Peterson et al. (1970), Horn et al. (1972), Kulm et al. (1972), Winterer et al. (1972), Meyer (1973), Skornyakova and Zenkevich (1976), Glasby (1978), Sorem et al. (1979), Shnyukov and Orlovskiy (1980), Bazilevskaya (1981), Skornyakova (1985), Svalnov and Novikova (1986) and many others.

MORPHOLOGY OF NODULES

Types of manganese formations at the oceanic bottom are extremely variable. They are semi-liquid ore mud of the Red Sea, loose or dense metalliferous deposits of the East Pacific Rise and of basal sedimentary layers revealed by deep-sea drilling, various micronodules and films in granulometric fractions of sediments, ore crusts and coatings on the surface of consolidated rocks and, finally, manganese nodules proper.

According to some concepts, nodule morphology is more or less a reflection of the conditions of nodule formation. Therefore, classification of nodules by their morphology is of particular importance.

The first classification of nodules was proposed by Murray and Renard (1891) who distinguished three morphological groups of nodules: irregular and pyramidal shape; spherical and ellipsoidal; and flat irregular.

Menard (1964) proposed another classification which recognized nodules, micronodules, plates and crusts.

Later, the classification began to consider such parameters as size and shape of nodules, composition and number of cores, thickness of ore crust, and mineral and chemical composition of ore material.

Grant (1967) classified nodules as follows: (1) nodules proper represented by rounded concretions with a single core the radius of which is less than one half of a nodule radius; (2) oval botryoidal poly-nodules; (3) crusts in which the lithogenic portion exceeds one half of the sample's volume; (4) agglomerates – clustering of cores of various form cemented by ore matter – their total volume exceeds one half of a sample's. Other classifications were also proposed (Goodell et al., 1971; Meyer, 1973; Meylan, 1974; Moritani et al., 1977; Halbach and Ozkara, 1979).

Nodules can be also classified according to their surface. It is a common belief that when nodules grow rapidly their surface is rough, whereas a smooth surface reflects slow growth of nodules or even erosion of the previously formed surface. In some oceanic regions, in particular in the Clarion-Clipperton zone, nodules have a ring-like swelling or equatorial belt that marks the zone of active growth at the bottom/water interface. In this case, a peculiar thing can be observed: nodules from the northern Pacific have a smooth upper and rough lower surface, whereas the inverse regularity is typical of nodules from the southern part of the ocean (Anikeeva et al., 1984), but this rule does have some exceptions.

Meylan's (1974) and Moritani's (Moritani et al., 1977) classification is the most common and convenient. It is based on the morphology and the type of nodule surface. According to this classification the major morphological types of nodules are: spheroidal (S), ellipsoidal (E), discoidal (D), tabular (T), polynodule (P), biomorphic (B), variegated, simulating the shape of clastic nuclei (V), irregular (I), crusts over bed rocks (C), and nodule pavement (NP).

To describe the surface texture of nodules the classification uses the following indices: smooth (s), rough (r), intensively rough (r'), botryoidal (b), upper surface is smooth, lower surface is rough (s/r), upper surface is rough, lower surface is smooth (r/s), cracked surface (c), fragments of nodules (f).

The following combinations are used to describe complicated forms of nodules: irregular-spheroidal (IS), spheroidal polinodule (SP). For instance, "Es" means that nodules are ellipsoidal with uniformly smooth surface, "PEcs/r" means polynodule ellipsoidal with cracked smooth upper surface and rough lower one.

This classification proved to be sufficiently effective when describing in detail the morphology of nodules. Most researchers use it widely (Raab and Meylan, 1977; Sorem et al., 1979; Halbach et al., 1981; Skornyakova, 1984).

The Meylan-Moritani classification was used to analyse the representative data on nodules collected in the Pacific by researchers from P.P. Shirshov Oceanology Institute, Ac. Sci. of the U.S.S.R., during the 28th cruise of R/V "Dmitry Mendeleev" (1982). The material was mainly sampled from three sites.

Site 2474 comprises 12x12 miles within the north-western periphery of the Clarion-Clipperton ore field (9°31.4°N, 152°40.3°W); the site embraces a series of latitudinal abyssal hills with minimum summit depths 4640-4900 m and gradual abyssal plain; 5200-5300 m deep surface sediments are semi-liquid radiolarian oozes and clayey radiolarian oozes that change for radiolarian marly oozes at a depth of less than 4900 m.

Site 2483 is 14x10 miles in the axial zone of the Clarion-Clipperton ore field (10°02.4°N, 146°29.4°W). The site includes a slightly-hilled plain with depths 5000-5250 m and a meridional depression with depths 5200-5250 m framed by a gradual swell-like uplifts with minimum depths 5050-5150 m: in the northwestern part of the site, an abyssal hill with minimum deptn 4885 m is located. Sediments are semi-liquid radiolarian clayey oozes rich in diatoms in depressions; below are Miocene mio- and eupelagic clays.

Site 2520 is 14x13 miles within the north-eastern part of the Central Basin (10°28°N, 175°07°W); bottom topography is rugged; sub-latitudinal abyssal hills lie at depths of 4760-5168 m; a flat abyssal plain has the depth of 5415-5456 m. Within the flat part of the site, at a depth over 5200 m the bottom is covered by radio-larian clayey oozes rich in diatoms underlain by mio- and eupelagic clays; in some places clays are covered by only a thin layer of ethmodiscus clays. These sediments change for siliceous-calcareous-argillaceous ones at a depth of 5100 m and then, in turn, for foraminiferal-coccolithic ooze.

The major morphological types of nodules spread over these polygons are given in Figure 27 (Skornyakova, 1984).

Within Site 2474 a similarity of morphogenetic types and of chemical and mineralogical composition was revealed for the nodules of abyssal hills and abyssal plain. Predominant on the surface of hills are clusters of flat, ellipsoidal, irregular in shape, discoidal and spheroidal nodules of 2-4 cm in diameter; in the plain part of the site, nodules of the same size (shape T, E, D, J, rare polynodules) occur. The greater part of the upper nodule surface, which contacts sea water, is s smooth, the lower part is microporous, rough, composed of globules to 1 mm or of microdendrites. Larger nodules (shapes E, T, D) are asymmetric: the thick upper shell has a convex surface, the lower surface is flat or concave. Nodules of these types are believed to be mainly hydrogenous.

Within site 2483, nodules with mainly globular surfaces are widespread. On an abyssal hill, clusters of small disco- and ellipsoidal nodules with fine-laminated structure and smooth upper and microrough lower surfaces were recognized as analoges of nodules from site 2474. At a slightly hilled plain, nodules of the shape E, D, T, rare S and B reaching 1 to 10 cm occur. The largest - 6-10 cm - are nodules of E, T and B shape. The majority of nodules have a large-globular surface. Large nodules - about 6 cm in diameter - have smooth upper and micro-rough globular lower surfaces and a distinct globular-dendritic equatorial belt that marks the position of a nodule at the water/sediment interface. The nodules are considered to be hydrogenous-diagenetic, being formed due to manganese supply from sea water and also from the underlying sediments. Within site 2520 two types of nodules were recognized, namely, those with a smooth upper and a micro-rough lower surface (hydrogenous) and those with a macro-globular surface (diagenetic). Mainly large (4-6 cm in diameter) spheroidal (S), ellipsoidal (E), discoidal (D), tabular (T). plate-like (P), isometric (V-crust like nodules with basalt cores), biomorphic (B) nodules cover the greater part of the site (Skornyakova, 1984, 1985).

The data on manganese nodules have also been classified by morphological types for the entire World Ocean (Anikeeva et al., 1984).

On the basis of this research, four major morphogenetic types of nodules can be distinguished in the northern near-equatorial part of the Pacific. They are: (1) irregular-spheroidal, polynucleus with rounded angles and facets; their surface is black, smooth, sometimes shagreen and uneven from below; the predominating size is 2 to 4 cm; (2) ellipsoidal and discoidal mainly single-cored, with equatorial belt, smooth upper and rough lower surface; their size is 2 to 10 cm; (3) plate-like with plate nucleus; rough upper and smooth lower coating, size being 0.5 to 8 cm in diameter; (4) crusts from tenths of a millimeter to 2 cm thick at outcrops and eluvium of bed rocks.

In the southern near-equatorial part of the Pacific, seven major nodule types are distinguished: (1) large ellipsoidal with a single nucleus; brown-gray with rough granular upper and smooth lower surface, size from 6 to 10 cm; (2) spheroidal brown-black with even rough surface, size 4 to 7 cm; (3) irregular spheroidal brown-black with even rough surface, size 2 to 4 cm; (4) discoidal brown-black nodules with equatorial belt that separates the upper rough and lower smooth surfaces, size 3 to 6 cm; (5) polynodules, often forming brown botryoidal agglomerations, evenly rough, size 1 to 2 cm; (6) nodule pavement formed due to coating of closely packed nodules by a common ore crust; the size of these plates varies from 40 to 50 cm in diameter, the thickness being 5-6 cm; (7) plates and tables at dense pelagic muds; brown or red-brown; their size is 4 to 10 cm in dia-meter, the thickness being 1 to 2 cm.

The density of nodules of various morphological types varies in this zone from 1.29 to 1.80 or more, often from 1.39 to 1.75 g/cm^2, their porosity being 32-40% on average. A correlation was established between porosity and density of nodules and their morphological type (Figure 28).

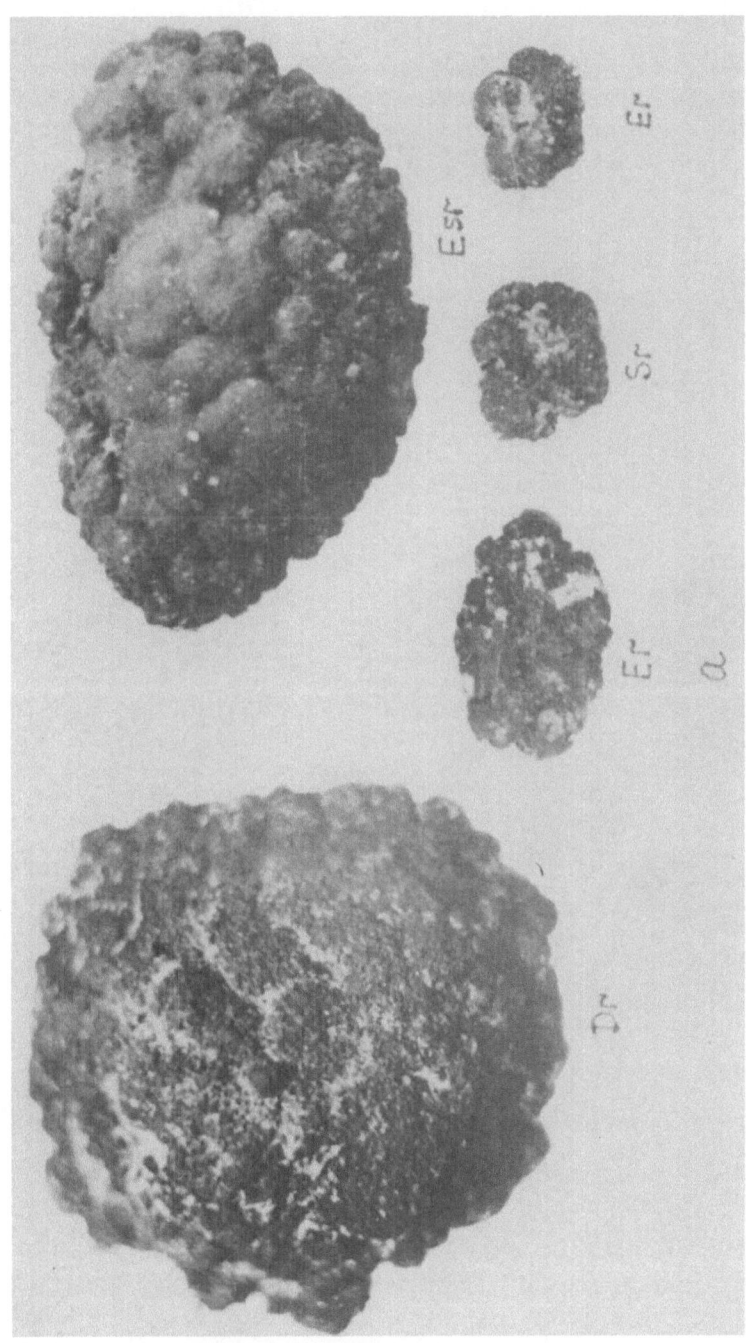

Fig. 27(a).

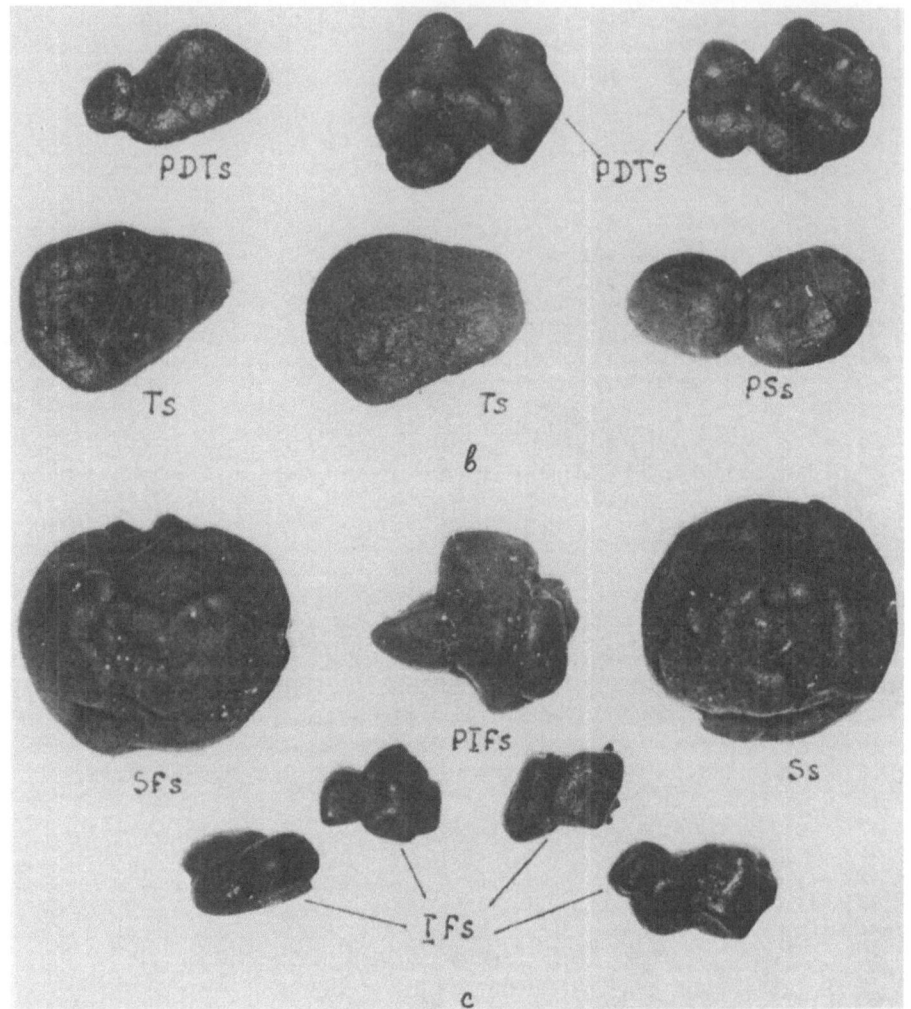

Fig. 27 (c)

Fig. 27. Morphological types of manganese nodules (Skornyakova, 1984): (a) Site 2483;
(b) Site 2474; (c) Site 2520.

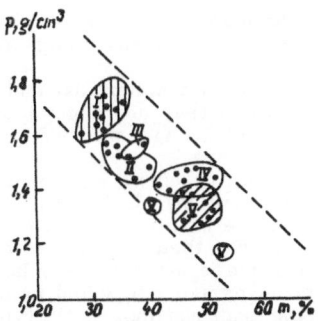

Fig. 28. Dependence of porosity (m) and density (p) of the South Pacific nodules
upon their morphological type (Anikeeva et al., 1984). (I) spheroidal (ideal
spherules); (II) ellipsoidal (potato-like); (III) irregular-spheroidal;
(IV) polynodules; (V) plates and crusts at dense clays.

In the Indian ocean, the following nodule types occur.
    In the Somali and Arabian basins, near the Arabian-Indian Ridge flanks,
spheroidal and pear-like or ellipsoidal nodules predominate; their size being 2-8 cm.
In the Somali field, in particular at the summits and slopes of sea mountains, ore
crusts often cover fragments of basalts and tuffs. The surface of these crusts is
usually loose and has a micro-nodular structure.
    In the Central Basin, nodules and crusts are widespread. Nodules have extreme-
ly variable sizes and shapes. Nodules of 2-4 cm with irregular knot-like shapes and
porous cavity-like surface predominate; they often form botryoidal clusters and
blocks. Crusts are well developed on the pieces of cemented radiolarian ooze (opalolite) coating them from one or several sides; sometimes thin layers of ore material
alternate with opalolite. Swells of rounded manganese aggregates up to 3.5 cm in
size are abundant on the upper surface of the crusts. These formations often reach
80-90 cm in diameter and can be classified as block nodules.
    In the Western Australia Basin, over 3000 manganese formations were analysed
in their morphometrical aspect during the site experiments; Figure 29 shows the

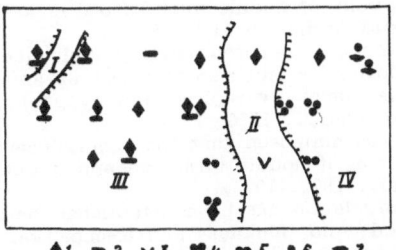

Fig. 29. Distribution of the major nodule morphotypes within the West Australia basin
(Anikeeva et al., 1984). (I) Cocos ridge extensions; (II) Ninety East Ridge
segment; (III) and (IV) deep oceanic basins. Nodules: (1) angular-rounded;
(2) tabular; (3) crustal; (4) botryoidal; (5) polynodules; (6) spheroidal;
(7) ellipsoidal.

morphological types recognized during the analysis: spheroidal, ellipsoidal, angular-
rounded, angular, clusters, botryoidal, tabular and crustal-types.
    Within the Diamantina ore field, mainly spheroidal nodules of two sub-types are
found: those with botryoidal and smooth surface. The size of the first sub-type is
3-10 cm, and of the second 0.5-2 cm.
    A morphometrical analysis of nodules was also made for the northern near-
equatorial regions of the Atlantic.

In the North American Basin, predominant types are rounded, ellipsoidal and elongated-ellipsoidal nodules with smooth or evenly rough surface. Their size is 2-4 cm.

Within the submarine Sierra Leone Rise spheroidal nodules occur. Sometimes they have knot-like swellings, sometimes they are elongated due to coupling of two or more nodules. Their surface is evenly rough and partly porous, the prevailing size is 1 to 3 cm.

In the Cape Verde Basin, besides the above mentioned types, one can find flat tabular nodules with knobby swells; their size is 5-7 cm.

Nodules from the Blake Plateau are mainly rounded or slightly elongated, rarely flat, with smooth lustered black surface; their size is 2-4 cm.

In general, nodules from the Atlantic are relatively homogeneous in their shape and size. Rounded, ellipsoidal and elongated ellipsoidal nodules with evenly rough and partly knobby surface form the main morphogenetic type. Figure 30 illustrates some typical examples of the upper, lower and side surface of nodules taken from the Clarion-Clipperton ore field.

INTERNAL STRUCTURE OF NODULES

Murray and Renard (1891) were the first to report on the inner structure of manganese nodules. They visually observed the concentric lamination or layering - the main feature of the nodules' texture. Nowadays, besides a macroscopic description of nodule texture, methods of microscopic and ultramicroscopic analysis are widely used.

Nuclei of Nodules

The occurrence of nuclei in nodules in many oceanic regions is one of the main features of their inner structure. Nuclei are represented by micro-nodules or fragments of ancient nodules in one case, and in another by lithogenic materials of various composition that occur in host deposits or in the underlying solid substrate.

Biogenic matter may form nodule nuclei as well. Within abyssal plains it might be teeth of sharks and ear bones of cetaceans; on the Blake Plateau, it is often molluscs' tests. If it is a sedimentary material that forms nodule nuclei, it may be dense or relatively loose biogenic calcareous ooze, siliceous ooze, or red clay, related to host sediments in their composition, but impregnated by ore matter to some degree. One of the descriptions of nodule nuclei from various regions of the Pacific is given by Skornyakova and Andruschenko (1976).

At the summits and slopes of sea mountains, nodule nuclei are formed from fragments of basalt, tuff, hyaloclastics and also of limestone and phosphorite. In the near-Antarctic zone, nodule nuclei are represented by pebbles carried by icebergs (Mero, 1965; Manganese nodules..., 1976, 1984).

In this connection it was supposed that the occurrence of a nucleation centre is one of the essential conditions of nodule formation and it determines their shape, size and composition (Horn et al., 1972, 1973).

As yet, no definite correlation has been established between the nucleus composition and size, on one hand, and thickness and composition of nodule ore coating on the other. In many cases, one cannot recognize any nucleus or its relict in a nodule.

Structure and Texture of Nodules

Many works are devoted to structural and textural properties of nodules determined by macro- and microscopic analyses (Andruschenko and Skornyakova, 1967, 1969; Sorem, 1967, 1973; Skornyakova and Andruschenko, 1968, 1970, 1971, 1976; Cronan and Tooms, 1968; Friedrich et al., 1969; Foster, 1970; Sorem and Foster, 1972, 1979; Durnham and Glasby, 1974; Heye, 1975; Andruschenko, 1976; Sorem and Fewkes, 1977, 1979; Skornyakova, 1979; Sorem et al., 1979; Usui, 1979; Halbach et al., 1981; Cronan, 1980; Glasby et al., 1982; Skornyakova, 1985, 1986; etc.). Since, in the World Ocean, nodule structure and textures are extremely variable, a uniform classification is still lacking. Some features of nodule structure and texture are similar in various parts of the ocean, though often they are called by different terms.

Fig. 30(b)

Fig. 30 (a)

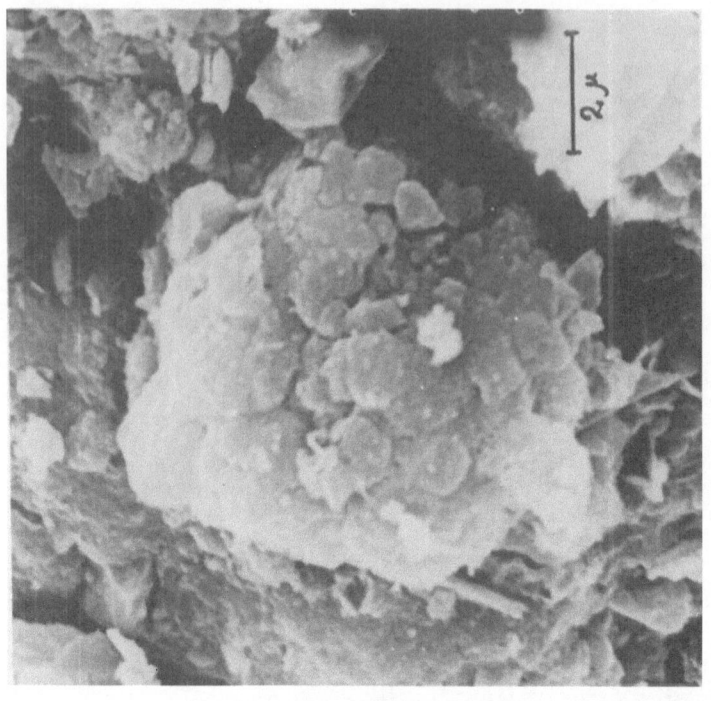

Fig. 30(d)

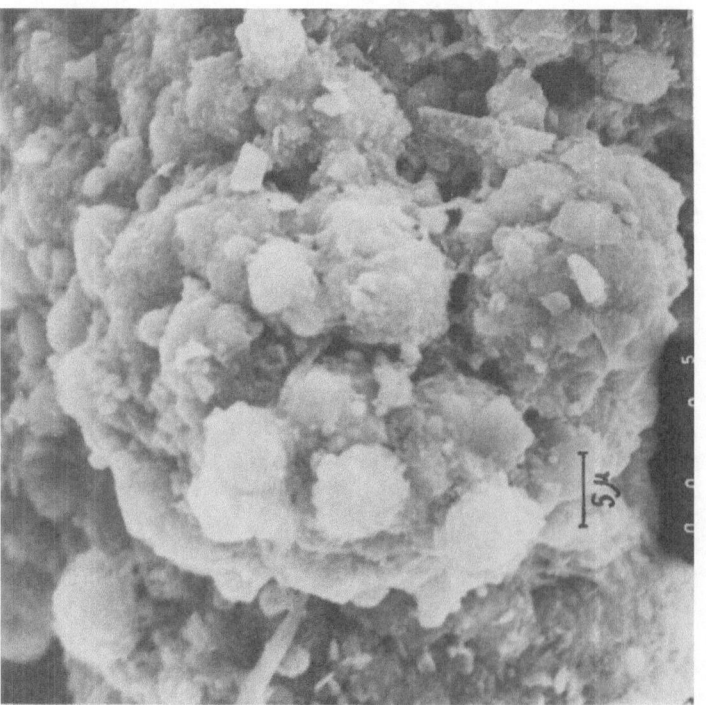

Fig. 30(c)

Fig. 30. Surficial structure of nodules from the Pacific radiolarian zone. (a) globular-
cavernous structure of nodule tops. Magnification: 90; (b) surface of an
individual globule. Magnification: 900; (c) same. Magnification: 1800;
(d) same. Magnification: 7000.

One of the first classifications of manganese nodule structure was attempted by Andruschenko and Skornyakova (1967, 1969) who described concentric, laminated, parallel laminated and shell laminated textures and globular assemblages.

Another group of researchers (Foster, 1970; Sorem and Foster, 1972; Sorem, 1973; Sorem and Fewkes, 1977) proposed distinguishing five zones according to structural-textural parameters determining the ore matter of nodules. They are: massive, mottled, compact, columnar and laminated.

Massive and mottled zones contain a considerable proportion of crystalline material. The massive zone is characterized by a diffuse lamination because of intergrowth of manganese minerals, and only minor admixture of clastic minerals. From the results of microprobe analysis, the concentration of major metals is as follows (in %): 32 Mn, 4 Fe, 2 Ni, 0.8 Cu. In some places, manganese concentration reaches 60%, nickel 7% and copper 2%.

The mottled zone has an irregular alternating pattern of crystallized manganese and amorphous iron phases, irregular and discontinuous lamination, chaotic distribution of structural elements, cavernous texture. This zone contains approximately 15% of argillaceous and amorphous material, 21% Mn, 12% Fe, 1% Ni and 0.5% Cu.

The structure of a compact zone is similar to that of the massive zone, but is composed of X-ray amorphous laminae with high reflectivity. Small lenses of crystallomorphic ore material can be also recorded there, but are rare. The compact zone contains approximately 19% Mn, 17% Fe, 0.6% Ni and 0.2% Cu.

The columnar zone consists of radially oriented columns of laminated X-ray amorphous material that forms a series of swells. Lighter laminae seem to be composed of manganese, the darker ones being of iron. The intervals between columns are filled with clayey material.

The laminated zone is considered as a modification of a columnar zone and originates when columns are shortened, coupled and intergrown, whereas thin interlayers are intergrated. Both zones have a similar composition: 19% Mn, 16% Fe, 0.4% Ni, 0.25% Cu.

Investigating the structure of manganese nodules in the Pacific, Andruschenko (1976) proposed distinguishing the following textural types: (1) coarse-layered, parallel layered, concentric layered and collomorphic; (2) dendrite forming tree-like and plumage aggregates; (3) globular and globular-collomorphic; (4) organogenic associated with biogenic remains in nodules (phyto- and zoo-plankton, fish teeth, etc.); (5) crustification typical of ore impregnated sediments; (6) cataclastic formed of nodule fragments cemented by ore material of a later generation.

Generalizing the vast information on the inner structure of manganese nodules, Cronan (1980) distinguished as major textures: laminated, collomorphic globular, shell laminated, columnar and mottled.

Analysing the material collected in several expeditions in the northern near-equatorial Pacific, Anikeeva and co-authors (1984) recognized four major types of nodules by their structural-textural parameters:

(1) Globular and globular massive structures and concentric laminated, rarely homogenous (isotropic) texture. The ore material consists of dense opaque black globules of 0.1-0.3 mm, condensed as a single mass; the globular surface is rough with mamillae entering the neighbouring globules; an inter-globular space is filled with clay making a small part of the total rock volume.

(2) Homogeneous structure and massive texture. Ore material represents a continuous opaque black mass with minor intrusions of clay matter and crypto-crystalline calcite that forms spherical concretions of 0.1-0.3 mm.

(3) Collomorphic and dendritic structure with coarse parallel laminated and concentric laminated texture. This is typical of iron-rich ore crusts composed of alternating layers of iron and manganese hydrooxides. Brown iron-rich layers are 0.1-1.5 mm thick, dark black manganese layers are less than 0.1-0.3 mm thick. In a transmitted light, manganese layers reveal a concentric zonal structure. Ore matter is crimped collomorphic plicated. The plicated layers of 0.5-1.5 mm alternate with linear laminated layers that have globular structure and are often composed of manganese hydrooxides. The layer-to-layer interfaces are powdered by cryptocrystalline carbonate or clay particles of 0.1-0.2 mm in diameter, brown because of iron hydrooxides.

(4) Relict hyaline structure and porous texture. This is typical of the nodules from the Shatsky Rise. A black amorphous ore mass impregnates a glassy matrix of

considerably porous hyalobasalt. The ore matter is evenly spread over a glassy mass and fills its cracks and cavities. These transformations are typical of highly altered nodules composed of porous glassy effusives, with plagioclase phenocrysts and grains of dark minerals being preserved.

According to Skornyakova (1985), the following types of structure and texture of manganese nodules are predominant in the central part of the Pacific, including the Clarion-Clipperton ore field.

In hydrogenous nodules: a fine-concentric laminated ore coating due to alternation of ore and clay layers. Beneath the outer shell, ore matter often has a globular structure composed by rounded or elongated globules of 0.2-0.4 mm in diameter which have fine-concentric-laminated microtexture. At the contact with the nucleus the ore matter has a collomorphic structure with slightly waved or plicated texture.

Dense ore crusts have a coarse-laminated, rarely fine-parallel-laminated or columnar texture, whereas loose ore crusts have a stalagmite porous texture.

Diagenetic nodules have a coarse-laminated texture due to alternation of layers of different thickness and composition that have a globular-dendritic or globular-collomorphic structure. The same structural elements may form a non-laminated mottled texture in a nodule body.

The cited review demonstrates that, despite the absence of a uniform nomenclature in describing structural and textural characteristics of nodules, a number of common features can be recognized in their structure with respect to different oceanic regions. Each of these features is determined by the aggregate state and mineral composition of ore, detrital and other phases determining both structure and chemistry of nodules.

Examples of the internal structure of nodules are given in Figures 31-34.

Ultramicroscopic Structure of Nodules

Results of the study of manganese nodules by electron and scanning microscopes at various magnifications are given in numerous works (Fewkes, 1973; Margolis and Glasby, 1973; Woo, 1973; Andruschenko, 1976; Fleischmann and Von Heimendal, 1977; Burns and Burns, 1978a, b, c, 1979a; Chukhrov et al., 1978a, b, 1979a, b, 1983a, b; etc.). The main goal of all these investigations was rather to decipher the mineral composition of nodules than to describe their structural and textural features. Yet, no attempt has been made to elaborate a classification of ultramicroscopic structures of nodules. It became more complicated as the resolution capacity of the microscopes grew and a greater variety of textural and structural types became revealed with respect to the same samples analysed at lower magnification.

The most common features of the ultramicroscopic structure of ore matter in nodules are globular, laminated, blocky collomorphic, fibric forms that, despite their outer similarity, might have different mineral composition. A set of these forms as individual features of macro- and microscopic structure of nodules may occur in different parts of the World Ocean.

A great variety of ultramicroscopic structures of ore matter is demonstrated by the nodules from the Antarctic part of the Pacific where a series of globular, laminated and fibric structures is recognized (Figures 33, 34).

MINERALOGY OF NODULES

The first fundamental works on the mineralogy of manganese nodules were the papers by Buser and Grütter (1956, 1957). They distinguished three minerals of manganese, namely, 10Å-manganite, 7Å-manganite and δ-MnO$_2$. X-ray diffraction patterns show the following reflections: 10Å-manganite -9.7; 4.8; 2.43; 1.42 Å; 7Å-manganite -7.2; 3.6; 2.4; 1.4 Å; δ-MnO$_2$ - 2.4 and 1.4 Å. In the following years, several investigations were performed that verified these results in principle and revealed a number of new mineral phases and confirmed the composition of those defined earlier (Manheim, 1965; Burns and Fuerstenau, 1966; Andruschenko and Skornyakova, 1967, 1969; Barnes, 1967; Cronan and Tooms, 1967, 1969; Sorem, 1967; Grill and Murray, 1968; Glasby, 1972; Ostwald and Frazer, 1973; Skornyakova et al., 1975; Halbach et al., 1975; Andruschenko, 1976; Bazilevskaya, 1976; Chukhrov et al., 1976; Burns and Burns, 1977; Glover, 1977; Sorem and Fewkes, 1977, 1980; Sterenberg, 1978;

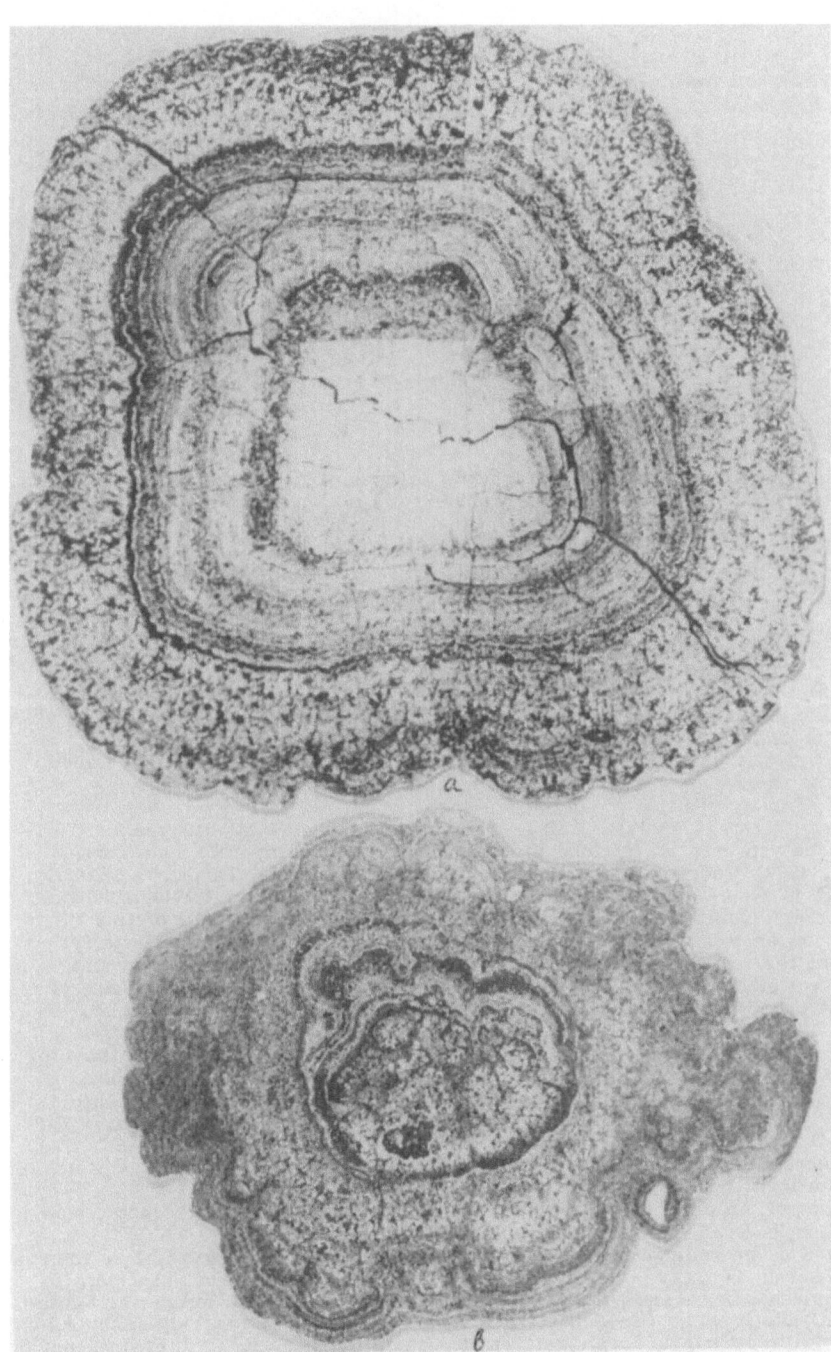

Fig. 31(a)

Fig. 31(b)

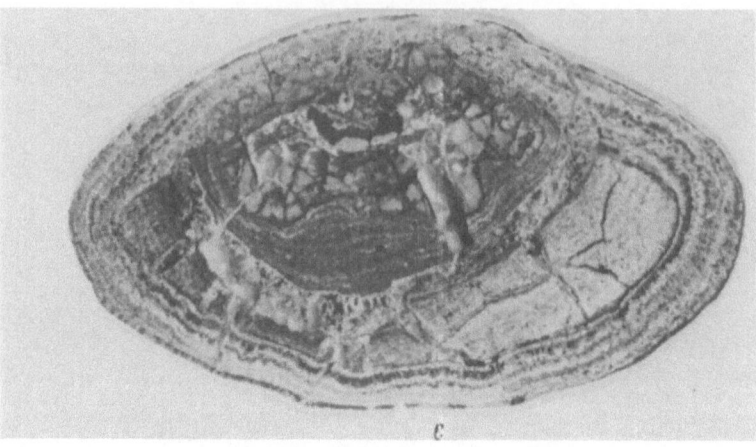

Fig. 31(c)

Fig. 31. Inner structure of nodules: fragments of ancient nodules in their nuclei,
        coarse and fine lamination of ore shells. (a) nodules from the Shatsky Rise.
        Magnification: 3; (b) nodules from the Pacific radiolarian zone, site 2483.
        Magnification: 1.8; (c) same, site 2474. Magnification: 3.7.

Chukhrov et al., 1978a, b, 1979a-d, 1980, 1981b, 1983a, b; Burns and Burns,
1978a, b, c, 1979a, b, c; Johnston and Glasby, 1978; Lonsdale et al., 1980; Potter
and Rossman, 1979; Usui, 1979a, b; Volkov et al., 1980; Friedrich and Schmitz,
1980; Halbach et al., 1981, 1982; Piper and Williamson, 1981; Turner and Buseck,
1981; Piper and Blueford, 1982; Turner et al., 1982; Cronan, 1980; Chudaev et al.,
1983; Burns et al., 1983; Lallier-Verdes and Clinard, 1983; Siegel and Turner,
1983; Dymond et al., 1984; Dritz et al., 1985; Chudaev, 1986). The most detailed
analyses of the mineralogy of nodules were made by R. and P. Burns and
Chukhrov and his co-authors.
        According to recent analyses, 10 Å -manganite, identified by Buser and
Grütter, can be represented by four diverse minerals: by manganite (or buserite
by Giovanoli, 1980), todorokite, asbolan and laminated mineral - asbolan - buserite;
7Å -manganite is assumed as birnessite analog identified earlier in fluvioglacial deposits in Scotland (Jones and Milne, 1956). However, lately, the right of this
mineral phase to an independent existence is doubted, as it might be an artefact
- a product of the transformation of todorokite with a low content of nickel and
copper (Dymond et al., 1984). δ - MnO_2 is recognized as an independent mineral
phase; it was proposed that it be called vernadite (Chukhrov et al., 1978a, b,
1979a, 1980, 1981b, 1983a, b; Burns et al., 1983).
        Besides these major mineral phases of manganese, nodules contain a number of
other minerals, though in subordinate amounts, namely, pyrolusite, ramsdellite,
nsutite, psilomelane, chalcophanite, rancieïte, manganosite (Andruschenko, 1976;
Burns and Burns, 1977, 1979; Chukhrov et al., 1979b; Potter and Rossman, 1979;
Baturin and Dubinchuk, 1984a; etc.). The basic parameters of the major and
accessory minerals of manganese in oceanic nodules are given in Table 52; results
of manganosite identification are given in Table 53.
        The bulk portion and, in most cases, the prevailing portion of the ore compo-
nent of nodules is represented by X-ray amorphous material.
        Iron minerals in manganese nodules have been studied considerably less than
manganese minerals (Murray, 1979; Chukhrov et al., 1980). The most important of
them are: maghemite (γ- Fe_2O_3)detected in nodules of the South Pacific, the Scotia
Sea and the Drake Passage (Goodell et al., 1971); magnetite (Fe_3O_4) determined by
thermomagnetic properties in fresh water and oceanic nodules; its amount does not
exceed 0.1-0.6% of the total rock mass (Carpenter et al., 1972); ferri-hydrite
(5 Fe_2O_3 x 9 H_2O) presumed as having been determined (Calvert, 1978); goethite

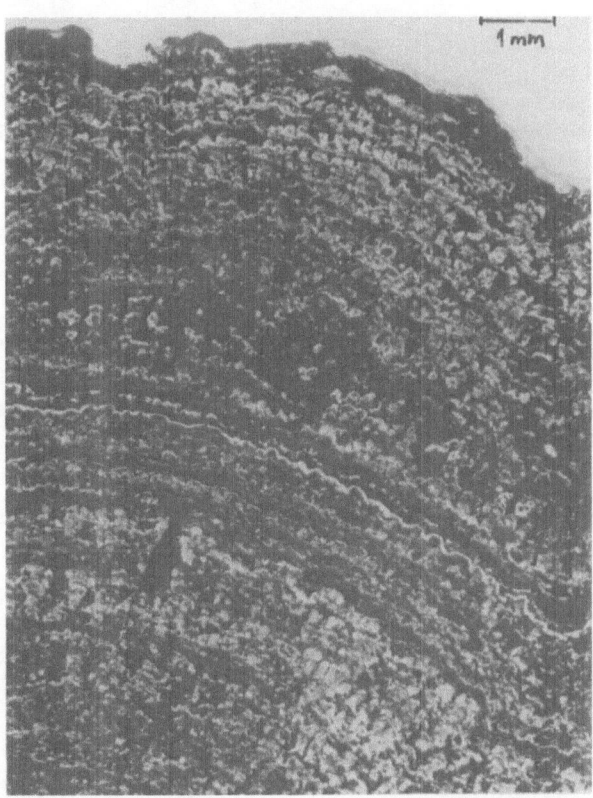

Fig. 32(a)

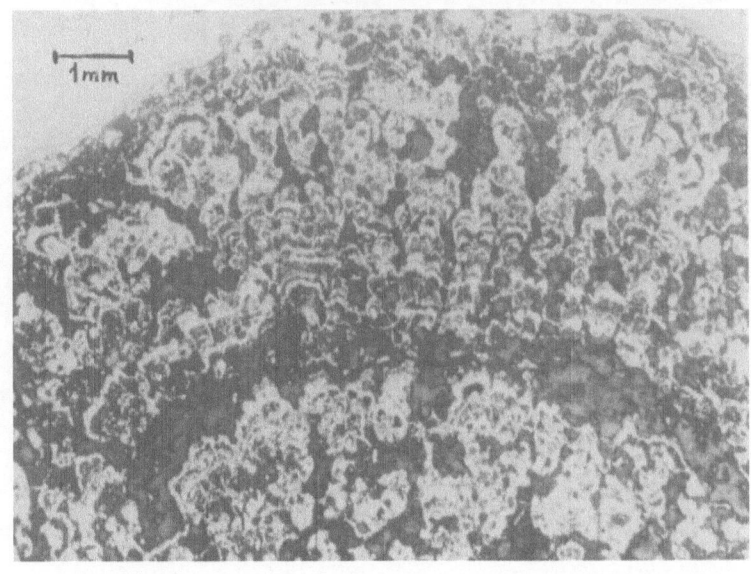

Fig. 32(b)

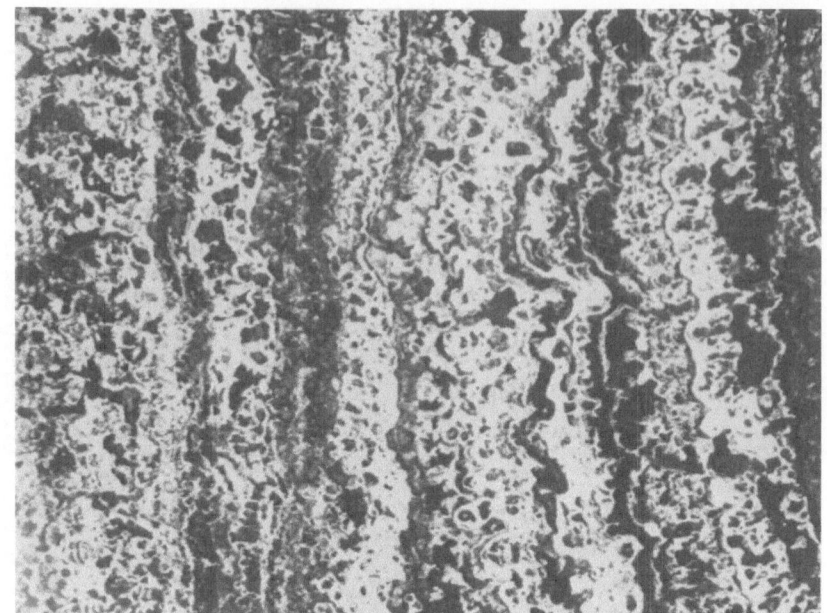

Fig. 32(c)

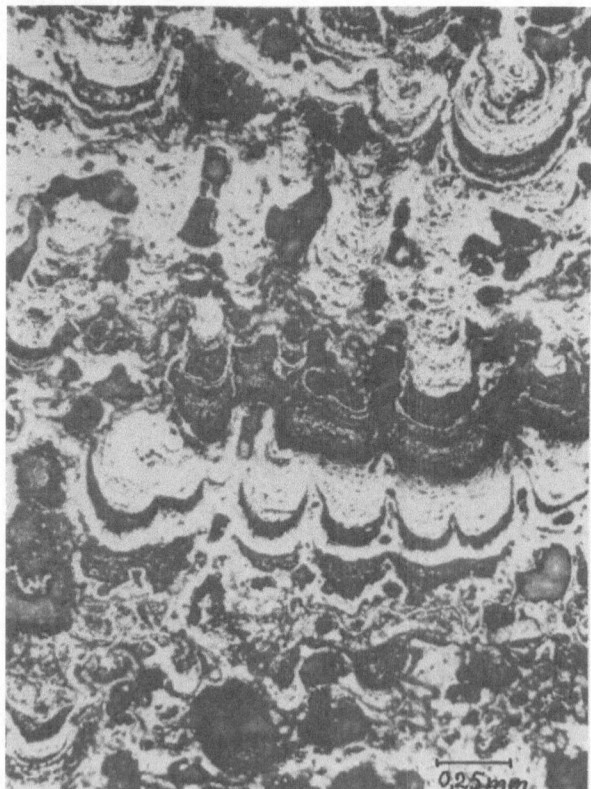

Fig. 32(d)

Fig. 32. Laminated structures and columnar texture of ore matter in nodules from
the Pacific radiolarian zone (polished section). Magnification: 10 (a)
12(b), 10 (c), 190 (d).

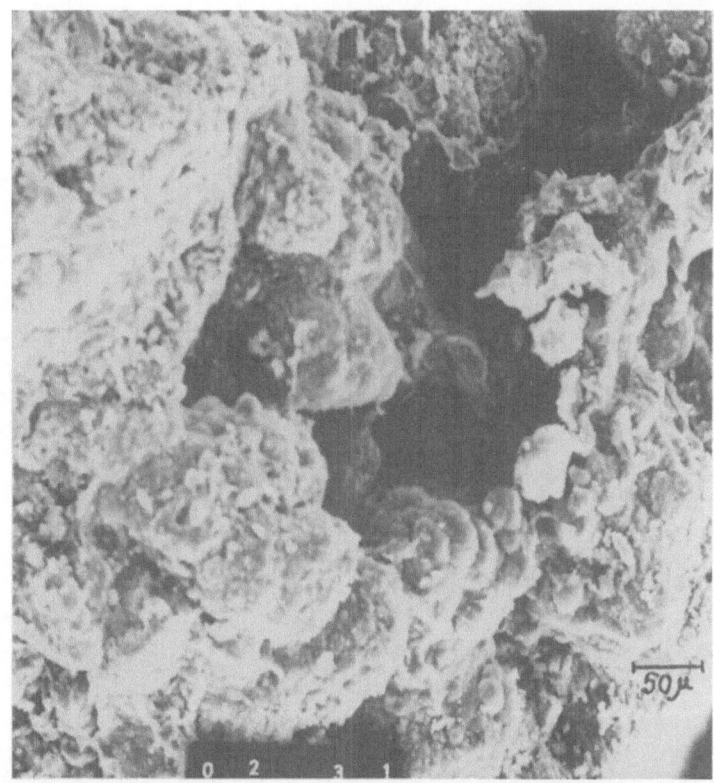

Fig. 33(a)

Fig. 33(b)

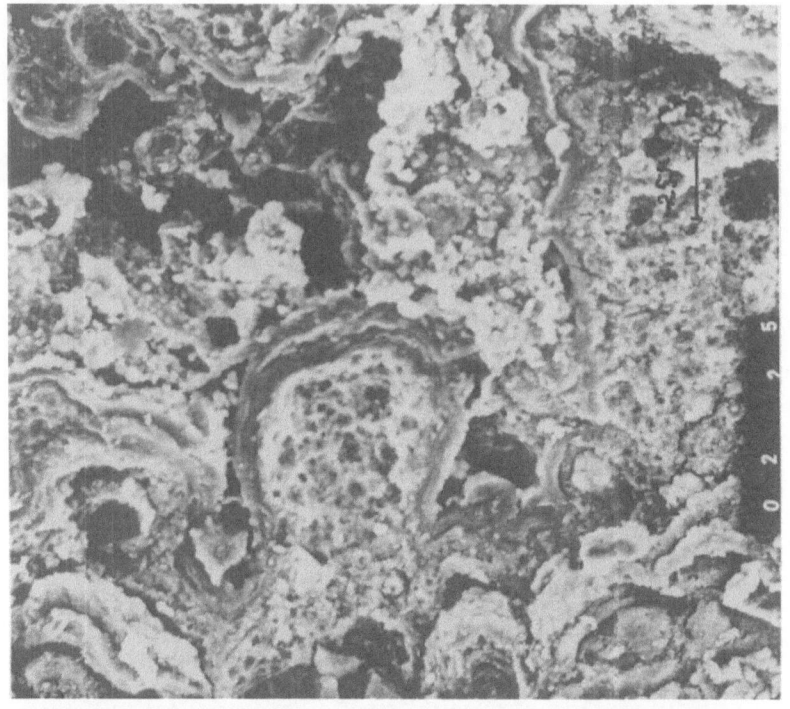

Fig. 33(d)

Fig. 33. Inner structure of nodules under scanning microscope at slight magnification. (a) Globular. Magnification: 150. (b) Globular-laminated. Magnification: 170. (c) Concentric-laminated. Magnification: 1000. (d) Loop-like. Magnification: 400.

Fig. 34(a)

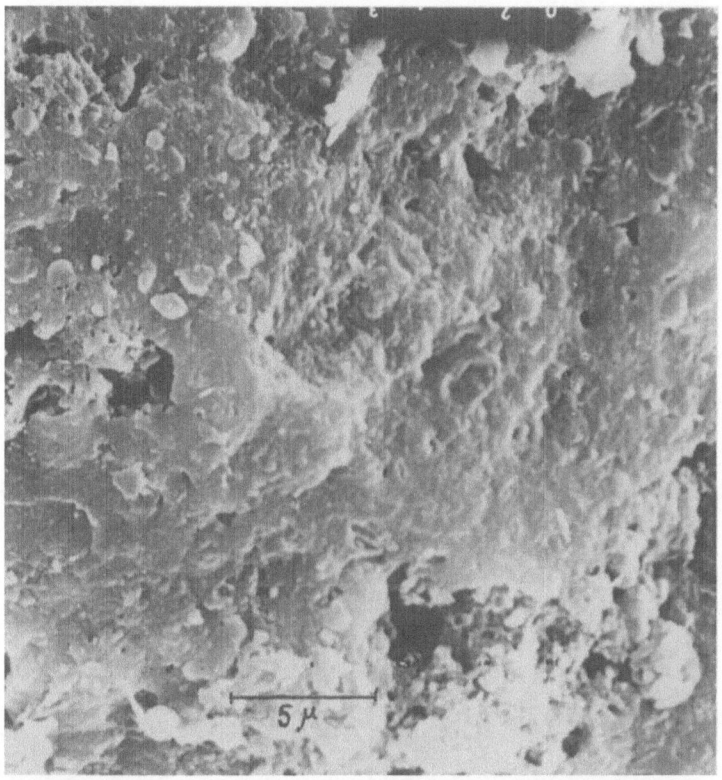

Fig. 34(b)

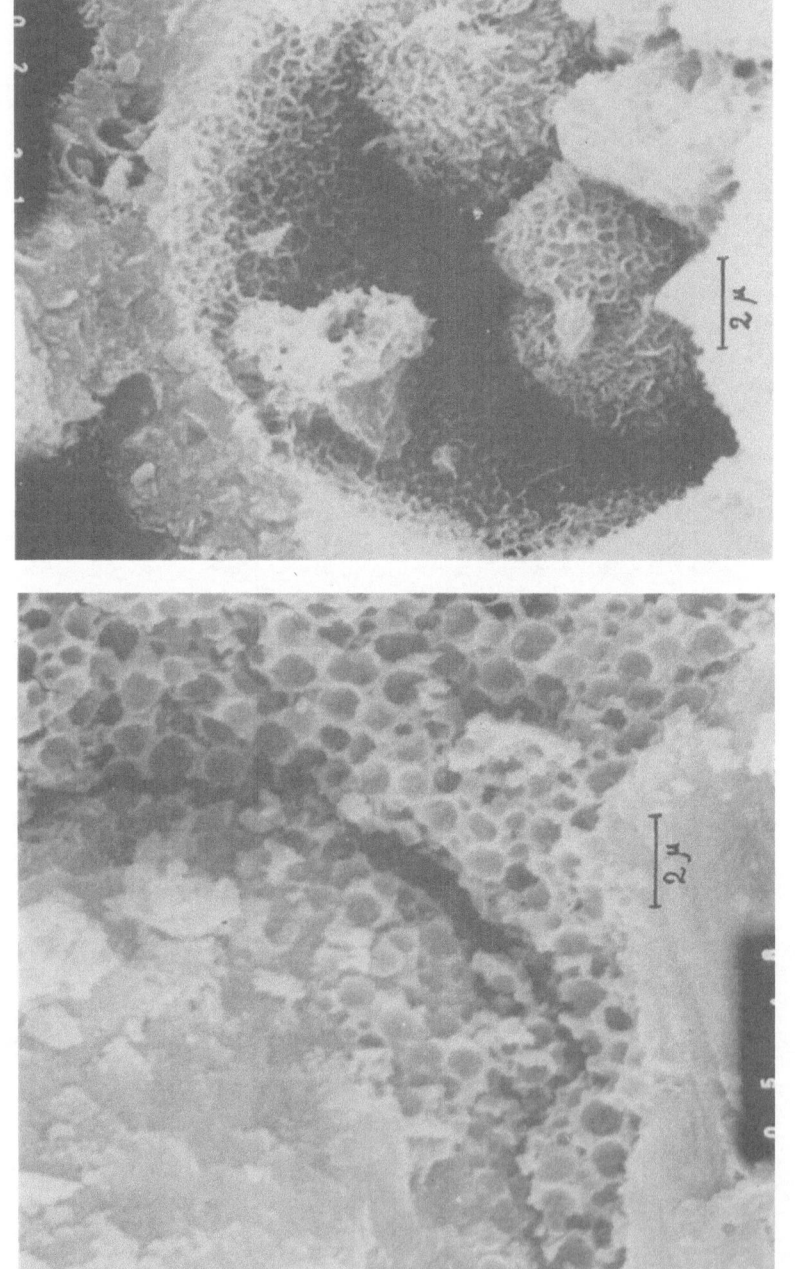

Fig. 34(d)

Fig. 34(c)

Fig. 34. Inner structure of nodules under scanning microscope at great Magnification.
(a) Globular. Magnification 3700. (b) Collomorphic. Magnification 4000.
(c) Cell-like. Magnification 6000. (d) Fibrous. Magnification: 6000.

TABLE 52

General characteristics of manganese minerals in nodules (Burns and Burns, 1979)

Mineral	Tentative composition	Crystal system	Parameters of elementary cell ($\overset{\circ}{A}$)
Todorokite	$(Ca,Na,K)(Mg,Mn^{2+})Mn_5O_{12} \cdot xH_2O$	Monoclinic	a=9.75; b=2.849; c=9.59
Buserite (10$\overset{\circ}{A}$-manganite)	Na Mn hydrooxide	Hexagonal	a=8.41; c=10.01
Birnessite	$(Na,Ca,K)(Mg,Mn)Mn_6O_{14} \cdot x5H_2O$	Hexagonal	a=2.85; c=7.08-7.31
Vernadite (δ-MnO_2)	$MnO_2 \cdot xnH_2Oxm(R_2O,RO,R_2O_3),R=Na,Ca,Co,Fe,Mn$	Hexagonal	a=2.86; c=4.7
Pyrolusite	MnO_2	Tetragonal	a=4.39; c=2.87
Ramsdellite	MnO_2	Orthorhombic	a=4.53; b=9.27; c=2.87
Nsutite (γ-MnO_2)	$(Mn^{2+},Mn^{3+},Mn^{4+})(O,OH)_2$	Hexagonal	a=0.65; c=4.43
Psilomelane	$(Ba,K,Mn^{2+},Co)_2Mn_5O_{10} \cdot xH_2O$	Monoclinic or Orthorhombic	a=9.56; b=2.88; c=13.8 a=8.254; b=13.40; c=2-86
Chalcophanite	$Zn_2Mn_6O_{14} \cdot x6H_2O$	Triclinic	a=7.54; b=7.54; c=8.22
Rancieite	$(Ca,Mn)Mn_4O_9 \cdot x3H_2O$	Hexagonal	a=2.84; c=7.07

(α - FeOOH) detected in the acid-insoluble residue (Buser and Grütter, 1956); acaganeite (β - FeOOH) identified by the Mössbauer method (Goncharov et al., 1973); lepidocrocite (γ- FeOOH) is in coupling with other minerals (Goodell et al., 1971; Glasby, 1972b; Okada et al., 1972; Chukhrov et al., 1980); feroxyhite (δ' - FeOOH) is in coupling with clay minerals and goethite, the plates of this mineral of 0.1-0.4μm in diameter can be transformed into wüstite under an electron beam (Chukhrov et al., 1976); wüstite, however, was detected in nodules as an independent mineral phase (Table 53).

Lately, a number of publications have appeared on the occurrence in nodules of inclusions of sulphide minerals that are authigenic according to some parameters. Among them are pyrite, troilite, pyrrhotite, chalcopyrite, bornite, covellite, violarite (Müller, 1979; Baturin and Dubinchuk, 1983a, 1984b) (Figure 35). Ostwald (1983) has noted that sulphide minerals occur in todorokite as microinclusions. Parameters typical of sulphide minerals that occur in nodules are given in Table 54.

In nodules, besides violarite (Ni_2FeS_4), other rare minerals of nickel were detected, namely, taenite (NiFe), bornite (NiO), and nickeline (NiAs) (Baturin and Dubinchuk, 1984b).

Inclusions of rock minerals in ore material of nodules are extremely variable in their size, shape and composition. Among them one can distinguish quartz, feldspar, mica, olivine, pyroxene, amphibole, prehnite, zeolite, apatite, calcite, aragonite, rutile, anatase, heavy spar (barite), spinel (Burns and Burns, 1977). Clay minerals from the Pacific nodules were analysed specially; to separate them manganese was removed out of a pelitic fraction. A smectite complex, i.e. laminated formations and minerals of the montmorillonite group, is most widely spread (25-85% of the total clay minerals), the second place is taken up by hydromica (10-60%) and the third is taken up by chlorite and caolinite (5-25%) (Sterenberg and Schurina, 1974; Sterenberg et al., 1977; Volkov et al., 1980).

TABLE 53

Standard and measured parameters of Manganosite and Wüstite
(Baturin and Dubinchuk, 1984a)

N of line	Manganosite $a_{o(st.)}$ = 4.435Å, $a_{meas.}$ = 4.428Å					Wüstite $a_{o(st.)}$ = 4.284Å, $a_{meas.}$ = 4.24Å				
	hkl	$I_{st.}$	$I_{meas.}$	$d_{st.}$	$d_{meas.}$	hkl	$I_{st.}$	$I_{meas.}$	$d_{st.}$	$d_{meas.}$
1	111	8	9	2.561	2.57	111	7	9	2.47	2.46
2	200	10	10	2.218	2.20	200	10	10	2.14	2.12
3	220	10	10	1.568	1.55	220	8	8	1.51	1.48
4	311	8	6	1.337	1.31	311	4	2	1.293	1.24
5	222	8	5	1.280	1.27	222	2	washed	1.23	1.10
6	400	7	6	1.108	1.09			out		
7	331	8	5	1.017	1.00					
8	420	10	4	0.991	0.98					
9	422	10	4	0.905	0.89					

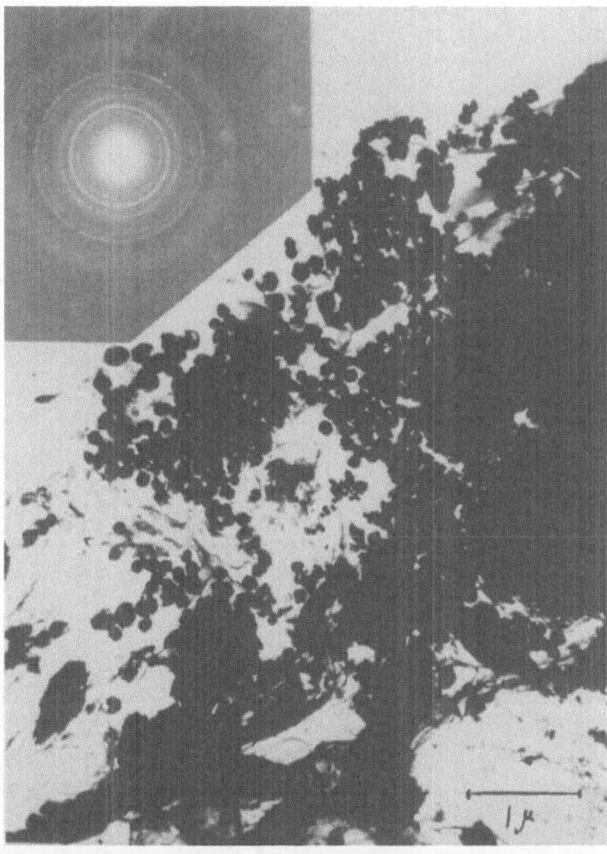

Fig. 35(a)

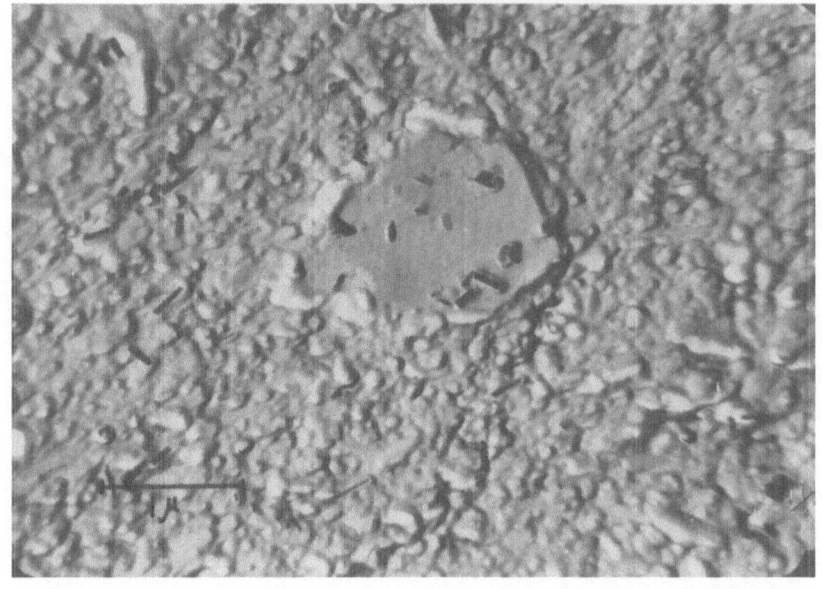

Fig. 35(b)

Fig. 35(c)

Fig. 35(d)

Fig. 35. Sulphide minerals in the Pacific nodules (Baturin and Dubinchuk, 1984b). (a) fine dispersed pyrite and its microdiffraction pattern; (b) relict of pyrite grains in globular mass of Mn and Fe hydrooxides; (c) crystalline pyrite and its microdiffraction pattern; (d) pyrite crystals in chalcopyrite matter.

TABLE 54

Standard and measured intensity of lines (I) and interplane distance (d)
of sulphide minerals in manganese nodules
(Baturin and Dubinchuk, 1983, 1984)

N of line	hkl (hkil	Pyrite				hkl (hkil)	Pyrrhotite			
		standard		measured			standard		measured	
		I	d	I	d		I	d	I	d
1	111	2	3.102	3	3.08	1120	6	2.97	3	2.96
2	200	8	2.696	10	2.67	1122	8	2.64	6	2.63
3	210	8	2.417	7	2.40	1014	1	2.45	0.5	2.42
4	211	7	2.206	2	2.18	1024	10	2.06	10	2.04
5	220	6	1.908	4	1.88	3030	7	1.72	3	1.70
6	221,300	4	1.796	1	1.77	1126	1	1.61	2	1.59
7	311	10	1.629	9	1.60					
8	222	3	1.560	2	1.57					
9	230	4	1.498	3	1.48					
10	321	6	1.444	4	1.42					

		Troilite					Chalcopyrite			
1	1010	2	5.11	–	–	112	10	3.03	10	3.02
2	0003	3	4.73	2	4.70	200,004	6	2.61	4	2.60
3	1120	4	3.82	8	3.82	220	10	1.855	8	1.83
						312	10	1.586	7	1.57
4	1122	6	2.96	9	2.96	224	6	0.515	4	1.50
5	1014	7	2.64	9	2.62	400	6	1.320	5	1.30
6	1124	1	2.53	0.5	2.50	008	6	1.294	4	1.26
7		10	2.085	10	2.06	332	8	1.205	6	1.18

		Bornite					Covellite			
1	311	8	3.304	8	3.31	1010	4	3.33	2	3.30
2	222	8	3.165	8	3.17	1011	6	3.24	4	3.22
3	400	8	2.737	8	2.72	1012	8	3.04	6	3.01
4	331	6	2.510	5	2.50	1013	10	2.81	10	2.78
5	511,333	4	2.103	4	2.08	0006	8	2.72	6	2.70
6	440	10	1.924	10	1.91	1015	6	2.30	4	2.30
7	531	4	1.849	3	1.83	1016	2	2.09	1	2.03
8	533	2	1.668	1	1.65	0008	4	2.03	2	2.00
9						1120,1017	10	1.890	10	1.86

MAJOR ORE ELEMENTS OF MANGANESE NODULES

Manganese and iron are the major elements of nodules determining their geochemical pattern. Ore and basic metals, namely, copper, nickel, cobalt, zinc, lead, are associated with them. Therefore, these five elements are considered in the first place when analysing nodules, whereas other components are usually determined selectively depending on the geochemical problems under consideration from time to time, that is why the scope of information about the major ore elements is considerably larger than about other elements. Thus, for mineral prospecting purposes, Soviet geologists analysed some 4000 samples of nodules and determined concentrations of manganese, nickel, copper and cobalt in them (Manganese nodules..., 1984). McKelvey et al. (1983) generalized the results of some 2400 analyses of nodules in search for the mentioned elements and iron, the number of analyses for zinc and lead being half of the total number of samples, major non-metallic minerals are analysed in 400 to 600 samples, and minor and trace elements in no more than 100 samples. The information on the latter is sometimes restricted to only few determinations, or is even missing at present.

This chapter is devoted to the data on distribution, forms of occurrence and behaviour of the seven major ore elements of nodules, namely, manganese, iron, nickel, copper, cobalt, zinc and lead.

CONCENTRATION AND DISTRIBUTION OF METALS IN NODULES FROM VARIOUS AREAS OF THE WORLD OCEAN

In chapter VII it was mentioned that the patterns of nodule distribution and structure in different oceanic areas is extremely variable. This is true for their chemical composition, which varies over a wide range and depends upon many parameters of the oceanic environment. The most thorough analysis was made of the nodules from the Pacific where major ore fields are located. The range within which their composition varies can be demonstrated by the following percentage values: manganese concentration in them varies from 0.07 to 50.3; iron from 0.3 to 41.9; nickel from 0.0008 to 2.48; copper from 0.003 to 1.90; cobalt from 0.001 to 2.53; zinc from 0.012 to 7.00; lead from 0.004 to 0.46%. Maximum-to-minimum ratios for each element will be: 700 for manganese, 140 for iron, 310 for nickel, 630 for copper, 2530 for cobalt, 580 for zinc and 115 for lead. Manganese, nickel, copper and zinc have similar values; iron and lead also have similar corresponding values. Cobalt shows the greatest variability.

When analysing average metal concentrations in nodules from various parts of the Pacific (except for the Clarion-Clipperton field), the variability range reduces by factors of ten. Average concentrations (in percent) are: manganese, 6.5 (the Drake Passage) to 38.8 (eastern equatorial zone); iron, 2.92 (eastern equatorial zone) to 25.85 (the Drake Passage); nickel, 0.11 (northwest) to 1% (the Californian field); copper, 0.04 (the Hawaiian field) to 0.74 (eastern part of the central zone); cobalt, 0.08 (northwest) to 1.127 (the Mid Pacific sea mountains); zinc, 0.047 (southwest) to 0.234 (eastern equatorial zone); lead, 0.017 (eastern zone) to 0.18 (sea mountains) (Table 55).

The composition of manganese nodules from the Clarion-Clipperton field are of special interest as the largest nodule deposits of economical value are accumulated

TABLE 55

Average concentration of ore metals in manganese nodules and crusts from various Pacific regions*

Element	Concen-tration extrema (1)	North (2)	North-west (2)	Hawaiian ore field (3)	Central zone Centre (3)	Central zone Centre (4)	Central zone East (5)	Central zone South (5)	Southern zone (6)	Southern zone (4)	Southern zone East (2)	Southern zone (7)
Mn	0.07–50.3	15.14	8.15	23.2	20.16	15.71	20.3	18.6	16.6	16.61	15.64	15.91
Fe	0.4–41.9	10.98	12.94	–	–	9.06	10.7	11.5	10.1	13.92	16.27	14.86
Ni	0.008–2.48	0.65	0.11	0.40	0.73	0.956	0.80	0.60	0.72	0.433	0.42	0.55
Cu	0.003–1.90	0.54	0.09	0.04	0.58	0.711	0.74	0.47	0.30	0.185	0.22	0.29
Co	0.001–2.53	0.33	0.08	0.86	0.29	0.213	0.21	0.25	0.22	0.595	0.37	0.28
Zn	0.012–7.00	0.084	–	–	–	–	–	–	0.08	–	0.071	–
Pb	0.004–0.46	0.12	–	–	–	0.049	–	–	0.047	0.073	0.130	–

Element	South-western zone (5)	(3)	(8)	(5)	(9)	South-eastern zone (8)	(10)	(11)	Drake Pass.- Scotia Sea (3)	Drake Pass.- Scotia Sea (4)	West (4)	East (2)	Calif. field (3)	Eastern equatorial zone (12)
Mn	15.2	14.78	16.3	16.2	14.75	19.81	30.3	18.37	6.45	12.98	16.87	34.8	25.54	38.6
Fe	15.4	–	21.1	18.6	15.77	10.20	7.8	13.9	–	25.85	13.30	1.49	–	2.92
Ni	0.41	0.34	0.40	0.43	0.269	0.961	1.09	0.406	0.28	0.244	0.564	0.15	1.04	0.742
Cu	0.22	0.20	0.22	0.24	0.140	0.311	0.51	0.229	0.14	0.105	0.393	0.082	0.61	0.371
Co	0.36	0.35	0.38	0.40	0.390	0.164	0.08	0.192	0.20	0.179	0.395	0.01	0.16	0.0195
Zn	–	–	0.078	–	0.0475	–	–	0.060	–	–	–	0.05	–	0.234
Pb	–	–	–	–	0.116	0.030	–	0.068	–	0.084	0.034	0.017	–	–

Element	Mid Pacific and Wake-Necker sea mountains (4)	(2)	(13)	(3)	Sea mountains of the southern borderland (4)
Mn	13.96	18.39	24.6	18.21	15.85
Fe	13.10	15.01	14.5	–	22.22
Ni	0.393	0.44	0.49	0.44	0.348
Cu	0.061	0.13	0.065	0.24	0.077
Co	1.127	0.61	0.79	0.44	0.514
Zn	–	0.066	0.072	–	–
Pb	0.174	0.18	0.18	–	0.085

* From: (1) Volkov, 1979; McKelvey et al., 1983; (3) Andreev et al., 1984; (4) Cronan, 1980; (5) Skornyakova, 1976; (6) Exon, 1983; (7) Murray and Renard, 1891; (8) Glasby, 1976; (9) Landsmesser et al., 1976; (10) Halbach et al., 1981; (11) Girin et al., 1977; (12) Dymond et al., 1984; (13) Halbach and Manheim, 1984.

MAJOR ORE ELEMENTS OF MANGANESE NODULES

TABLE 56

Average concentration of ore metals in nodules from the Clarion-Clipperton province*

Element	Concentration extrema (1)	Northern zone (2)	Southern zone (2)	(3)	(4)	(5)	(6)	(7)	(1´)	(8)	(9)
Mn	2.50–37.50	27.8	24.6	23.33	23.1	21.89	22.0	24.64	28.80	25.24	23.53
Fe	0.18–15.31	5.8	11.5	9.44	.798	7.14	5.91	6.77	6.64	6.92	–
Ni	0.11–1.95	1.40	0.94	1.080	1.1	1.04	1.29	1.0768	1.22	1.18	1.14
Cu	0.11–1.66	1.15	0.48	0.627	0.89	0.88	1.16	1.1353	0.99	0.94	0.86
Co	0.05–0.91	0.28	0.25	0.192	0.18	0.21	0.209	0.1600	0.23	0.20	0.22
Zn	0.04–0.25	0.13	0.07	–	0.11	0.108	0.121	0.0869	0.13	0.18	–
Pb	0.015–0.17	–	–	0.028	0.056	0.050	0.022	0.0483	0.048	–	–

Element	Site A		Site B		Site C		Reference samples			
	(10)	(11)	(10)	(11)	(10)	(11)	(12)	(13)	(14)	(15)
Mn	23.3	20.7	26.4	28.1	25.4	27.0	31.69	29.14	27.00	30.02
Fe	8.17	9.2	4.0	5.5	8.8	8.5	10.77	5.78	5.00	5.81
Ni	1.12	1.00	1.42	1.54	1.20	1.48	1.65	1.337	1.45	1.40
Cu	0.85	0.79	1.16	1.20	0.97	0.99	1.39	1.151	1.25	1.03
Co	0.248	0.24	0.206	0.29	0.264	0.27	0.25	0.224	0.25	0.22
Zn	0.095	–	0.125	–	0.119	–	0.19	0.1595	0.15	0.114
Pb	–	–	–	–	–	–	0.06	0.0555	0.05	0.040

* From: (1) McKelvey et al., 1979; (2) Friedrich and Plüger, 1974; (3) Cronan, 1982; (4) Skornyakova, 1976; (5) Skornyakova et al., 1981; (6) Marchig and Gundlach, 1976; (7) Calvert, 1978; (8) Halbach et al., 1981; (9) Andreev et al., 1974; (10) Piper et al., 1979; (11) Monget et al., 1976; (12) Description...., 1977; (13) Flanagan and Gottfried, 1980; (14) King and Pasho, 1979; (15) IOAN reference sample.

in it. Nodules of the zone in question are characterized by considerable
variability of their composition in general, however, average concentrations of
metals in its various parts are relatively regular: 20.7 to 28.1% of manganese; 4.0
to 11.5 of iron; 0.94 to 1.54 of nickel; 0.48 to 1.26 of copper; 0.16 to 0.29 of
cobalt; 0.07 to 0.18 of zinc; 0.022 to 0.05% of lead (Table 56).

Table 56 also gives the data on the composition of reference samples of
manganese nodules in the Clarion-Clipperton ore field that was prepared for
standardization of various geological and geochemical analyses of nodules within
and beyond the ore field. The composition of reference samples by the U.S.
Geological Service, prepared, by the Canadian Kennecott company and by Shirshov
Institute of Oceanology, Academy of Sciences of the USSR, in general, appeared
to be extremely close for all major components: 27 to 30% of manganese; 5.0 to 5.8
of iron; 1.337 to 1.45 of nickel; 1.03 to 1.25 of copper; 0.22 to 0.25 of cobalt;
0.114 to 0.1595 of zinc; 0.040 to 0.0555% of lead.

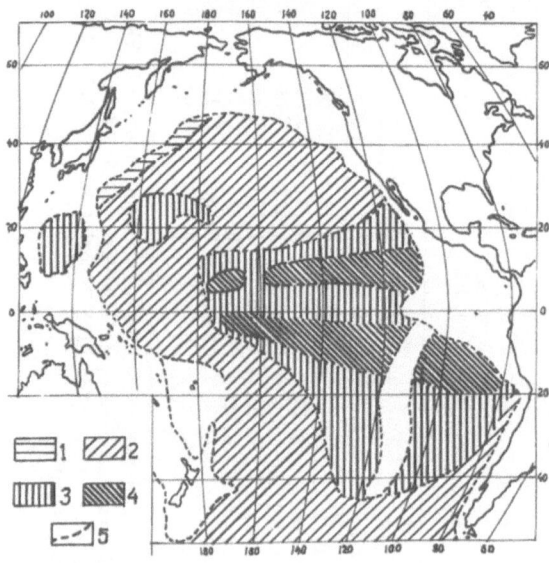

Fig. 36. Mn distribution in the Pacific nodules (Skornyakova, 1976b). Mn concentra-
        tion, %: (1) less 10; (2) 10 to 20; (3) 15 to 25; (4) 20 to 35; (5) boundary
        of the oxidized sediment layer over 1 m.

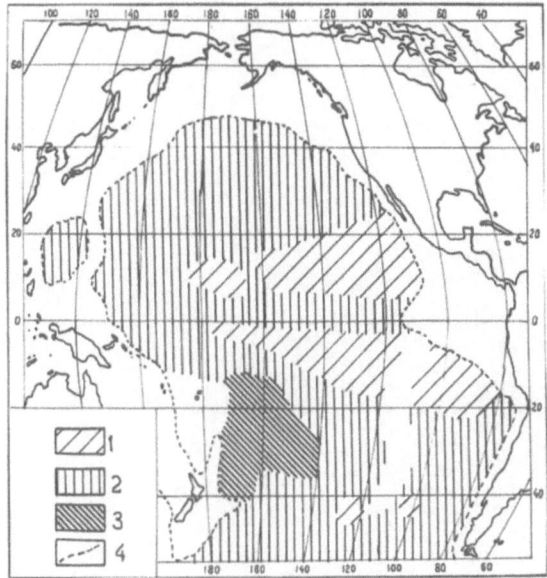

Fig. 37 Fe concentration (%) in the Pacific nodules (Skornyakova, 1976b). (1) 5 to 15; (2) 10 to 20; (3) 15 to 25; (4) boundary of the oxidized sediment layer over 1 m.

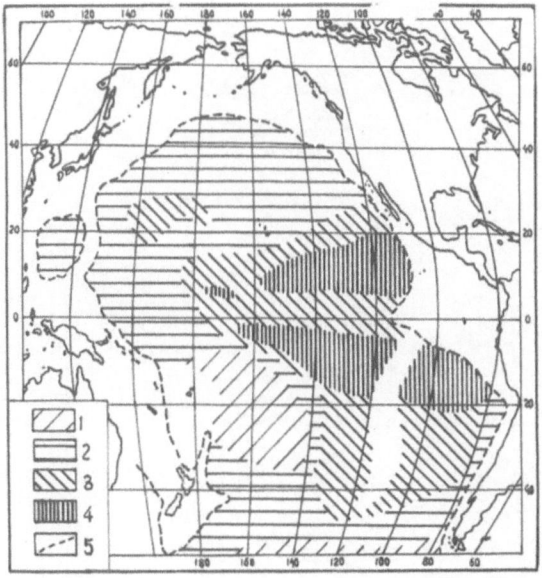

Fig. 38. Ni concentration (%) in the Pacific nodules (Skornyakova, 1976b); (1) less 0.4; (2) 0.2 to 0.8; (3) 0.4 to 1.2; (4) 0.8 to 2.0; (5) boundary of the oxidized sediment layer over 1 m.

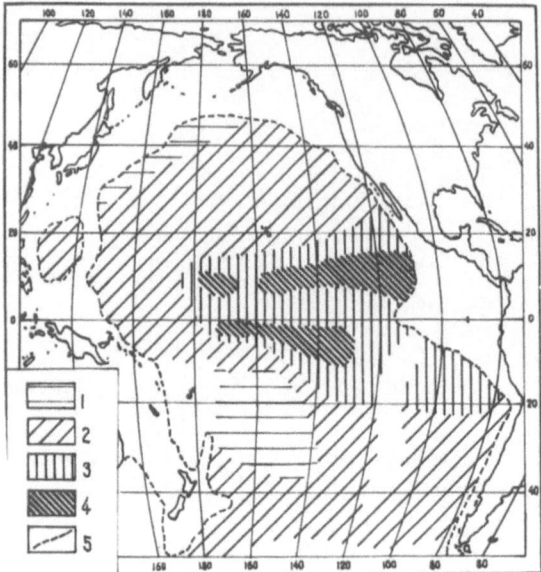

Fig. 39. Cu concentration (%) in the Pacific nodules (Skornyakova, 1976b); (1) less 0.4; (2) 0.2 to 0.6; (3) 0.4 to 1.0; (4) 0.8 to 2.0; (5) boundary of the oxidized sediment layer over 1 m.

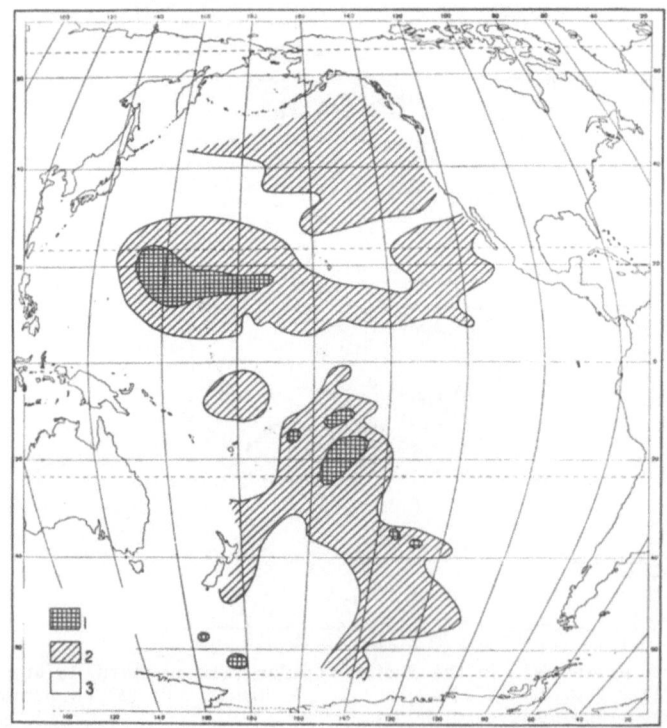

Fig. 40. Co concentration (%) in the Pacific nodules (Calvert, 1978): (1) over 0.5; (2) 0.25 to 0.5; (3) less 0.25.

Figures 36 to 40 illustrate the pattern of manganese, iron, nickel, copper and cobalt distribution in nodules from the Pacific.

Minimum manganese concentrations (less than 10%) were recorded in nodules of the northwestern basin periphery. On the whole, a moderate manganese concentration (10 to 20%) is typical of the entire northern, western and southern parts of the ocean. In the pelagic zone of the ocean, two tendencies in manganese distribution in nodules are distinct: increase of its concentration from the periphery towards the central regions and occurrence of two latitudinal belts of maximum concentration north and south of the Equator (Figure 36).

Opposite tendencies were detected for iron distribution. The latitudinal near-equatorial belts of manganese maximum coincides in shape with the belts of minimum iron. The zone of maximum iron is localized in the South Pacific (Figure 37).

In general, nickel and copper distribution is close to that of manganese, though it has some shades of difference. Thus, nodules in the South Basin have the lowest concentration of the metals considered (Figures 38, 39), whereas manganese does not show this tendency (Figure 37). Besides, latitudinal near-equatorial zones of copper maximum (0.8 to 2%) are more restricted spatially as compared to similar zones of manganese and nickel maximum concentrations.

The scheme of cobalt distribution in nodules compiled by Calvert (1978) (Figure 40) differs considerably from the scheme compiled by Skornyakova for iron (Figure 37). The most pronounced feature of cobalt is the localization of its maxima (over 0.5%) in nodules and ore crusts at sea mountains. Fields of high cobalt concentration in nodules (0.25 to 0.5%) are confined to the northern and southern peripheries of the ocean and to the sub-latitudinal northern near-equatorial zone.

The composition of nodules in the Indian ocean is as variable as in the Pacific (Table 57). Average minimum concentrations of manganese (11-13%), nickel (0.2 –0.3%), copper (0.11-0.17%) and zinc (0.03-0.07%) are typical of nodules in the Crozet depression, Madagascar, Mozambique basins, in the Antarctic zone, at the South African submarine plateau and sea mountains. Maximum concentrations of these metals are typical of the Western Australian field and especially of the Central Basin where manganese reaches 26%, nickel 1.43%, copper 0.84% and zinc 0.12%. Maximum concentrations of cobalt (0.443%) and lead (0.161%) are registered in ore formations at sea mountains.

In the Atlantic, one can also delineate a number of regions with lower and higher content of metals in nodules. The northwestern Atlantic and the northern part of the Mid-Atlantic Ridge are characterized by a low content of metals, namely: manganese, 9.5-15%; nickel, 0.14-0.30; copper, 0.07-0.15; zinc, 0.06%. Regions of high ore matter concentrations are the northern part of the Blake Plateau, Argentina and Cape basins and the Walvis Ridge where manganese is 19.8 to 23.7, nickel is 0.48 to 0.95, copper is 0.09 to 0.50, and zinc is 0.06 to 0.18%. Higher concentrations of cobalt (0.41 to 0.68%) are detected in nodules and ore crusts at some sea mountains and hills: within the Walvis Ridge, northeastern Atlantic, central part of the Mid-Atlantic Ridge, in some samples from the Blake Plateau and also in nodules from the Cape Verde basin (Table 58).

Despite the irregular distribution of nodules, the variability of their composition and different progress in their analysis within various oceanic areas, average concentrations of elements have been estimated for individual oceans and for the World Ocean as a whole. Such estimates were first performed by D. Mero (1965) for the Pacific; however, the values were presented on the ore basis, free from the alumo-silicate material that often forms a considerable portion of nodules. This resulted in overestimated values of some elements, in particular, 24.2% of manganese, 14% of iron, 0.99% of nickel. The corresponding averages are lower: 17.9 to 20.1, 11.4 to 11.96, 0.59 to 0.634%. (Volkov, 1979; Cronan, 1980; McKelvey et al., 1983). Averages of other major ore elements in nodules of the Pacific vary within the following ranges: 0.39 to 0.54% of copper; 0.27 to 0.35% of cobalt; 0.047 to 0.16% of zinc; 0.083 to 0.11% of lead (Table 59).

For the Indian ocean, estimates are: 14.17 to 15.25% of manganese; 13.05 to 14.7% of iron; 0.43 to 0.464% of nickel; 0.173 to 0.295% of copper; 0.21 to 0.254% of cobalt; 0.061 to 0.149% of zinc; 0.070 to 0.101% of lead. For the Atlantic the estimates are: 13.25 to 15.78% of manganese; 13.08 to 20.78% of iron; 0.32 to 0.484 of nickel; 0.116 to 0.155 of copper; 0.27 to 0.323 of cobalt; 0.066 to 0.123 of zinc; 0.127 to 0.14% of lead.

TABLE 57

Average concentrations of ore metals in nodules from various Indian ocean regions*

Element	Concentration extrema (1)	Central Basin			Warton Basin		Southern basin (3)	Diamantina field (4)	Grozet Basin		Western Australian field (4)
		(2)	(3)	(4)	(2)	(3)			(2)	(3)	
Mn	0.6-37.0	22.1	20.64	26.0	15.1	15.97	18.44	21.77	12.8	13.00	21.4
Fe	1.20-39.63	7.6	10.82	-	11.4	11.67	13.93	-	15.5	15.76	-
Ni	0.01-1.58	-	0.68	1.43	-	0.39	0.51	0.72	-	0.30	0.86
Cu	0.09-1.66	0.991	0.54	0.84	0.355	0.22	0.26	0.43	0.150	0.14	0.55
Co	0.014-0.998	0.113	0.15	0.15	0.173	0.16	0.16	0.20	0.190	0.17	0.19
Zn	0.016-9.00	0.119	0.12	-	0.071	0.07	0.07	-	0.066	0.037	-
Pb	0.005-0.75	0.065	0.054	-	0.087	0.043	0.075	-	0.097	0.076	-

Element	Madagascar Basin		Mozambique Basin		Somali Basin			Antarctic zone (3)	Mozambique Strait (2)	South-Africa Plateau (2)	Sea mountains (2)	Mid-oceanic ridges (2)
	(2)	(3)	(2)	(3)	(2)	(3)	(4)					
Mn	12.6	13.31	11.6	12.96	15.3	17.57	16.0	13.02	16.5	17.5	13.4	12.5
Fe	16.1	18.10	13.0	12.09	12.00	14.73	-	13.02	16.5	14.7	17.4	19.4
Ni	-	0.20	-	0.26	-	0.50	0.53	0.22	0.800	0.285	0.275	0.119
Cu	0.115	0.11	0.166	0.17	0.241	0.16	0.22	0.17	0.204	0.94	0.065	0.088
Co	0.256	0.29	0.150	0.12	0.139	0.20	0.14	0.12	0.120	0.264	0.443	0.205
Zn	0.049	0.055	0.041	0.11	0.064	0.053	-	0.067	0.112	0.056	0.053	0.051
Pb	0.104	0.11	0.120	0.10	0.071	0.074	-	0.095	0.051	0.128	0.161	0.097

* From: (1) Volkov, 1979; McKelvey et al., 1983; (2) Cronan, 1980; (3) Skornyakova and Vanshtein, 1983; (4) Andreev et al., 1984

TABLE 58

Average concentrations of ore metals in nodules from various Atlantic regions*

Element	Concentration extrema (1)	North-western Atlantic (2)	(3)	North-eastern Atlantic (2)	Cape Verde Basin (2)	Blake Plateau North (2)	South (2)	(3)	Argentina Basin (2)	Brazil Basin (2)
Mn	0.04–40.3	15.19	13.4	19.75	16.65	19.34	15.26	10.44	23.67	16.07
Fe	1.54–50.0	23.17	–	18.53	22.93	12.95	24.44	–	17.88	19.95
Ni	0.013–1.56	0.270	0.30	0.361	0.249	0.637	0.227	0.46	0.549	0.266
Cu	0.01–0.88	0.13	0.15	0.126	0.105	0.095	0.200	0.14	0.182	0.118
Co	0.01–1.44	0.287	0.27	0.488	0.438	0.381	0.095	0.41	0.138	0.327
Zn	0.035–8.00	0.098	–	0.071	0.064	0.098	0.065	–	0.106	0.109

Element	Cape Basin (2)	(3)	Sea mountains (2)	Mid-Atlantic Ridge (2) 45°N	Centre	North	Walvis Ridge (2)	Region west of South Africa (2)
Mn	23.45	15.6	16.22	13.54	15.54	9.55	23.05	17.18
Fe	13.68	–	22.41	21.84	20.37	20.27	18.10	20.26
Ni	0.952	0.63	0.245	0.188	0.225	0.140	0.482	0.258
Cu	0.500	0.60	0.047	0.048	0.047	0.070	0.078	0.058
Co	0.208	0.28	0.539	0.372	0.439	0.301	0.676	0.406
Zn	0.178	–	0.058	0.056	0.059	0.062	0.116	0.156

* From: (1) Volkov, 1979; McKelvey et al., 1983; (2) Cronan, 1980; (3) Andreev et al., 1984.

TABLE 59

Average concentrations of ore metals in nodules from various regions of the World Ocean*

Element	Concentration extrema (1)	The Pacific					The Indian Ocean			The Atlantic			South Basin	The World Ocean		
		(2)	(3)	(4)	(5)	(6)	(3)	(4)	(5)	(3)	(4)	(5)	(4)	(4)	(5)	(6)
Mn	0.04–50.3	24.2	17.94	19.78	20.1	18.3	14.74	15.10	15.25	14.93	15.78	13.25	14.69	16.02	18.60	17.4
Fe	0.3–50.0	14.0	11.72	11.96	11.4	12.77	13.05	14.74	14.23	13.08	20.78	16.97	15.78	15.55	12.40	13.6
Ni	0.08–2.48	0.99	0.59	0.634	0.76	0.63	0.441	0.484	0.43	0.484	0.328	0.32	0.450	0.480	0.66	0.55
Cu	0.003–1.90	0.53	0.39	0.392	0.54	0.41	0.173	0.294	0.25	0.155	0.116	0.13	0.210	0.259	0.45	0.34
Co	0.001–2.53	0.35	0.33	0.335	0.27	0.29	0.254	0.230	0.21	0.323	0.318	0.27	0.240	0.284	0.27	0.27
Zn	0.01–9.00	0.047	0.084	0.068	0.16	–	0.061	0.069	0.149	0.066	0.084	0.123	0.060	0.078	0.12	–
Pb	0.01–0.75	0.090	0.11	0.0846	0.083	–	0.070	0.093	0.101	0.134	0.127	0.14	–	0.0900	0.093	–

* From: (1) Volkov, 1979; McKelvey et al., 1983; (2) Mero, 1965; (3) Volkov, 1979; (4) Cronan, 1980; (5, 6) McKelvey et al., 1983; (5) for the entire World Ocean, (6) but for the Clarion–Clipperton zone.

Two estimates of average concentrations of metals in nodules were made for the World Ocean by Cronan (1980) and McKelvey et al. (1983). The values obtained by them are respectively: 16.02 and 18.60% of manganese; 15.55 and 12.47 of iron; 0.48 and 0.66 of nickel; 0.259 and 0.45 of copper; 0.284 and 0.27 of cobalt; 0.078 and 0.12 of zinc; 0.090 and 0.093% of lead. These estimates are identical for cobalt and lead, but for the rest of the elements the concentration ratio is from 0.58 (copper) to 1.25 (iron). On one hand, this is a result of the different quantity of initial data analysed and, on the other, from the various techniques used in their statistical processing which, at present, do not seem to be universal when considering the system of sea-floor geological features.

Manganese nodules spread over the seas, inlets, bays and fjords differ considerably in their geochemistry from the oceanic ones; though on the whole, the manganese and iron content in them varies within the same range, from 1 to 30% (manganese) and 4 to 26% (iron) (Table 60).

Sea nodules are characterized by an extremely low concentration of those metals which, in oceanic nodules, can be considered as ores. The concentration of nickel in sea nodules is 0.0035 to 0.071, of copper 0.0009 to 0.0067, of cobalt 0.003 to 0.023, of zinc 0.0023 to 0.0157, of lead 0.0008 to 0.0043%. This can be explained by sharp differences in sedimentary and ore-formation environment in sea and oceanic basins (Strakhov et al., 1968).

COMPOSITION OF MANGANESE NODULES AND OCEANIC BOTTOM ENVIRONMENT

When considering the compositional varieties of manganese nodules one should pay attention to the great variety of oceanic environments that are controlled to a considerable extent by such global factors as climate, and vertical and circum-continental zoning of the ocean. These factors alone can explain some features of oceanic sedimentation with which ore formation is associated (Bezrukov, 1970; Lisitsin, 1974, 1978).

Table 61 demonstrates the correlation between nodule composition and climatic or latitudinal zoning.

The maximum concentration (over 32%) of ore elements (in total) in nodules is recorded within the 40° N - to - 40° S interval in the Pacific, in the ranges from 60 to 40° N and from 0 to 20° S in the Atlantic, from 0 to 20 ° S in the Indian ocean and from 40° N to 40° S in the World Ocean. Thus, latitudinal distribution of nodules in the Pacific controles the general pattern of their composition in the World Ocean. This tendency is traced when considering the manganese, nickel and copper content in nodules. In the Pacific, nodules with a maximum concentration of these metals are localized within the interval of 0 to 20° N; in the Atlantic they are located in the interval from 20 to 40° S; in the Indian ocean, from 0 to 20° S; in the World Ocean, from 0 to 20 ° N (18.2% of Mn; 0.55% of Ni; 0.37% of Cu, on average).

Areas of iron maximum are localized in the Pacific within the interval of 0 to 40° S (16.93%); in the Atlantic, it falls within 60 to 40 ° N (19.58%), which determines the general pattern of iron occurrence in nodules of the World Ocean within the same latitudes.

Cobalt distribution turned out to be more complex. The maxima are confined to the intervals of 20 to 40° S in the Pacific (0.35%), 0 to 20° S and 60 to 40 ° N in the Atlantic (0.49% and 0.39%, respectively), and 0 to 20 ° N in the Indian ocean (0.29%). So, two belts of high cobalt concentrations in nodules of the World Ocean are formed, namely 40 to 60° N and 0 to 40 ° S. One of them coincides completely and another only partly with iron maxima.

Many researchers described a certain dependence of the chemical composition of nodules upon the oceanic depth. Analysis of the data on the Pacific nodules (Skornyakova, 1976b) showed that manganese maxima are confined to the depths down to 2000 m, iron maxima fall within the depth 3000-4000m, nickel and copper extends to under 5000 m, cobalt only to 2000 m (Table 62).

In the ore crusts of the Mid-Pacific sea mountains maximum concentrations of manganese, nickel and cobalt are confined to the depths of 1100 to 1500 m; to 2400-3000 m for iron; to 3000-4000 m for copper (Halbach and Puteanus, 1984). On the whole, manganese content in nodules of the World Ocean does not essentially

TABLE 60

Average concentrations of metals (%) in nodules of seas and bays*

Element	Baltic Sea			White Sea	Barents Sea	Kara Sea	Black Sea		Jervis Inlet	Loch Fyne
	Main sea	Gulf of Finland	Riga Bay				(1)	(2)		
Mn	12.40	14.65	10.05	1.06–26.3	16.50	13.39	6.79	14.10	32.76	30.19
Fe	19.55	19.32	22.80	5.1–17.6	14.36	14.22	26.62	18.20	6.16	3.92
Ni	0.071	0.0035	0.0047	0.0035	0.0311	0.0062	0.0283	0.010	0.0314	0.0077
Cu	0.0045	0.0009	0.0017	0.0016	0.0041	0.0015	0.0037	0.001	0.0067	0.0017
Co	0.0130	0.0096	0.0064	0.0030	0.0187	0.0030	0.0083	0.003	0.0157	0.0230
Zn	0.0105	0.0113	0.0135	0.0051	0.0157	0.0070	–	–	0.0023	0.0060
Pb	0.0023	0.0009	0.0025	0.0008	–	0.0011	0.0016	–	–	0.0043

* From: Volkov, 1979 – Baltic, White, Kara, Black Seas (1); Calvert and Price, 1977 – Baltic Sea, Jervis Inlet (Britisch Columbia), Loch Fyne (Scotland); Georgescu and Lupan, 1971 – the Black Sea (2); Shterenberg et al., 1985 – the White Sea; Ingri, 1985 – Bartents Sea.

TABLE 61

Average and Maximum (in brackets, %) concentrations of metals in manganese nodules with respect to geographical latitude (after McKelvey et al., 1983)

Metal	60°–40° N	40°–20° N	20° N –0°	0°–20° S	20–40° S	40–60° S	South of 60 ° S
				The Pacific			
Mn	15.71 (33.9)	19.72 (39.56)	22.39 (50.30)	18.30 (42.3)	16.46 (38.02)	14.89 (36.0)	10.11 (22.40)
Fe	11.33 (26.0)	12.28 (41.9)	8.90 (26.7)	14.17 (34.0)	16.93 (32.20)	14.13 (28.20)	15.57 (25.64)
Ni	0.35 (0.72)	0.66 (1.69)	0.96 (1,95)	0.54 (1.80)	0.47 (1.73)	0.54 (1.65)	0.32 (1.00)
Cu	0.19 (0.49)	0.38 (1.10)	0.76 (1,90)	0.34 (1.54)	0.22 (0.65)	0.24 (0.84)	0.20 (0.58)
Co	0.24 (0.86)	0.28 (1.64)	0.26 (1,50)	0.31 (1.88)	0.35 (2.23)	0.22 (1.06)	0.19 (0.57)
In total	27.82	33.32	33.27	33.66	34.43	30.02	26.39
Number of sites	30	355	855	244	146	98	32
				The Atlantic			
Mn	15.03 (27.39)	13.25 (40.90)	14.06 (34.80)	15.13 (24.00)	14.51 (26.78)	11.16 (26.64)	1.00
Fe	19.58 (24.28)	17.62 (50.00)	15.84 (23.70)	19.54 (22.4)	14.67 (26.40)	16.56 (34.55)	7.10
Ni	0.27 (0.68)	0.30 (1.56)	0.24 (0.73)	0.25 (0.43)	0.42 (1.42)	0.30 (0.99)	0.14
Cu	0.08 (0.34)	0.13 (0.62)	0.11 (0.48)	0.09 (0.15)	0.15 (0.88)	0.13 (0.41)	0.10
Co	0.39 (1.01)	0.29 (1.04)	0.25 (0.51)	0.49 (1.44)	0.30 (1.01)	0.16 (0.76)	0.10
In total	35.35	31.59	30.50	30.50	30,05	28.31	8.44
Number of sites	30	120	12	7	56	72	1
				The Indian ocean			
Mn			16.52 (32.30)	18.05 (32.25)	14.36 (26.50)	13.53 (24.57)	3,07 (3,93)
Fe			16.64 (22.80)	11.92 (19.70)	15.47 (39.63)	12.25 (21.60)	8,94 (12.22)
Ni			0.38 (0.81)	0.58 (1.58)	0.37 (1.19)	0.48 (1.43)	0,04
Cu			0.10 (0.30)	0.47 (1.66)	0.18 (0.84)	0.25 (0.89)	0,02
Co			0.29 (0.93)	0.20 (0.94)	0.22 (0.62)	0.13 (0.37)	-
In total			33.93	31.22	30.6	26.64	12.07
Number of sites			28	73	159	42	2
				The World Ocean			
Mn	15.67 (33.9)	18.10 (40.9)	22,10 (50.3)	18.90 (42.3)	15.20 (38.0)	13.40 (36.0)	9.40 (25.6)
Fe	15.46 (26.0)	13.64 (50.0)	9,20 (26.7)	13.80 (34.0)	15.90 (39.6)	14.60 (30.6)	14.90 (25.6)
Ni	0.31 (0.72)	0.57 (1.69)	0,93 (1.95)	0.55 (1.80)	0.42 (1.73)	0.45 (1.65)	0.31 (1.00)
Cu	0.14 (0.49)	0.32 (1.10)	0,73 (1.90)	0.37 (1.66)	0.20 (0.88)	0.20 (0.88)	0.19 (0.58)
Co	0.32 (1.01)	0.28 (1.64)	0,26 (1.50)	0.29 (1.88)	0.29 (2.23)	0.19 (1.06)	0.19 (0.57)
In total	31.90	32.91	33.22	33.21	32.01	28.84	24.99

TABLE 62

Average concentration of metals in manganese nodules and ore crusts with respect to oceanic depth
(in brackets: number of samples)

Ele- ment	Pacific nodules*					Ore crusts from the Central Pacific sea mountains**						Nodules of the World Ocean***			
	to 2000 m	2000- 3000 m	3000- 4000 m	4000- 5000 m	over 5000 m	1100- 1500 m	1500- 1900 m	1900- 2400 m	2400- 3000 m	3000- 4000 m	4000- 4400 m	to 2000 m	to 3000 m	over 3000 m	over 4000 m
Mn	19.22 (27)	16.17 (14)	15.61 (44)	17.11 (196)	17.77 (128)	28.4	24.7	25.5	20.5	25.5	19.7	18.00 (186)	17.16 (324)	18.60 (1931)	18.6 (1561)
Fe	13.36 (27)	15.74 (14)	17.50 (44)	10.54 (196)	12.36 (128)	14.3	15.3	16.1	19.5	18.0	16.7	14.81 (186)	15.63 (324)	12.17 (1926)	11.7 (1357)
Ni	0.50 (27)	0.36 (14)	0.42 (44)	0.61 (196)	0.62 (128)	0.50	0.42	0.41	0.18	0.35	0.24	0.41 (185)	0.39 (323)	0.69 (1936)	0.71 (1565)
Cu	0.12 (27)	0.13 (14)	0.21 (44)	0.39 (196)	0.47 (128)	0.03	0.06	0.07	0.39	0.13	0.12	0.08 (174)	0.10 (309)	0.49 (1935)	0.52 (15.65)
Co	0.79 (27)	0.53 (14)	0.37 (44)	0.31 (196)	0.28 (128)	1.18	0.90	0.88	0.69	0.63	0.67	0.53 (170)	0.46 (307)	0.23 (1855)	0.24 (1530)
Mn/Fe	1.44	1.03	0.89	1.62	1.44	1.99	1.65	1.58	1.05	1.41	1.17	2.17	-	-	2.32

From: *Skornyakova, 1976b; **Halbach and Puteanus, 1984; ***McKelvey et al., 1983.

depend upon the depth, whereas other elements show definite trends in their distribution with depth. For instance, iron accumulates in nodules at a depth to 3000 m (15.63 on average), nickel and copper tend to concentrate at a depth below 4000 m (on average, 0.71 and 0.52% respectively), cobalt accumulates at a depth to 2000 m (0.53% on average) (Table 62).

The systematization of the data on metal occurence in nodules with respect to facial conditions in the Pacific and Indian oceans revealed that nodules of deep depressions, sea mountains and hemipelagic regions usually differ considerably in their composition (Table 63).

In pelagic zones and deep basins, manganese is abundant in nodules of radiolarian belts, whereas iron tends to accumulate in nodules of the deep basins beyond radiolarian belts. As a result, the average Mn:Fe ratio in nodules of the Pacific radiolarian belt is 2.9; it is 1.7 beyond the belt considered. For the Indian ocean this ratio is 1.8 within and 1.0 outside the belt. The behaviour of other metals seems to be controlled by the major ore-forming elements. Thus nickel, copper and zinc accompany manganese, while cobalt and lead accompany iron.

The manganese average in nodules and ore crusts at sea mountains is almost the same as in deep basins, whereas iron concentration grows considerably: in the Pacific it reaches 15%; in the Indian ocean it is as great as 16% at sea mountains and 18.7% at the mid-oceanic ridge. The concentration of other metals varies as deep basins change over to sea mountains; nickel reduces by 1.5 times in the Pacific and by 1.2 - 2 times in the Indian ocean; copper reduces by 4 and 2 times, respectively; zinc concentration lowers by 1.4 times in both (being constant at the mid-oceanic ridge); cobalt increases by 2 times in the Pacific and by 1.6 times in the Indian ocean (being constant at the mid-oceanic ridge); lead increases by 1.5 and by 1.1 - 1.3 times, respectively.

In hemipelagic zones, the average concentration of iron and manganese varies over a wider range than in pelagic zones, as is demonstrated by Table 64.

The composition of the host and underlying sediments is the determining factor for a geochemical environment for the formation of manganese nodules. Table 64 presents the data on the composition of nodules resting over sediments of various types in the Pacific and Indian oceans. This table shows once again that nodules with high concentrations of nickel, copper and zinc are confined to pelagic radiolarian ooze in both the oceans considered. The second place, according to this parameter, is taken up by eupelagic clays, the third by miopelagic sediments of the Pacific and Indian oceans, and the fourth by diatomaceous and calcareous ooze of the Indian ocean.

Hemipelagic sediments, which are extremely variable in composition and in the physical-chemical environment of their diagenesis, produce nodules that are also extremely variable in composition. The average concentration of manganese varies in them from 8.15 to 38.8%, and the iron average varies from 1.49 to 16.99%. This became clear from a comparison of the composition of nodules from hemipelagic sediments of the northwestern Pacific, from some regions of the eastern Pacific and from hemipelagic sediments of the Indian ocean. Concentrations of other metals in nodules of hemipelagic sediments vary within the following ranges: nickel, 0.11 to 0.74%; copper, 0.09 to 0.48; cobalt, 0.01 to 0.2; zinc, 0.05 to 0.23%. According to their manganese and iron concentrations (8.1 to 18.7 and 12.9 to 14.3%, respectively), nodules from hemipelagic sediments of the northwestern Pacific are similar to those of the north-eastern tropical part of the Ocean, whereas nodules in the vicinity of the Californian province (34.8 to 38.8% of Mn; 1.5 to 2.9% of Fe) are similar to those of the eastern equatorial part of the Pacific. However, the concentration of nickel, cobalt and copper revealed another association: nodules of the northwestern Pacific are similar to those of the near Californian region and nodules of the north-eastern tropical part of the ocean resemble those of the eastern equatorial part of the ocean. The former contain 0.11 to 0.15% of nickel, 0.08 to 0.09% of copper, 0.01 to 0.08% of cobalt; the latter have 0.655 to 0.742% of nickel, 0.371 to 0.484% of copper and 0.195 to 0.258% of cobalt. Such geochemical associations, differing sharply from those observed in the nodules of pelagic zone, result from the variability of the process of diagenesis in hemipelagic sediments.

According to the various compositions of manganese nodules and the bottom environment, one can distinguish geochemical provinces on a regional scale.

TABLE 63

Concentration of metals in the Pacific and Indian ocean nodules with respect to facial conditions

Element	Pelagic region							Hemipelagic region		
	Deep oceanic basins		Radiolarian belt		Sea mountains		Mid-oceanic ridge	The Pacific		Indian ocean**
	In total									
	The North Pacific*	Indian ocean**	The Pacific*	Indian ocean**	The Pacific*	Indian ocean**	Indian ocean**	***	****	
Mn	19.14	16.60	23.1	18.86	18.39	15.23	15.03	18.74	38.8	15.70
Fe	10.98	14.49	7.98	10.60	15.01	16.21	18.70	14.28	2.92	12.74
Ni	0.65	0.37	1.1	0.58	0.44	0.30	0.18	0.655	0.742	0.26
Cu	0.54	0.17	0.89	0.42	0.13	0.0,8	0.10	0.484	0.371	0.10
Co	0.33	0.19	0.18	0.15	0.61	0.31	0.21	0.258	0.195	0.14
Zn	0.084	0.06	0.11	0.097	0.066	0.043	0.060	-	0.234	0.060
Pb	0.12	0.075	0.056	0.054	0.18	0.082	0.096	-	-	0.070
Mn/Fe	1.74	1.05	2.9	1.79	1.18	0.93	0.85	1.31	19.29	1.23

From: *Skornyakova, 1976b; **Skornyakova and Vanshtein, 1983; ***Volkov et al., 1980 (Transoceanic profile from Japan to California); ****Dymond et al., 1984 (Site H, eastern equatorial zone).

TABLE 64

Concentration of metals in the Pacific and Indian ocean nodules with respect to the type of host sediments*

Element	Eupelagic clays			Miopelagic sediments			Radiolarian ooze		Diatomaceous ooze	Calcareous ooze	Hemipelagic sediments				
	P(N)(1)	P(S)(1)	I(2)	P(N)(1)	I**(2)	I***(2)	P(1)	I(2)	I(2)	I(2)	P(NW)(1)	P(3)	P(C)(1)	P(H)(4)	I(2)
Mn	19.14	15.64	14.18	14.55	16.01	17.65	23.1	21.76	12.14	15.84	8.15	18.74	34.8	38.8	14.60
Fe	10.98	16.27	13.42	12.38	14.02	11.07	7.98	9.79	13.16	14.48	12.94	14.28	1.49	2.92	16.99
Ni	0.65	0.42	0.31	0.47	0.43	0.55	1.1	0.77	0.22	0.40	0.11	0.655	0.15	0.742	0.30
Cu	0.54	0.22	0.17	0.29	0.21	0.34	0.89	0.65	0.15	0.14	0.09	0.484	0.082	0.371	0.15
Co	0.33	0.37	0.15	0.24	0.19	0.16	0.18	0.14	0.13	0.21	0.08	0.258	0.01	0.195	0.10
Zn	0.084	0.071	0.065	0.074	-	0.093	0.11	0.12	0.067	0.051	-	-	0.05	0.234	0.06
Pb	0.12	0.13	0.061	0.12	0.062	-	0.056	0.050	0.10	0.096	-	-	0.017	-	0.05
Mn/Fe	1.74	0.96	1.06	1.17	1.14	1.59	2.9	2.12	0.92	1.09	0.48	1.31	27.6	19.29	0.85

(P) The Pacific, (I) Indian ocean, (N) North, (S) South, (NW) North-West, (C) California region, (H) Site H, DOMES
*From: (1) Skornyakova, 1976; (2) Skornyakova and Vanshtein, 1983; (3) Volkov, 1980 (eastern part of the Transoceanic profile); (4) Dymond et al., 1984.
**Marlaceous clays;
*** Marlaceous radiolarian clays.

In the Pacific, Mero (1965) distinguished four provinces that differ in composition of manganese ore formations: an iron-rich province (southern and western zones of the ocean, the region near Central America); a manganese-rich province (eastern part of the ocean); a province of high concentration of nickel and copper (central and eastern parts of the ocean); and a cobalt-rich province (submarine hills of the central, southern and western parts of the ocean).

Skornyakova (1976b) proposed to distinguish geochemical provinces according to the Mn:Fe ratio in nodules. When this ratio is less than 0.25, nodules are considered as being iron-rich, from 0.25 to 1 as manganese-iron, from 1 to 4 as iron-manganese, and over 4 as manganese-rich.

As Mn:Fe ratio in the nodules of the peripheral regions of the ocean varies from 0.06 to 51, three sub-provinces were additionally delineated within the eastern periphery of the Pacific (province I), namely, Ia- iron-rich and manganese-iron; Ib-iron-manganese; Ic-manganese-rich nodules (Figure 41).

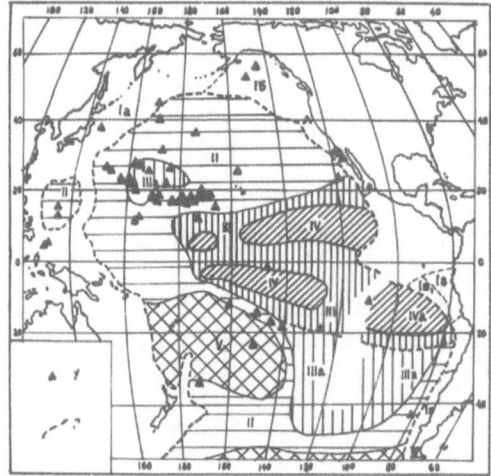

Fig. 41. Geochemical provinces of the Pacific nodules (Skornyakova, 1976b), (1) sea mountains; (2) boundary between pelagic and hemipelagic zones; (I) peripheral ((a) iron; (b) iron-manganese; (c) manganese nodules); (II) province with 10 to 20% Mn, 10 to 20% Fe, 0.2 to 0.8% N, 0.2 to 0.6% Cu; (III) province with 15 to 25% Mn, 0.4 to 1% Cu, 0.4 to 1.2% Ni; (IV) province with 20 to 35% Ni and 0.8 to 2% Cu; (IVa) province of the Peru basin; (V) province of the Southern basin (15 to 25% Fe, 0.2 to 0.6% Ni).

In pelagic zone, provinces II (with moderate content of Mn, Ni, Cu) and III (with their high concentrations) are delineated. Province IV is characterized by the maximum concentration of nickel and copper, also of zinc and molybdenum. Province IV is represented by two latitudinal belts extending north and south of the Equator. A sub-province of the Peru Basin can be also recognized (IVa). Province V (manganese-iron nodules) is localized in the South Basin.

COMPOSITIONAL VARIABILITY OF INDIVIDUAL NODULES AND RELATIONS BETWEEN METALS

Manganese nodules demonstrate great variability in their composition, not only from region to region of the World Ocean, but also within one ore field and in seperate features of bottom topography. The chemical composition of nodules depends upon their morphogenetic type, as was revealed by detailed investigations

TABLE 65
Concentration of metals in manganese nodules from
the Pacific radiolarian zone with respect to morphogenetic
types of nodules (Skornyakova, 1984)

Ele-ment	Types of nodules; sites; number of samples			
	D and HD; 2483; 48 samples	H; 2474; 43 samples	H; 2520; 30 samples	Ore crusts of faults; 31 samples
	from-to; average	from-to; average	from-to; average	from-to; average
Mn	18.0 - 28.4; 25.5	14.6 - 24.3; 19.8	13.8 - 19.3; 17.3	11.7 - 19.4; 15.2
Fe	4.9 - 9.4; 6.23	8.0 - 14.0; 11.5	9.1 - 16.3; 12.7	15.0 - 22.2; 18.5
Ni	0.96- 1.45; 1.14	0.41- 0.90; 0.69	0.39- 0.64; 0.49	0.12- 0.40; 0.22
Cu	0.70- 1.35; 0.98	0.30- 0.59; 0.46	0.25- 0.52; 0.36	0.04- 0.28; 0.16
Co	0.13- 0.29; 0.20	0.20- 0.36; 0.29	0.24- 0.40; 0.32	0.08- 0.50; 0.28
Zn	0.08- 0.16; 0.11	0.046-0.096; 0.071	0.043-0.076; 0.063	0.032-0.062; 0.047
Pb	0.013-0.029; 0.022	0.045-0.095; 0.068	-	0.06- 0.11; 0.084
Mn/Fe	2.5 - 6.0; 4.22	1.34- 2.43; 1.74	0.98- 1.96; 1.36	0.54- 1.19; 0.82
Mineral composition of nodules	todorokite	slightly crystal-lized todorokite	slightly crystal-lized todorokite and vernadite	vernadite

*Nodules: (D) diagenetic; (DH) diagenetic-hydrogeneous; (H) hydrogeneous

within several sites in the Central Pacific, the radiolarian belt included (Skornyakova, 1984). From the results obtained, diagenetic and hydrogenous-diagenetic nodules (their morphology was described in Chapter VII) are mainly characterized by high concentrations of manganese, nickel, copper and zinc and low concentration of iron, cobalt and lead (Table 65, Figure 42). In hydrogenous nodules, the concentration of metals of the first group is lower, while of the second it is higher.

Ore crusts from the fault zones within the same sites have an even lower content of manganese, nickel, copper and zinc, but they are rich in iron and lead at a moderate concentration of cobalt. Considerable fluctuations of chemical compositions are observed in each of these ore formation groups; however, average concentrations are distinct and individual and agree well with the described morphogenic types of nodules and their mineral composition.

As is shown in Chapter VII, the inner structure, texture and mineralogy of nodules are also variable, as is reflected in their chemical composition.

The analysis of the top, bottom and side parts of nodules from various regions of the ocean showed that the nodule tops are usually richer in iron, cobalt and lead with respect to the bottom parts, whereas the nodule bottom is richer in manganese, nickel, copper and zinc. The equatorial parts are of intermediate composition. Such a situation is typical of nodules of all morphogenetic types; however, in some diagenetic nodules of the Pacific radiolarian belt, the distribution of manganese and zinc can be inverse or regular, though this does not affect the distribution of nickel and copper, which relatively enrich the bottom nodules and do not follow manganese behaviour in this case (sample 3 in Table 66).

Another example of a non-traditional relation between manganese and other ore elements is demonstrated by diagenetic nodules from the eastern equatorial hemipelagic zone of the Pacific (Site H, MANOP Program). The tops of these nodules contain, on average, 34.7% of manganese and 3.8% of iron, the bottoms contain 43.8% of manganese and 1.1% of iron. However, the remainder of the ore elements in these nodules are accumulated only in their tops, with relative concentration coefficients (top/bottom) being 1.5, nickel; 2.5, copper; 2.7, cobalt; 1.2, zinc (Table 66).

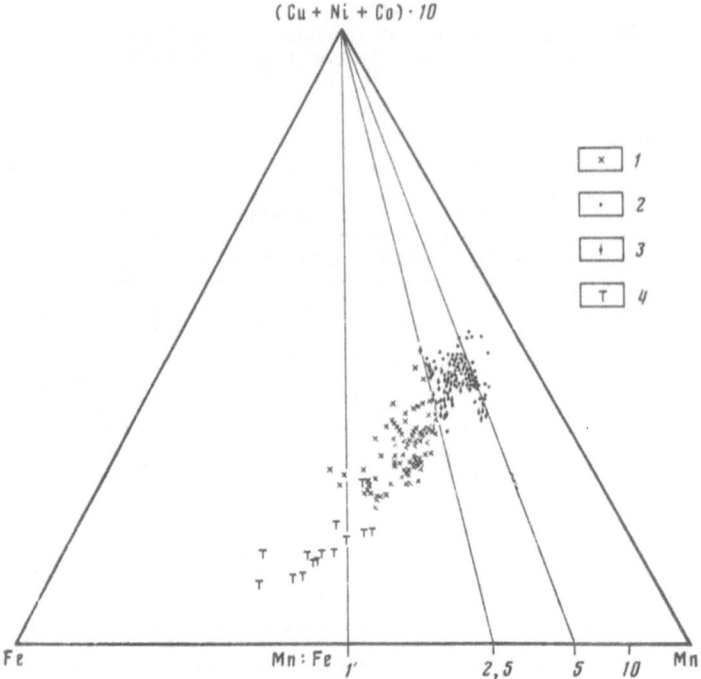

Fig. 42. Ternary diagram of chemical composition of various genetic nodule types
(Skornyakova, 1984). (1) mainly hydrogenous nodules; (2) diagenetic;
(3) hydrogenous-diagenetic; (4) ore crusts at bed rock surface.

The ratios of metal concentrations in the tops and bottoms within the Site
considered, and two more Sites, are given in Figure 43.

The statistical processing of the results obtained from the analyses of total
samples of oceanic nodules performed by various authors revealed an inverse
correlation between manganese and iron in all cases. It is the strongest in the
Atlantic (-0.74), moderate in the Pacific (-0.43) and weak in the Indian ocean
(-0.20) (Tables 67-69). The positive correlation between manganese and other
metals has the following coefficients in the Pacific, Indian and Atlantic oceans,
respectively: Mn : Ni, 0.83, 0.63; 0.78; Mn : Cu, 0.59, 0.41, 0.51; Mn : Zn,
0.32, 0.24, 0.20; Mn : Co, 0.21, 0.12, 0.35. In all three oceans iron shows a
strong negative correlation with nickel and copper, and a weak correlation with
zinc. It has a positive correlation with lead. The Fe : Co ratio is controversial,
since it is 0.25 in the Pacific, 0.40 in the Indian ocean and -0.27 in the Atlantic.

These results are independent statistical indices of the relation between the
major ore elements in nodules, as was demonstrated earlier when describing
variations in their composition with respect to the geographical latitude, oceanic
depth, facial setting, sediment lithology and the types of nodules proper.

The mineralogical composition of nodules plays a special role in the accumulation of ore elements in them. In particular, as is shown in Table 65, concentrations
of manganese, nickel, copper and zinc in the nodules from the Pacific radiolarian
belt gradually increase as the ore matter in the nodules changes from vernadite
to a mixture of vernadite and poorly crystallized todorokite, then to poorly
crystallized todorokite and well crystallized todorokite. Therefore it is expedient to
consider the chemical composition of the major manganese minerals in nodules,
namely, of todorokite, birnessite and vernadite, the mineralogical parameters of
which were given in Table 52.

TABLE 66

Chemical composition of the Pacific nodules: tops, bottoms, equatorial belts

Ele-ment	Station 2520 (H)			Station 2483-2 (HD)				St. 2483-41 (D)			
	T	B	Bulk	T	B	E	Bulk	T	B	E	Bulk
Mn	16.4	25.4	17.6	22.0	34.1	29.9	27.4	28.4	25.2	30.0	28.4
Fe	13.9	4.9	7.1	7.9	2.6	4.2	6.0	8.1	3.6	4.1	5.75
Ni	0.45	1.40	0.65	0.80	2.2	1.45	1.48	0.84	1.10	1.34	1.10
Cu	0.36	1.45	0.45	0.50	1.52	1.05	1.12	0.75	1.15	1.25	1.03
Co	0.30	0.07	0.25	0.36	0.06	0.17	–	0.28	0.13	0.16	0.16
Zn	0.074	0.11	0.065	0.12	0.17	0.19	0.12	0.15	0.10	0.15	0.13
Mn/Fe	1.15	5.2	2.5	2.8	13.1	7.1	4.5	3.5	7.0	7.3	4.9

Ele-ment	Site S (5 samples) (Radiolarian zone)			Site R (7 samples) (Red clays)			Site H (27 samples) (Hemipelagic sediments)		
	T	B	Bulk	T	B	Bulk	T	B	Bulk
Mn	27.12	31.2	29.1	17.5	22.4	20.6	34.7	43.8	38.8
Fe	7.76	4.47	4.80	15.6	10.8	11.5	3.8	1.1	2.9
Ni	1.30	1.77	1.52	0.56	1.01	0.86	0.806	0.531	0.742
Cu	0.786	1.270	1.187	0.303	0.672	0.498	0.437	0.172	0.371
Co	0.364	0.243	0.214	0.328	0.240	0.304	0.0216	0.0081	0.0195
Zn	0.136	0.211	0.169	0.056	0.107	0.076	0.243	0.203	0.234
Mn/Fe	3.5	7.3	6.1	1.1	2.1	1.8	9.1	39.8	13.4

Stations 2520 and 2483 by Skornyakova, 1984; Sites by Dymond et al., 1984.
2520-17 (H) – hydrogenous, 2483-2 (HD) – hydrogenous – diagenetic;
2483-41 (D) – diagenetic.
(T) tops; (B) bottoms; (E) equatorial belt; (Bulk) bulk sample

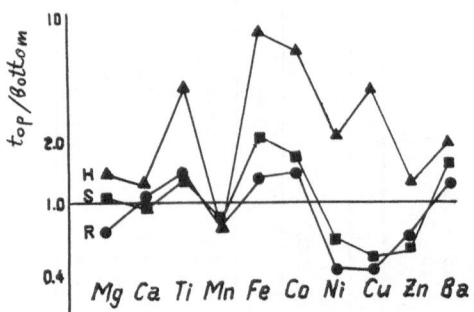

Fig. 43. Element concentration ratios in nodule tops and bottoms at Sites H, S and R in the eastern tropical Pacific (Dymond et al., 1984).

According to data from various authors, the total content of manganese oxides in todorokite from various sea and land environment is 71.1 to 80.86%. In this case, concentration of bivalent manganese in todorokite is 6.81 to 12.37%.

A considerable role in todorokite composition, besides a bivalent manganese, is played by other bivalent metals, in particular ore metals that are believed to stabilize its crystalline structure (Dymond et al., 1984). The occurrence of todorokite with 6% of Ni, about 4% of Cu, 5% of zinc in various natural environments, in particular in manganese nodules, can be explained by this fact (Table 70). In contrast to the three metals mentioned, nickel does not accumulate in todorokite; its maximum concentration does not exceed 0.2%. The share of iron oxides in todorokite is also inconsiderable: 0.07 to 2.22%.

In birnessi⁺e, the total content of manganese oxides is 50 to 80%, similar to todorokite, the share of bivalent manganese is rather large, 4.66 to 16%. Maximum concentrations of oxides of other ore elements are: 3.81 of nickel, 0.63 of copper, 0.25 of cobalt and 0.12 to 2.88% of iron (Table 71).

Vernadite or δ - MnO_2 differs considerably in its chemical composition from both todorokite and birnessite. It has a lower content of manganese oxides (35.7 to 61.2%) and a higher content of iron oxides (7 to 25.6%), which possibly form ultramicroscopic intergrowths. Maximum contents of oxides of other ore metals are 0.95 of nickel, 0.25 of copper, 3.41% of cobalt (Table 72).

Thus, the main fraction of nickel, copper and zinc accumulated in nodules is concentrated in todorokite and birnessite, whereas the major part of cobalt is accumulated in vernadite.

This is verified by special analysis of thoroughly sampled nodules in the Pacific, Atlantic and Indian oceans composed mainly of todorokite or mainly of vernadite (Table 73). Such an analysis revealed a number of regional features typical of the composition of two major manganese minerals which form nodules.

Todorokites of all three oceans have an almost similar manganese concentration (19.53 to 20.0%); todorokites of the Pacific and Indian oceans are close in concentrations of iron (9.0 to 9.5%), nickel (0.935 and 0.844%), copper (0.562 and 0.689%) and cobalt (0.172 and 0.130%). Todorokites in the Atlantic are characterized by a higher content of iron (16.4%) and cobalt (0.29%) and a lower content of nickel (0.62%), copper (0.17%) and zinc (0.09%). The content of lead in todorokite increases from the Pacific (0.033) to the Indian ocean (0.076) and to the Atlantic (0.133%).

Vernadites (δ MnO_2) of the Pacific and Atlantic have an almost similar content of manganese (15.3 and 14.7%), zinc (0.057 and 0.054%) and lead (0.115 and 0.142%) vernadites of the Indian and Atlantic oceans are close in concentration of iron (19.3 and 21.4%), nickel (0.25 and 0.25%), copper (0.69 and 0.78%) and cobalt

TABLE 67

Correlation coefficients of elements in nodules from the Pacific pelagic zones (Skornyakova, 1976)

Element	Mn	Fe	Ti	Ni	Co	Cu	Zn	Pb	Ba	Mo	V
Fe	-0.43										
Ti	-0.20	0.58									
Ni	0.83	-0.66	-0.47								
Co	0.21	0.25	0.44	-0.26							
Cu	0.59	-0.74	-0.54	0.94	-0.33						
Zn	0.32	-0.29	-0.20	0.23	-0.17	0.32					
Pb	-0.21	0.32	0.44	-0.37	0.56	-0.47	0.16				
Ba	0.26	-0.15	0.099	0.13	0.53	0.051	-	0.45			
Mo	0.51	-0.28	-0.095	0.45	0.25	0.26	-	0.24	0.42		
V	-0.14	0.41	0.43	-0.30	0.48	-0.40	-	0.34	0.36	0.32	
Siliceous part	-0.32	-0.33	0.20	0.026	-0.51	0.14	-	-0.38	-0.29	-0.33	0.39
	-	-	-0.44	0.38	0.032	0.73	0.30	-0.37	-	-	-
Depth	-0.016	-0.21	-0.11	0.36	-0.47	0.48	0.14	-0.45	-0.44	-0.18	0.46

TABLE 68

Correlation coefficients of elements in the Indian ocean nodules
(Skornyakova and Vanshtein, 1983)

Element	Mn	Ti	Co	Ni	Cu	Zn	Mo	V	Cr	Pb
Fe	-0.20	0.36	0.40	-0.39	-0.33	-0.10	-0.14	0.45	0.30	0.46
Mn		-0.39	0.12	0.63	0.41	0.24	0.41	0.24	0.42	0.02
Ti			0.65	-0.39	-0.45	-0.20	-0.14	0.06	0.29	0.04
Co				0.08	-0.24	-0.11	0	0.51	0.06	0.48
Ni					0.60	0.34	0.26	-0.04	-0.29	-0.11
Cu						0.47	-0.02	-0.39	-0.13	-0.16
Zn							0.12	-	-	0.16
Mo								0.61	-0.03	-0.31
V									0.09	0.41
Cr										-0.17

TABLE 69

Correlation coefficients of elements in the Atlantic nodules and ore crusts
(Cronan, 1975)

Element	Fe	Mn	Na	K	Ca	Mg	Cu	Zn	Ni	Co
Fe	1.00									
Mn	-0.74	1.00								
Na	-0.02	0.06	1.00							
K	0.01	-0.14	0.56	1.00						
Ca	-0.42	-0.11	-0.28	-0.10	1.00					
Mg	-0.16	-0.21	-0.11	-0.08	-0.18	1.00				
Cu	-0.41	0.51	0.22	0.27	-0.18	0.02	1.00			
Zn	-0.17	0.20	0.11	0.24	0.02	-0.10	0.28	1.00		
Ni	-0.70	0.78	0.06	0.10	0.03	0.06	0.75	0.27	-1.00	
Co	-0.27	0.35	-0.35	-0.47	0.05	-0.20	-0.26	-0.11	0.07	1.00

TABLE 70
Chemical composition of todorokite, % (Burns and Burns, 1979)

Component	1	2	3	4	5	6	7	8
MnO_2	65.39	68.46	72.98	71.1	78.03	80.86	74.91	67.98
MnO	12.37	10.70	6.81					
CaO	3.28	2.13	1.24	1.3	-	1.67	1.21	5.99
SrO	-	0.13	0.11	-	-	-	-	-
BaO	2.05	0.40	0.19	0.3	-	0.28	1.51	0.50
Na_2O	0.21	1.44	0.16	2.7	-	3.09	1.10	0.19
K_2O	0.54	0.75	0.92	0.6	-	0.66	1.55	0.19
MgO	1.01	3.22	2.26	3.4	-	3.27	3.12	1.51
CoO	-	0.18	-	0.2	0.5	-	0.01	0.21
NiO	-	-	-	6.4	6.76	0.06	0.32	-
CuO	-	0.44	0.77	3.8	2.05	0.31	0.55	-
ZnO	-	-	4.99	0.3	-	0.04	0.07	-
Al_2O_3	0.28	0.19	0.02	0.2	-	0.38	0.35	-
Fe_2O_3	0.20	0.07	0.59	0.1	2.22	0.35	1.64	0.08
SiO_2	0.45	0.41	0.55	0.2	1.67	-	0.96	0.20
H_2O	11.28	10.99	8.34	(9.4)	(9.07)	(8.51)	(12.46)	(14.31)
Other components	1.98	-	-	-	0.15	-	0.24	0.24
Total	99.24	99.51	100.74	(100.0)	(100.0)	(100.0)	(100.0)	99.29

(1) todorokite from Todoroki mine, Hokkaido (Yoshimura, 1934);
(2) todorokite from Charko Redondo mine, Cuba (Straczek et al., 1960);
(3) zinc todorokite from Montana, Philippsbourgh, contains 0.55% PbO (Larsen, 1962);
(4) todorokite from healed crack of manganese nodule, Site A,DOMES, 8° N,151° W (Burns and Burns, 1978a);
(5) todorokite layer in the North Pacific nodules (Stevenson and Stevenson, 1970);
(6) almost pure todorokite, well crystallized with slight admixture of birnessite from hydrothermal hill at the southern flank of the Galapagos Rift, 00°36'N, 86° 04'W (Corliss et al., 1978);
(7) todorokite from micronodules of metalliferous sediments from the Bauer depression, 9°S, 102°W (after Dymond and Ecklund);
(8) radial-crystalline todorokite in cavity of high phosphatized zeolithic hyaloclastic tuff from the Pacific, Site 6333, 22 °41'09''S, 160°50'084''W (Andruschenko et al., 1975)

TABLE 71
Chemical composition of birnessite, % (Burns and Burns, 1979)

Component	1	2	3	4	5	6	7
MnO_2	54.24	66.66	80.03	75.80	77.62	78.40	48.98
MnO	4.66	16.07					
CaO	1.65	1.05	-	0.39	2.38	2.29	1.03
Na_2O	2.17	0.16	8.67	1.90	2.82	2.04	-
K_2O	-	0.09	-	1.80	0.76	1.68	1.44
MgO	-	0.23	-	6.20	3.06	4.60	3.48
CoO	-	-	-	0.14	0.05	0.13	0.25
NiO	-	-	-	0.80	-	0.22	3.81
CuO	-	-	-	0.33	-	0.25	0.63
Al_2O_3	3.32	0.83	-	-	-	-	5.67
Fe_2O_3	2.88	0.86	-	0.62	0.12	0.24	2.86
SiO_2	18.92	2.62	-	0.90	0.06	1.56	1.07
TiO_2	0.28	-	-	-	-	-	0.03
H_2O	10.87	10.83	10.02	(10.80)	(13.10)	(8.6)	(30.4)
Other components	-	1.47	-	0.30	-	-	0.37
Total	98.99	100.87	98.72	(100.0)	(100.0)	(100.0)	(100.0)

(1) birnessite from fluvial-glacial deposites of Birness, Scotland
    (Jones and Milne, 1956);
(2) birnessite films on siliceous and calcareous rocks from Cummington,
    Massachussetts (Frondel et al., 1960);
(3) synthetic birnessite (Giovanoli et al., 1970);
(4) birnessite from micronodules of foraminiferal sediments, Gulf of Mexico,
    21°32'N, 85°45'W (Glover, 1977);
(5) birnessite from the Pacific nodules (Chukhrov et al., 1978B);
(6) birnessite in nodules from submarine volcano caldera, 8°48'2''N, 103°53'8''W
    (Lonsdale et al., 1980);
(7) birnessite in nodules from sea mountain, 16°29'S, 145°33'W (Burns and
    Fuerstenau, 1966).

TABLE 72
Chemical composition of vernadite, % (Burns and Burns, 1979)

Component	1	2	3	4	5
MnO_2	46.47	56.45	39.3	36.34	35.71
MnO	1.09	4.81			
CaO	2.15	5.17	3.4	3.08	3.44
Na_2O	2.29	0.12	1.1	-	-
K_2O	0.60	0.23	0.24	0.36	-
MgO	2.62	0.28	1.9	1.16	-
CoO	3.41	-	-	1.53	-
NiO	0.95	-	-	0.76	-
CuO	-	-	-	-	0.25
Al_2O_3	1.00	1.00	1.7	2.08	2.51
Fe_2O_3	10.47	7.00	25.6	17.16	24.74
SiO_2	0.80	1.30	4.35	5.35	11.06
TiO_2	1.50	-	1.5	2.00	-
H_2O	25.44	16.53	(20.09)	(29.8)	(22.5)
Other components	1.53	7.26	-	0.15	-
Total	100.32	100.15	(100.0)	(100.0)	(100.0)

(1) vernadite from hydrothermal sediments in Kurchatov fault zone, the Pacific, (Chukhrov et al., 1978a);
(2) vernadite from Lovozero (Chukhrov et al., 1978a);
(3) vernadite from nodules, the Pacific (Chukhrov et al., 1978a);
(4) vernadite from sea mountain, the Pacific, 16°29'S, 145°33'W (Burns and Fuerstenau, 1966);
(5) vernadite from hydrothermal hill, the Galapagos Rift, 00°36'N, 86°04'W (Corliss et al., 1978).

TABLE 73

Mean composition of manganese nodules formed of various minerals
(Cronan and Tooms, 1969; Cronan, 1975; Cronan and Moorby, 1981; Cronan, 1980)

Predominating mineral	Mn			Fe			Ni			Cu		
	P	I	A	P	I	A	P	I	A	P	I	A
Todorokite	19.53	20.9	19.85	9.07	9.5	16.39	0.935	0.844	0.62	0.562	0.689	0.17
Vernadite	15.30	14.7	17.35	13.78	19.3	21.45	0.390	0.250	0.26	0.107	0.079	0.078

Predominating mineral	Co			Zn			Pb		
	P	I	A	P*	I	A	P	I	A
Todorokite	0.172	0.130	0.29	0.132*	0.103	0.09	0.033	0.076	0.133
Vernadite	0.790	0.360	0.47	0.057*	0.054	0.073	0.115	0.142	0.260

*From data by Skornyakova, 1984; 1986; Gordeev, 1986.
(P) the Pacific;
(I) Indian ocean
(A) the Atlantic.

(0.36 and 0.47%). Vernadites of the Pacific are characterized by high concentration of cobalt (0.79%), whereas vernadites of the Atlantic are rich in lead (0.26%).

However, the general problem of the chemical composition of manganese minerals in nodules has been insufficiently studied. In particular, a disagreement exists on the degree of manganese oxidation in nodules.

According to Glagoleva (1972), the degree of manganese oxidation in the nodules of the Trans-Pacific profile is 1.80 to 1.95 (Figure 44).

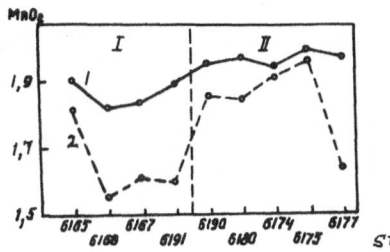

Fig. 44. Oxidation degree of Mn in nodules and host sediments along the profile trans the northwestern Pacific (Glagoleva, 1972). (I) hemipelagic sediments; (II) pelagic red clays; (1) nodules; (2) host sediments.

In the Central Pacific, the manganese oxide concentration in nodules is 0.25 to 1.86%, i.e. 1.3 to 9.4% of the total manganese content (Skornyakova et al., 1975; Bazilevskaya, 1985). However, from the data presented by American researchers, a relative concentration of bivalent manganese does not exceed 1%, even in todorokites that are rich in manganese (Murray et al., 1984). It was suggested that manganese oxidation may occur during the period from the raising of samples on board the vessel and the time of their processing (Dymond et al., 1984), similar to the bivalent iron oxidation in smectite (Lyle, 1983).

FORMS OF METALS IN NODULES AND RELATIONS BETWEEN THE COMPOSITION OF NODULES AND SEDIMENTS

When investigating the forms of metal occurrence in nodules, the technique of selective leaching was used, similar to the one applied to sediments (Chapter V). The aim of this investigation is to reveal the relation between mobile and inert forms of metals. Three major forms are recognized: sorbed; oxide and hydrooxide with various degree of crystallization; and lithogenic; the latter occurs in minerals of terrigenous or volcanic origin resistant to submarine weathering.

Sixty-two samples from the Pacific, Atlantic and Indian oceans were analysed (Moorby and Cronan, 1981; Gordeev, 1986); the results are given in Table 74 and Figure 45.

Despite a certain difference in the techniques applied by the authors of the papers cited, the results appeared to be consistent.

A sorbed form of manganese (fraction 1) in nodules plays a subordinate role in all cases, its relative concentration being less than 1% (Moorby and Cronan, 1981) or even 0.01 to 0.02% (Gordeev, 1986). The share of sorbed iron forms is larger, 0.01-0.03% in the Pacific nodules, <0.5 to 3% in the Indian ocean, and <1 to 6% in the Atlantic. A share of sorbed nickel is even larger, 0.5 to 0.8% in the Pacific, <0.5 to 5% in the Indian ocean and 1 to 18% in the Atlantic. For copper the

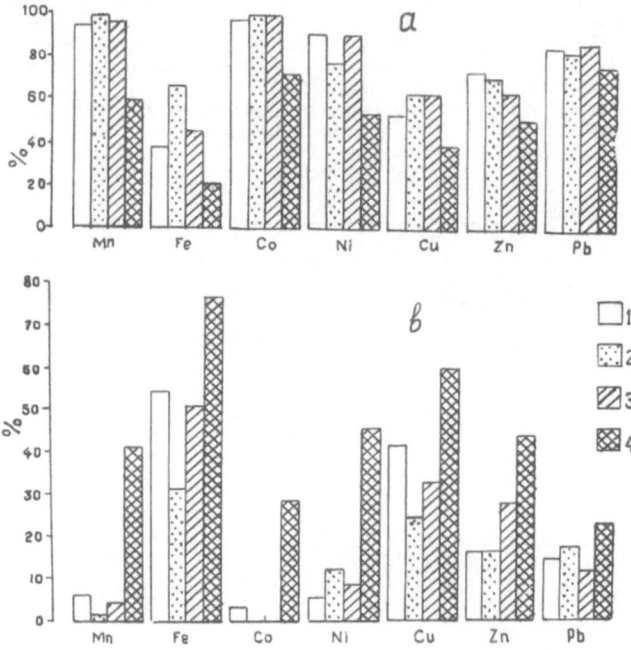

Fig. 45. Forms of metals in the Atlantic nodules (Moorby and Cronan, 1981).
(a) acid-reducing-agent leach; (b) hydrochloric acid leach. Phases:
(1) amorphous; (2) vernadite; (3) todorokite; (4) goethite.

corresponding values are: 2.0 to 2.2%, 1 to 12%, 1 to 19%; for cobalt, 0.08 to
0.12%, 0.5 to 1%; for zinc, 3.3 to 3.5%, 2 to 15%, 2 to 26%; for lead, 11.4% in the
Pacific nodules and 1 to 2% in the Atlantic nodules.

Oxide and hydrooxide in readily leachable forms (fraction 2) are predominant
for manganese. Their relative concentration in nodules of the Pacific is 98.8 to
99.7%; in the Indian ocean, it is 93 to 100%; in the Atlantic, 74 to 100%. For iron,
the corresponding values are considerably lower: 17.3 to 32.6; 65 to 85; 5 to 73%.
For other metals, the share of this form varies within a wide range. For nickel, it
is 92 to 97; 88 to 100; 52 to 98%; for copper, 59.3 to 87.5; 12 to 99; 39 to 82%;
for cobalt, 81.2 to 92.5; 88 to 100; 81 to 100%; for zinc, 34.7 to 83.8; 75 to 96;
32 to 94%; for lead, 15.5; 72 to 100; 70 to 97%.

Oxide and hydrooxide refractory forms (fraction 3), resistant to reducing
reagents, have no considerable share in the total balance of manganese in nodules.
Their relative content in nodules of the Pacific and Indian oceans does not exceed
1.5%, and only in some nodules of the Atlantic, it increases to 26%. In iron balance,
on the contrary, they can play the main role, being 54 to 78% in the Pacific;
13 to 37% in the Indian Ocean; 21 to 92% in the Atlantic. For nickel, as for manganese, their share is not great: 1 to 9% in the Pacific and Indian oceans, and to 45%
in some nodules of the Atlantic. A similar distribution is typical of copper, zinc
and cobalt; the behaviour of the latter in the selective leaching process resembles
the behaviour of manganese, not iron.

The role of a residual fraction (not dissoluable in hydrochloric acid) in the
balance of metals in nodules is subordinate. Its maximum share in iron is 13 to 18;
12 in lead; 10 in zinc; 7 in cobalt; 5 in copper; 2 in nickel; less 1% in manganese.
The results of the research show that the bulk of ore metals in nodules is in the

TABLE 74

Metal forms in manganese nodules as resulted from selective leaching analyses
(in relative % of total content)

Region and material properties	Forms	Mn	Fe	Ni	Cu	Co	Zn	Pb
The Pacific, Site 2474, nodules of todorokite composition, 8 samples*	1	0.02	0.03	0.7	2.0	0.12	3.3	11.4
	2	98.8	17.3	92.3	68.8	81.2	34.7	15.5
	3	1.2	77.8	7.0	28.0	18.5	59.3	72.6
	4	0.02	4.9	0.04	1.2	0.2	2.7	0.5
	5	11.7	0.43	0.63	0.43	0.30	0.065	0.066
The Pacific, Site 2483, nodules of todorokite composition, 8 samples*	1	0.01	0.01	0.5	2.0	0.08	3.4	
	2	99.7	32.6	97.1	87.5	92.5	83.8	
	3	0.3	54.2	2.1	9.2	6.7	9.5	
	4	0.02	13.2	0.3	1.3	0.7	3.3	
	5	24.7	6.15	1.17	0.99	0.19	0.12	
The Pacific, Sites 2492 and 2520, nodules of vernadite composition, 5 samples*	1	0.01	0.03	0.8	2.2	0.08	3.5	
	2	98.5	25.2	92.0	59.3	89.5	68.2	
	3	1.5	71.1	6.9	37.7	10.0	26.6	
	4	0.01	3.6	0.3	0.8	0.4	1.7	
	5	19.4	12.3	0.53	0.41	0.34	0.057	
Indian ocean, Madagascar basin, nodules of vernadite composition, 11 samples*	1	1	0.5-3	2 - 5	7 - 12	0.5	9 - 15	-
	2	93-100	66 -83	88 -97	72 - 93	88-100	75 - 85	72-91
	3	0-6	15 -35	3 - 8	3 - 17	2- 12	2 - 33	4-17
	4	1	1	1	<4 - 5	1	5 - 10	5-12
	5	6.04-15.3	13.6-19.3	0.157 - 0.347	0.0845 - 0.704	0.094 - 0.390	0.039 - 0.060	0.082 - 0.131
Indian ocean, Central Basin, nodules of todorokite composition, 16 samples**	1	1	0.5- 3	0.5- 3	1- 6	1	2-14	-
	2	94-100	65 -85	90 -100	81-99	89-100	80-96	87-100
	3	0- 6	13 -37	1 - 9	3-11	4- 8	2- 5	4- 10
	4	1	2 -18	-	-	0- 7	2- 5	4- 7
	5	11.4-31.0	3.39-12.7	0.465 - 1.57	0.415 - 1.77	0.057 - 0.164	0.074 - 0.174	0.023 - 0.111

Table 74 (continued)

Region and material properties	Forms	Mn	Fe	Ni	Cu	Co	Zn	Pb
The Atlantic, sea mountains southwest of Portugal, 10 nodules and 4 ore crusts of various composition**	1	1	1-6	1-18	1-19	1	2-26	1-2
	2	74-100	5-73	52-98	39-82	81-100	32-94	70-97
	3	1-26	21-92	2-45	16-58	2-19	7-45	2-27
	4	1	2-8 22	< 2	2-4	-	3-7	2-3
	5	2.6-13.9	12.7-27.9	0.11-0.46	0.042 - 0.0330	0.029 - 0.0790	0.042 - 0.066	0.092 - 0.265

*after Gordeev, 1985: (1) fraction leached by acetic solution; (2) fraction leached by 0.5M hydroxilamine; (3) fraction leached by 5M hydroxilamine and 3M HCl; (4) residual fraction; (5) concentration of metals in initial samples.
** after Cronan and Moorby, 1981: (1) fraction leached by acetic solution; (2) fraction leached by 1M hydroxilamine-hydrochloride; (3) fraction leached by 6M HCl; (4) residual fraction; (5) concentration of metals in initial samples.

form of geochemically mobile oxides and hydrooxides (fractions 2 and 3) which are authigenic in origin.

From the results of other experimental works using the ion-exchange technique, a part of the ore matter in nodules is in the form of interlayer cations of chalcophanite type (Mn^{2+}, Cu, Zn, Cd), part of them is in asbolan layers (Ni, Co^{2+}), whereas $Co^{3+}$is associated with Mn^{4+} (Chelischev and Gribanova, 1985).

To clarify the role that the underlaying and host sediments play in the supply of ore material to nodules the total composition of both and mobile forms of metals were compared.

Along the profile through the North Pacific, Volkov (1979) found that Mn:Fe ratio in nodules and in leachable fractions of sediments are close, being about 1.1 on average. That permitted the conclusion that there was a genetic link between nodules and host sediments. However, other comparisons have shown that the problem is more complicated.

The coefficients of metal concentration in nodules with respect to host sediments on the scale of the World Ocean are: 3.3 for iron; 9 for zinc; 19 for copper; 41 for cobalt; 60 to 68 for manganese and nickel, the average Mn : Fe ratio being 2.26 (Table 75). Within the Central Pacific sites, all average concentration coefficients were close to those mentioned, except for iron (17 to 24 instead of 3.3); the Mn : Fe ratio was 2 to 2.5 times lower than its average for the entire ocean and 2.4 times higher than in host sediments (1.36 to 4.22 in nodules and 0.83 to 1.05 in sediments) (Table 76).

The correlation coefficients of metal concentration in nodules and their mobile forms in sediments calculated for the three Sites showed a direct correlation between them (except for manganese and iron) only within Site 2483, where diagenetic and hydrogenous-diagenetic nodules occur. Within the other two sites where hydrogenous nodules (according to their morphological properties) are spread, such a correlation is distinct only for cobalt.

Besides such a comparative analysis, it is also possible to use ratios of various metals to manganese, as has been done by Volkov (1979). The data obtained by Volkov, and the results from the three sites considered, are given in Table 76.

As we can see, an average ratio of nickel, copper and cobalt to manganese in nodules and in leachable sediment fractions are really close in the most cases. This is true for both diagenetic (Site 2483) and for hydrogenous nodules (Sites 2474 and 2520). So different techniques of statistical processing of analytical results give various answers to the question whether a genetic link between nodules and host sediments exists.

The problem of the ore matter supply to nodules and the formation mechanism of the latter will be considered in turn, after we review the data on element concentrations and distribution in nodules.

TABLE 75

Average concentrations of metals in nodules and host sediments; their enrichment and correlation coefficients
(Skornyakova et al., 1984)

Parameter	Mn	Fe	Ni	Cu	Co	Zn	Mn/Fe
Average concentration in oceanic nodules*	18.60	12.47	0.66	0.45	0.27	0.12	2.26
Average concentration in oceanic sediments**	0.31	3.79	0.0097	0.0237	0.0055	0.013	0.08
Enrichment coefficient in nodules	60	3.3	68	19	41	9	28
Average concentrations in nodules at Sites							
2483	25.5	6.23	1.14	0.98	0.20	0.11	4.22
2474	19.8	11.5	0.69	0.46	0.29	0.071	1.74
2520	17.3	12.7	0.49	0.36	0.32	0.063	1.36
Average concentration in sediments at Sites (mobile forms)							
2483	0.39	0.37	0.0151	0.0199	0.0044	0.0043	1.05
2474	0.41	0.47	0.0134	0.0194	0.0052	0.0041	0.88
2520	0.42	0.52	0.0127	0.0195	0.0063	0.0042	0.83
Enrichment coefficients in nodules							
2483	65	17	75	40	45	26	4
2474	48	24	51	24	56	17	2
2520	41	24	38	18	51	15	1.6
Correlation coefficients (nodules/sediments)							
2483	0.02	-0.14	0.33	0.42	0.34	0.23	0.08
2474	-0.18	-0.21	0.01	0.29	0.10	-0.1	-0.07
2520	0.16	-0.27	0.02	-0.25	0.17	-0.54	0.13

* From McKelvey et al., 1983;
** From Lisitsin, 1978.

TABLE 76

Average Me/Mn ratios in the Pacific nodules and host sediments (mobile forms)*

Ratio of Me/Mn × 100	Transoceanic profile		Site 2483		Site 2484		Site 2520	
	Nodules	Sediments	Nodules	Sediments	Nodules	Sediments	Nodules	Sediments
Ni/Mn	3.11	2.31	4.5	3.9	3.8	5.1	0.8	1.1
Cu/Mn	2.33	2.42	3.5	3.3	2.3	4.8	1.5	1.3
Co/Mn	1.23	1.42	2.8	3.0	2.1	4.6	1.8	1.5

*Transoceanic profile after Volkov, 1979; Sites: data from Table 75.

MAJOR LITHIC* ELEMENTS IN NODULES

The most widespread lithic elements in manganese nodules are elements of the group of alkaline and alkaline-earth elements, namely, potassium, sodium, calcium, magnesium, barium; elements of the lithogenic group or hydrolysate elements: aluminium and titanium; silicon which is of a polygenic nature (litho- and biogenic); and elements of the biogenic group, namely, carbon, phosphorus and sulphur (we shall also consider nitrogen as being related to this group). We shall also consider pore water, which plays an important role but which is sometimes ignored.

CONCENTRATION OF LITHIC ELEMENTS IN NODULES

The generalized data obtained by various authors on major lithic elements in nodules are given in Table 77. The range within which their concentration can vary is as wide as for the major ore elements considered in the previous chapter. The ratio of maximum-to-minimum concentrations varies from tens (as for potassium, sodium, aluminium, titanium, sulphur, carbon) to hundreds, and even thousands (other elements). For alkaline and alkaline-earth elements the variability range widens as their atomic weight rises.

An average potassium concentration in nodules decreases from the Pacific (0.74 to 0.82%) to the Atlantic (0.56 to 0.57%) and to the Indian ocean (0.48%). According to the majority of the available data, the same pattern is typical of sodium, the average concentration of which decreases from the Pacific (about 2%) to the Atlantic (about 1.9%) and to the Indian ocean (1.7%). The data on average magnesium concentration in nodules of the Pacific and Atlantic is not completely consistent (1.5 - 1.9%), but its concentration in the Indian ocean nodules seems to be lower. An average calcium concentration in the Pacific nodules is about 2%, in the Atlantic it is about 3%, in the Indian ocean it seems to reach 2%. The data on average barium concentration are controversial. From Volkov (1979), the barium maximum was registered in the Indian ocean nodules, and its minimum in the Atlantic nodules. However, from McKelvey et al. (1983) barium concentrations are similar in the three oceans and are 0.21 to 0.23%.

An average silicon concentration seems to be at a maximum in the nodules of the Indian ocean (9.4% from McKelvey et al., 1983) and it is lower in the Pacific and Atlantic nodules (6.3 to 8.3%).

According to the above mentioned authors, an average aluminium concentration in nodules of all three oceans is approximately the same (2.4 to 2.7%). It is slightly higher in the Pacific nodules which follows from the previous estimates based on a smaller number of samples. An average titanium concentration is maximum in the Pacific nodules (about 0.7%), slightly lower in the nodules of the Indian ocean and are minimum (0.42 to 0.43%) in the Atlantic nodules.

The estimates of the average concentration of phosphorus in the Pacific

---

*In this work this term is relative, to a certain extent, since it is a label for the macroelements of nodules that do not belong to the group of "useful elements" of the latter.

TABLE 77

Concentration of major lithic elements in oceanic nodules*

Ele-ment	Concentration extrema	Average concentration in the oceans									Average concentrations in the World Ocean			
		The Pacific			The Atlantic			Indian ocean						
	1	2	3	1	2	3	1	2	3	1	4	3	1	***
K	0.11 – 3.70	0.74	0.753	0.82	0.56	0.567	0.57	–	–	0.48	0.8	0.6427	0.73	490
Na	0.30 – 6.55	2.06	2.054	2.05	2.31	1.88	1.86	–	–	1.70	2.6	1.9409	1.97	410
Mg	0.09 – 5.39	1.76	1.710	1.50	1.7	1.89	1.75	1.51	–	1.43	1.7	1.8234	1.57	531
Ca	0.02 –28.73	1.98	1.960	1.96	2.84	2.96	3.72	4.10	2.37	1.97	1.9	2.47	2.23	1083
Ba	0.000– 2.14	0.39	0.276	0.235	0.498	0.498	0.228	0.158	0.182	0.21	0.18	0.2012	0.23	463
Si	0.24 –28.80	8.27	8.320	7.62	7.36	9.58	6.34	7.46	11.40	9.39	9.4	8.624	7.69	509
Al	0.11 – 8.00	3.27	3.060	2.75	3.20	3.27	2.37	2.41	2.49	2.67	2.9	2.82	2.70	592
Ti	0.01 – 8.90	0.80	0.674	0.73	0.434	0.421	0.42	0.63	0.662	0.62	0.67	0.647	0.69	900
P	0.02 – 6.03	0.197	0.235	0.28	0.299	0.098	0.91	1.34	–	0.37	0.13	0.2244	0.37	321
S	0.05 – 3.50	0.188	–	0.32	–	–	1.34	0.14	–	0.83	–	–	0.51	106
C$_{total}$	0.033– 1.50	0.15**	–	0.342	–	–	0.77	–	–	0.212	–	–	0.327	75

*For description of the term see the text.
*After: *McKelvey et al., 1983; (2) Volkov, 1979; (3) Cronan, 1980; (4) Mero, 1965.
**G$_{org}$;   ***number of sites.

nodules made by various researches are almost the same, 0.20 to 0.28%. For the Atlantic they differ approximately by ten times (0.098% from Cronan, 1980; and 0.91 from McKelvey et al., 1983). An average phosphorus concentration in the Indian ocean nodules is 1.34% by Volkov's estimates (44 samples), whereas it is 0.37% from McKelvey's data (the total number of samples analysed is 321 regardless of the ocean they belong to).

The estimates of average sulphur concentration also differ considerably: in the Pacific it is 0.19% according to Volkov (1979) and 0.32% from McKelvey et al. (1983); in the Indian ocean it is 0.24% and 0.83%, respectively. However, both give an average of 1.34% for the Atlantic.

Volkov (1979) estimated the average concentration of organic carbon in the Pacific nodules as being 0.15%. McKelvey et al. estimate carbon concentration as 0.342% in the Pacific, 0.77% in the Atlantic and 0.212% in the Indian ocean, but do not mention whether it is total or organic; such high values lead us to presume that most probably it is total.

The estimates of average concentration of lithic elements in the World Ocean nodules differ (Table 77); however, for the majority of individual elements this discrepancy does not exceed 20-30%. According to the number of samples considered, from 75 samples for carbon to over 1000 samples for calcium, the most representative data are given by McKelvey et al. (1983).

Silicon is the most widespread metalloid. Its average concentration in manganese nodules of the ocean is about 7.7%. Then follows aluminium (2.7), calcium (2.3), sodium (about 2), magnesium (about 1.6%). The concentration of the other lithic elements varies from 0.2 to 0.7%: potassium and titanium, 0.7; sulphur, 0.5; phosphorus and total carbon, about 0.3; barium, about 0.2%.

As sulphur and carbon are not so thoroughly studied contrast to other elements, we shall give some additional data on them.

From the work of Glasby et al. (1978) who analysed 4 samples, we find that the sulphur content in nodules of the Pacific and Indian oceans is 0.09 to 0.33%, being 0.2% on average.

Japanese researchers (Terashima et al., 1982) analysed 111 samples of nodules from the Central Pacific and found extremely low concentrations of sulphur in them, 0.012 to 0.159%; 0.065% on average (Table 78).

In the nodules from the Pacific radiolarian zone, the sulphur content as determined by American researchers (McKelvey et al., 1979) is many times higher: 0.05 to 1.00%; 0.22% on average (39 samples).

The reference nodule samples from the Pacific radiolarian belt and from the Atlantic ocean contain 0.47 and 0.86% of SO_3 or 0.19 and 0.34% of S (Neil, 1980), respectively.

The concentration of total carbon in nodules from the Central Pacific varies from 0.02 to 0.418%, and concentration of the carbonate carbon varies from less than 0.001 to 0.368 and of organic carbon from 0.02 to 0.14%. Average concentrations of carbonate and organic carbon are 0.039 and 0.521%, respectively (Terashima et al., 1982).

Romankevich (1977) studied the organic components of nodules in the Pacific and Indian oceans. According to his data, the concentration of organic carbon in nodules varies from 0.04 to 0.27% and of organic nitrogen from less than 0.001 to 0.21%; the N_{org}/C_{org} weight ratio being 1 to 37%.

From various authors (Mero, 1965; McKelvey et al., 1979), we find that the concentration of pore water in nodules varies from 5.3 to 47.15%. When both the water concentration and loss on ignition are determined, the latter is usually several percent higher, due to dehydration of hydrooxides (Mero, 1965; Glasby et al., 1978).

The content of HCl-insoluble residues in nodules and ore crusts varies over a wide range, from 0.2 to 68.8% thus reflecting a variety of proportions of ore, lithic and biogenic components in nodules.

TABLE 78

Concentrations of total sulphur, carbon and water-soluble sodium in the Central Pacific nodules, %
(Terashima et al., 1982)

Region	S	C	Na	Mn	Number of samples
Mid-Pacific sea mountains	0.0598	0.0663	0.24	19.97	16
Northern part of the basin	0.0535	0.0742	0.23	20.76	37
Central part of the basin	0.0582	0.0650	0.25	22.68	14
Southern part of the Central and Takelau basins	0.0678	0.0542	0.26	19.81	12
Manihiki Plateau and north-eastern part of the Manihiki basin	0.0704	0.0867	0.27	11.94	7
Penrhyn basin	0.0864	0.0854	0.26	17.56	25
Average	0.0650	0.0730	0.25	19.53	111

DISTRIBUTION OF LITHIC ELEMENTS IN NODULES

To recognize the distribution pattern of lithic elements in nodules, bulk samples from various oceanic regions and depths were analysed as well as individual parts of nodules. The results are given in Table 79.

Minimum potassium content was determined in the northernmost Arctic zone, and its maximum in the southernmost Antarctic zone. On the whole, deep-sea nodules are twice as rich in potassium as compared to relatively shallow-water nodules, i.e. those from a depth less than 3000 m.

Sodium concentration decreases slightly towards the Arctic zone and has two maxima within 0 to 20°S and 40 to 60°S, irrespective of depth.

Magnesium shows its minimum within 0 to 40 °S and its maximum in the Antarctic zone, irrespective of depth.

Calcium shows two minima, within 0 to 20°N and 40 to 60 °S, and one maximum, within 20 to 40 °N. Its concentration depends strongly upon the depth, being 2.7 times higher at a depth down to 3000 m, in contrast to deeper nodules.

Barium shows its minimum (0.109%) in the Antarctic zone and maximum in the northern temperate zone, 20-40°N. Its behaviour resembles that of calcium. At a depth less than 3000 m, its concentration is 1.5 times higher, on average, than at greater depth.

A minimum (7%) in the northern temperate zone and two maxima (12-13.8%) in the Arctic and Antarctic zones are typical of silicon. In deep-sea nodules its content is 1.6 times higher than in shallow water ones (5 and 8.3%, respectively).

The latitudinal distribution of aluminium is relatively even but its concentration in deep-sea nodules is 1.7 times higher than in shallow-water ones (this is similar to the silicon distribution with depth).

Titanium shows a slight minimum (0.53%) in the Arctic zone and a maximum (0.88%) within 20 to 40 °S, and does not seem to depend upon the depth.

Phosphorus shows a distinct minimum, 0.18% in the southern tropical zone of minimum biological productivity (0 to 20 °S), and a maximum, 0.86%, in the near Antarctic zone (40 to 60°S). In shallow-water nodules its concentration is some 7 times higher than in deep-sea nodules (1.15% and 0.22%, respectively).

The sulphur minimum (0.26%) is detected within 0 to 20°N and its maximum (0.82%) is registered within 20 to 40°S. Shallow-water nodules are twice as rich in sulphur than are deep-sea nodules.

The total carbon minimum (0.22%) coincides with the sulphur minimum, whereas its maximum (0.68%) coincides with the phosphorus maximum. In shallow-water nodules its concentration is three times higher than in deep-sea ones (0.91 and 0.31%, respectively).

Thus, the distribution pattern of lithic elements in nodules with respect to their latitudinal occurrence allows their subdivision into groups according to their minima and maxima: alkaline elements, potassium and sodium; alkaline-earth elements, calcium and barium (magnesium distribution differs from the mentioned two considerably); hydrolisating elements, aluminium and titanium; and biogenic elements, phosphorus, sulphur and carbon. Silicon has an individual pattern of distribution.

The distribution of lithic elements in the nodule tops and bottoms was analysed in detail by Dymond et al. (1984) who studied the material from H, S and R MANOP Sites in the eastern tropical Pacific. Hemipelagic sediments are spread over Site H, siliceous radiolarian and siliceous clayey ooze occurs within Site S, whereas red deep sea clay is typical of Site R. Distribution of lithic elements in nodule tops and bottoms and also in the bulk samples of nodules is shown in Table 80.

The analysis of the data on the basis of calculated averages shows that within Site H, all lithic elements, but for sodium, tend to accumulate in the tops of nodules with respect to their bottoms. This pattern is the most pronounced in the case of silicon, aluminium, titanium (concentration differs by a factor of 2) and barium (1.5 times). The difference in the concentration of other elements in the nodule tops and bottoms is not considerable.

Within Site S, only aluminium (by 2 times), titanium and barium (by 1.3 times) show higher concentration in the nodule tops with respect to their bottoms. Sodium and silicon tend to accumulate in the bottoms, the difference of average

TABLE 79

Average concentrations of lithic elements in nodules with respect to geographical latitude and oceanic depth
(after McKelvey et al., 1983)

Element	Geographical latitude							Depth	
	60-40°N	40-20°N	20°N-0	0-20°S	20-40°S	40-60°S	60-80°S	to 3000m	over 3000m
K	0.57	0.71	0.74	0.83	0.64	0.86	0.974	0.47	0.81
Na	1.64	1.87	1.84	2.30	1.96	2.37	1.96	1.83	2.02
Mg	1.72	1.61	1.57	1.40	1.42	1.58	2.64	1.61	1.553
Ca	2.19	3.30	1.87	2.07	2.12	1.82	2.25	4.65	0.22
Ba	0.291	0.370	0.267	0.173	0.234	0.136	0.109	0.34	0.22
Si	12.15	7.04	7.42	7.58	8.07	8.48	13.83	5.07	8.29
Al	2.81	2.61	2.58	2.56	3.28	2.46	3.18	1.72	2.95
Ti	0.53	0.66	0.64	0.76	0.88	0.62	0.64	0.76	0.68
P	0.26	0.50	0.28	0.18	0.47	0.86	-	1.148	0.22
S	0.73	0.51	0.26	0.64	0.82	0.51	-	0.85	0.43
C	0.26	0.29	0.22	0.30	0.412	0.682	-	0.91	0.311

Note: solid line stands for maxima and dashed line stands for minima.

TABLE 80

Concentrations of lithic elements in nodule tops and bottoms: within three sites of the Eastern Pacific*
(Dymond et al., 1984)

Element	Site H, 27 samples (hemipelagic sediments)			Site S, 5 samples (radiolarian sediments)			Site R, 7 samples (red clays)		
	Tops	Bottoms	Bulk sample	Tops	Bottoms	Bulk samples	Tops	Bottoms	Bulk samples
K	0.527-0.820 / 0.733	0.398-0.782 / 0.588	0.622-0.866 / 0.763	0.805-0.936 / 0.880	0.848-0.900 / 0.878	0.954-0.985 / 0.970	0.696-0.821 / 0.738	0.914-1.009 / 0.977	1.020-1.286 / 1.133
Na	2.45-4.31 / 3.16	3.03-5.05 / 3.68	2.53-3.95 / 3.05	1.91-2.12 / 1.99	2.09-2.44 / 2.27	2.03-2.24 / 2.12	1.57-1.79 / 1.71	1.90-1.92 / 1.91	1.84-2.16 / 1.95
Mg	1.31-2.52 / 2.16	1.24-1.93 / 1.55	1.38-2.14 / 1.89	1.55-1.74 / 1.65	1.50-1.66 / 1.56	1.66-1.74 / 1.75	1.14-1.71 / 1.37	1.66-2.23 / 1.86	1.40-1.87 / 1.70
Ca	1.30-1.78 / 1.45	1.13-1.40 / 1.29	1.25-1.76 / 1.46	1.54-1.83 / 1.73	1.49-1.79 / 1.71	1.75-1.88 / 1.81	1.36-2.08 / 1.52	1.30-4.69 / 1.79	1.09-1.55 / 1.38
Ba	0.155-0.445 / 0.326	0.110-0.308 / 0.199	0.230-0.477 / 0.320	0.160-0.355 / 0.265	0.159-0.255 / 0.210	0.222-0.522 / 0.390	0.119-0.147 / 0.139	0.104-0.135 / 0.115	0.177-0.208 / 0.192
Si	2.32-5.80 / 4.27	1.46-3.72 / 2.08	1.86-5.93 / 3.84	9.01-10.24 / 8.62	6.66-11.00 / 9.22	6.12-10.75 / 9.77	7.91-10.48 / 9.24	8.22-10.00 / 9.08	8.51-11.11 / 9.39
Al	0.871-2.587 / 2.057	0.631-1.745 / 1.012	0.998-2.458 / 1.725	1.400-2.335 / 1.908	1.755-2.300 / 0.953	2.183-2.528 / 2.336	2.139-3.690 / 2.757	2.909-3.280 / 3.115	2.950-4.150 / 3.340
Ti	0.053-0.208 / 0.138	0.028-0.111 / 0.063	0.067-0.137 / 0.109	0.345-0.399 / 0.365	0.211-0.312 / 0.294	0.288-0.346 / 0.323	0.811-0.867 / 0.835	0.535-0.620 / 0.592	0.619-0.743 / 0.765

*Numerator: extrema; denominator: average.

concentrations being 2 to 2.3% for sodium and 8.6 to 9.2% for silicon. Other elements behave indifferently.

Most of lithic elements (potassium, sodium, magnesium, calcium, aluminium) within Site R tend to accumulate in the bottoms being 1.1 to 1.3 times higher there. Barium and titanium slightly accumulate in the nodule tops, silicon is indifferent.

These results can be interpreted as follows. Under conditions of hemipelagic sedimentation and ore formation, lithic elements arrived to nodules from above, i.e. out of sea water. On the contrary, under oxic conditions of pelagic zone, in red clay, these elements, but for barium and titanium, are supplied to nodules from the underlying sediments more actively than out of sea water. Under suboxic conditions of deposition and diagenesis, in siliceous pelagic ooze, aluminium and titanium are mainly supplied from sea water, whereas silicon is supplied from below, out of underlying siliceous sediments.

Distribution of organic carbon and nitrogen in various parts of the Pacific and Indian ocean nodules was studied by Romankevich (1977). According to his data (Table 81), concentration of organic carbon in the upper ore shell of nodules is 0.04 to 0.12% (0.074% on average), in the bulk ore body it is 0.06 to 0.1 (0.10% on average), in the nuclei it is 0.05 to 0.27% (0.128% on average). In host sediments, concentration of organic carbon is considerably higher, from 0.13 to 0.35%; 0.234% on average.

Concentration of organic nitrogen in the same samples is: in ore shell, 0.001 to 0.007% (0.003% on average); in the bulk ore body, less 0.001 to 0.004% (0.002% on average); in nuclei, 0.005 to 0.021% (0.012% on average); in host sediments, 0.035 to 0.092% (0.061% on average).

It is possible to conclude from these results that the concentration of organic components is at a maximum in the nodule nuclei, somewhat lower in the bulk ore body, and a minimum in the ore coating. But in some nodules from the Pacific and Atlantic oceans the author found different patterns of organic carbon distribution.

FORMS OF LITHIC ELEMENTS IN NODULES

Lithic elements can occur in manganese nodules in various forms, namely, as a component of ore minerals, in a sorbed form, as a component of detrital and bio- genic minerals and also as a dry sea-salt residue (their natural water content may exceed 40%). Well-shaped cubic crystals of sodium chloride are sometimes recognized in nodules under the scanning electron microscope.

A number of lithic elements are permanent components of the chemical compo- sition of the major manganese minerals in nodules (Table 82).

Thus, the maximum concentration of K_2O in todorokite is 1.55%, in birnessite, 1.80%, in vernadite, 0.60%. For MgO the corresponding values are: 3.4; 6.20 and 2.62%; for CaO, 5.99; 2.38 and 5.17%; for SiO_2, 1.67; 1.56 and 11.06%; for Al_2O_3, 0.38; 5.67 and 2.51%. Concentration of BaO in marine todorokite is from 0.3 to 1.51%, concentration of TiO_2 in birnessite is to 0.28%, and 1.5 to 2.0% in vernadite.

The forms of lithic elements in nodules are studied by the same technique of selective leaching used to study the forms of ore elements (see the previous chapter).

Moorby and Cronan (1981), besides ore elements, studied the forms of calcium, aluminium and titanium in nodules of the Indian and Atlantic oceans. From their data, calcium is the most mobile lithic element in nodules (Table 83). The relative share of calcium (percent of its total content) leachable by acetic acid out of the Indian ocean nodules is from 36 to 83%; it is 8 to 96% in the Atlantic nodules, and 64%, on average, for the bulk of samples.

A share of calcium which can be leached by mixed acetic acid and hydroxyla- mine-hydrochloride varies from traces to 76%, and is 7.5 to 25% on average for nodules of various regions. On average, from 2 to 4% of calcium occur in 6 M- hydrochloric acid soluble form, and from 10 to 21% in the residue.

Aluminium forms in nodules are essentially different. On average, only 6% of the element considered is transferred to an acetic leach, 15 to 27% is transferred to a hydroxylamine leach and 18 to 31% to a hydrochloric one (in the Atlantic and Indian oceans, respectively). The residue fraction is the most rich in aluminium containing 36 to 62% of the element on average (in the Indian and Atlantic oceans, respectively).

TABLE 81

Concentration of organic carbon and nitrogen in the Pacific and Indian ocean nodules*

(Romankevich, 1927)

Sample	C_{org}	N_{org}	N/C, %
Upper crust	$\dfrac{0.04 - 0.12}{0.074 \ (7)}$	$\dfrac{0.001 - 0.007}{0.003 \ (10)}$	$\dfrac{1.7 - 11.6}{3.1 \ (6)}$
Ore matter	$\dfrac{0.06 - 0.13}{0.10 \ (4)}$	$C_{\pi.} \ \dfrac{- 0.004}{0.002 \ (8)}$	$\dfrac{1.0 - 3.8}{2.1 \ (4)}$
Nucleus	$\dfrac{0.05 - 0.27}{0.128 \ (4)}$	$\dfrac{0.005 - 0.021}{0.012 \ (4)}$	$\dfrac{6.3 - 30.3}{8.7 \ (4)}$
Host sediments	$\dfrac{0.13 - 0.35}{0.234 \ (9)}$	$\dfrac{0.035 - 0.092}{0.061 \ (12)}$	$\dfrac{18.9 - 37.0}{24.4 \ (9)}$

*Numerator: concentration extrema; denominator: average concentration; in brackets: number of samples.

TABLE 82
Concentration of lithic elements (%) in manganese minerals of nodules and ore
crusts (after Burns and Burns, 1979)

Compound	Todorokite	Birnessite	Vernadite
K_2O	0.19 - 1.55	0.76 - 1.80	0.23 - 0.60
Na_2O	0.19 - 3.09	1.90 - 2.82	0.12 - 2.29
MgO	1.51 - 3.4	3.06 - 6.20	0.28 - 2.62
CaO	1.21 - 5.99	0.39 - 2.38	2.15 - 5.17
BaO	0.3 - 1.51	-	-
SiO_2	0.2 - 1.67	0.06 - 1.56	0.80 - 11.06
Al_2O_3	0.2 - 0.38	3.32 - 5.67	1.00 - 2.51
TiO_2	-	0.03 - 0.28	1.50 - 2.00
H_2O	9.4 - 14.31	8.6 - 30.4	16.5 - 29.8

Titanium forms differ from both calcium and aluminium. Its average concentration in nodules from various regions is 8.5 to 21% in a hydroxylamine fraction; 46 to 87% in a hydrochloric fraction and 5 to 33% in a residue. Iron forms have similar pattern of distribution (Moorby and Cronan, 1981).

The data presented and many other discussed earlier (Calvert and Price, 1977; Bischoff et al., 1981; Glasby and Thijssen, 1982; Dymond et al., 1984) testify that both the concentration of lithic elements and forms of their occurrence in nodules are controlled by the regional and local sedimentation regimes and possibly by a micro-environment within different localities.

Some authors believe that alumosilicate minerals are the main carrier of potassium, sodium, barium, silicon and aluminium to manganese nodules (Bischoff et al., 1981). A fraction of these elements enters clinoptilolite and phillipsite which are usually components of nodules; the formula of clinoptilolite is

$$K_{2.3} Na_{0.8} Al_{3.1} Si_{14.9} O_{36} \cdot 12 H_2O; \text{ for phillipsite it is}$$

$$K_{2.8} Na_{1.6} Al_{4.4} Si_{11.6} O_{32} \cdot 10 H_2O \text{ (Kastner, 1979).}$$

However, in nodules of the eastern Pacific, studied by Dymond et al. (1984), silicon and aluminium are almost completely incorporated in the ore phase, or so the authors believe. A correlation between magnesium and manganese is also found in the Central Pacific nodules (Calvert and Price, 1977).

A biogenic factor may play a definite role in the composition of lithic matter of nodules, affecting the behaviour of such biogenic elements as silicon, calcium, phosphorus, sulphur and carbon.

Table 79 shows that the silicon concentration in nodules from high latitudes is almost twice as great as in other regions. No doubt it is associated with the high biological productivity of diatomaceous plankton that influences both sediments and nodule compositions. The latter is verified by direct observations of the inner structure of nodules from the Antarctic zone of the Pacific.

The data of selective leaching of nodules from the Indian and Atlantic oceans (Table 83) show that the greater portion of calcium in them occurs in carbonate or sorbed forms. This is also verified by other data.

The results of chemical analysis of calcium carbonate in the Pacific, Atlantic and Indian ocean nodules were first reported by Murray and Renard (1891). They estimated $CaCO_3$ concentration in nodules as 2.2 to 7.0% (4.1% on average). Later estimates (Calvert, 1978) of CO_2 concentration in deep nodules of the Pacific are 0.04 to 0.37%, and 0.14% (on average) in the southern part of the Central Pacific (Terashima et al., 1982). However, in other parts of the ocean, the calcium carbonate content in nodules may be considerably higher, which is verified by the results of electron microscopic analysis and by other data. The concentra-

TABLE 83

Forms of lithic elements (in relative %) in manganese nodules*
(after Moorby and Cronan, 1981)

Element	1 fraction	2 fraction	3 fraction	Residue
		Somali Basin, 11 samples		
Ca	36 - 82 (67)	4 - 13 (7.5)	2 - 13 (4)	10 - 47 (21)
Al	3 - 8 (6)	16 - 37 (30)	19 - 48 (29)	30 - 44 (36)
Ti	-	5 - 11 (8.5)	76 - 97 (87)	2 - 20 (5)
		Central basin of the Indian ocean, 16 samples		
Ca	48 - 80 (61)	16 - 32 (25)	traces - 8 (4)	3 - 54 (12)
Al	3 - 8 (6)	4 - 35 (24)	14 - 89 (34)	3 - 64 (37)
Ti	-	6 - 32 (21)	14 - 63 (46)	13 - 52 (33)
		The Atlantic		
Ca	8 - 96 (65)	traces - 76 (24)	traces - 17 (2)	traces - 25 (10)
Al	21 - 25 (7)	3 - 54 (15)	14 - 30 (18)	4 - 82 (62)
Ti	-	2 - 29 (13)	41 - 84 (62)	10 - 40 (19)

*from-to (average). (1) fraction leached by acetic acid, (2) fraction leached by reducting reagent,
(3) fraction leached by HCl, (4) residual fraction.

tion of CO_2 in a reference nodule sample from the Pacific radiolarian zone was determined as being 0.76 to 1.39%, and 11.62% from the Atlantic (the Blake Plateau) (Flanagan and Gottfried, 1980).

Phosphorus, which is sometimes anomalously high in nodules and especially in ore crusts at sea mountains, can form its own mineral phase that corresponds in its composition to fluorocarbonate-apatite (Figure 46). A small portion of sulphur in nodules can be associated with sulphide minerals, inclusions of which were recognized in several regions of the Pacific (Müller, 1979; Ostwald, 1983; Baturin and Dubinchuk, 1983, 1984). A wide occurrence of barite in pelagic sediments (Goldberg and Arrhenius, 1958) gave birth to a supposition that barium in nodules is mainly in the form of barite. However, barium in nodules (its maximum concentration may be 2.14% according to McKelvey et al., 1983) is also a component of todorokite which may contain to 3.87% of barium (Burns and Burns, 1979). The Ba : Mn ratio in the Central Pacific nodules also verifies this fact (Calvert and Price, 1977).

In the concluding section of this chapter we shall compare average concentrations of lithic elements in manganese nodules and in pelagic sediments of the World Ocean, thus illustrating a possible participation of sedimentary matter in nodule formation. We shall consider the data by Lisitsin (1978) on sediments and the data by McKelvey et al. (1983) on nodules (Table 84).

TABLE 84

Average concentration coefficients of major lithic elements in nodules with respect to pelagic sediments of the ocean*

Element	Average concentration in nodules (1)	Average concentration in sediments (2)	Concentration coefficient
K	0.73	1.33	0.5
Na	1.97	1.82	1.1
Mg	1.57	1.42	1.1
Ca	2.23	13.60	0.16
Ba	0.23	0.26	0.9
Si	7.69	19.65	0.4
Al	2.70	5.35	0.5
Ti	0.69	0.26	2.6
P	0.37	0.11**	3.3
S	0.51	0.30***	1.7

*Nodules: from McKelvey et al., 1983.
 Sediments: from Lisitsin, 1978.
 **from Baturin, 1978.
***from Volkov, 1984.

One can see from this comparison that the least active role in nodule formation is played by calcium (concentration coefficient is 0.16). Then follows a group of elements composed of potassium, silicon and aluminium (coefficient being 0.4 to 0.5). The concentration of sodium, magnesium and barium in nodules and sediments is similar. Sulphur tends to certain accumulation in nodules (the coefficient is 1.7). This tendency is more distinct with titanium (coefficient 2.6, which is close to iron) and phosphorus (3.3). The ability of phosphorus to accumulate in association with iron under various marine conditions is a typical feature of its geochemistry (Baturin, 1978). A considerable correlation between iron, titanium and phosphorus in nodules reveals a certain similarity in their behaviour (Calvert and Price, 1977; Calvert and Piper, 1984).

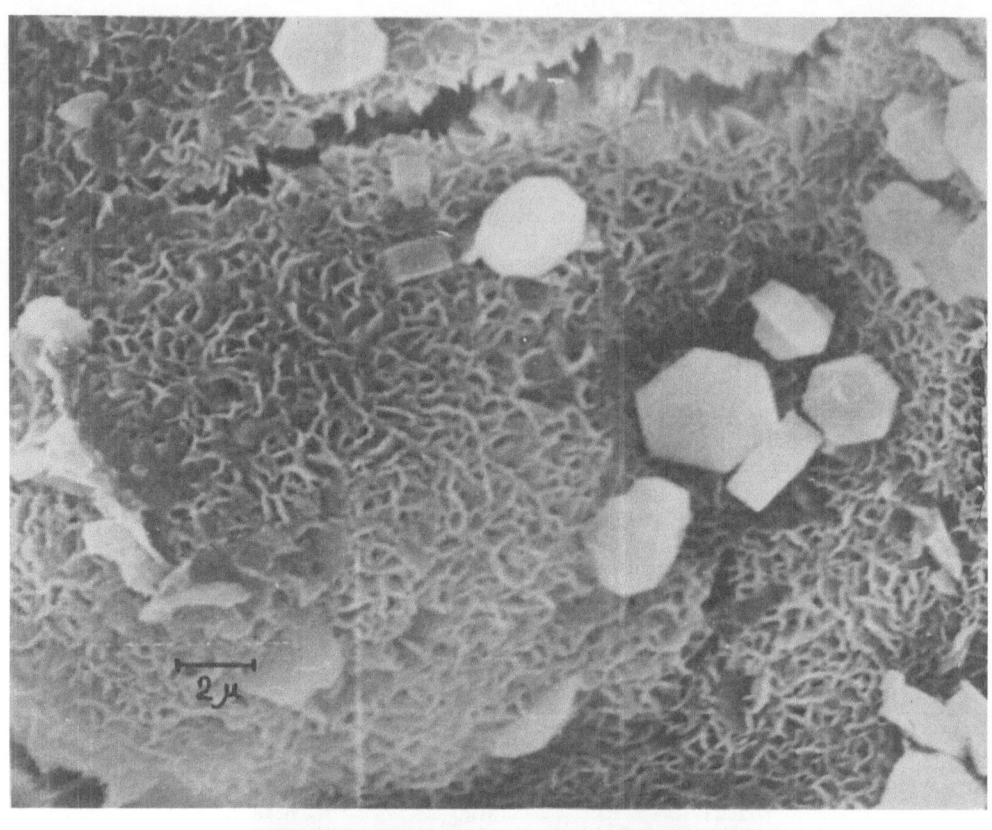

(a)

(b)

Fig. 46. Crystals of fluor carbonate-apatite in ore crusts from the Mid Pacific sea mountains: (a) Site 6352, depth 1630 m, magnification by 3500; (b) Site 6358-2, depth 2450 m, magnification by 1200.

MICROELEMENTS IN MANGANESE NODULES

The concentration of microelements in manganese nodules is less than 0.1% and for many of them is as low as $10^{-4} - 10^{-7}\%$. All of them are poorly studied, in contrast to macroelements, due to analytical difficulties and lack of economical interest in most of them. However, from a scientific standpoint, all elements of such interesting natural objects as nodules are worth being considered and analysed. In the course of these studies, some of the microelements would probably be recognized as valuable ones and would be processed along with the major ore elements. Besides it would be possible to distinguish between useful elements and contaminants that could be injurious to the industrial processing of nodules. So the knowledge of their behaviour and distribution in nodules would be of practical value.

Considering the data on the geochemistry of microelements we shall first touch upon the rare metals according to the groups of the Periodic Table, then we shall dwell, in turn, upon radioactive and precious metals and rare metalloids, halogens included.

RARE ALKALINE ELEMENTS

Rare alkaline elements, namely, lithium, rubidium and cesium belong to the first group of the Periodic Table.

Lithium is the lightest of the alkaline metals. It was studied in the Pacific, Indian and Atlantic manganese nodules. The results of these analyses are given in Table 85.

Dvoretskaya and Boiko (1979) analysed eight nodules from the Pacific (the cross-section from Japan to California) and determined that the bulk samples contain 36 to 120 ppm of lithium (80 ppm on average), nodule shells contain 27 to 91 ppm (44 ppm on average), nuclei contain 38 to 75 ppb (57 ppb on average). However, these results do not seem to be representative as the shells and nuclei analysed were taken from different samples. The lithium content was determined by McKelvey et al. (1983) in 19 nodule samples from various parts of the Pacific. Lithium concentration varied from less than 100 to 600 ppm, the detection limit of the analyses being 0.01%.

From one sample the lithium content in the bulk samples of nodules from the Pacific radiolarian belt (the Clarion-Clipperton zone) was 22 to 200 ppm; 38 to 240 ppm in nodule shells and 22 to 180 ppm in their nuclei, averages for bulk samples of nodules, shells and nuclei are 100, 117 and 70 ppm, respectively. This testifies to the predominant though relatively insignificant accumulation of lithium in the outer parts of nodules.

In two nodule samples analysed layer-by-layer the lithium concentration was higher in their bottoms, which were in contact with sediments (radiolarian ooze), 200-240 ppm in contrast to 90-100 ppm in their tops.

A reference nodule sample from a radiolarian zone prepared by the U.S. Geological Survey contains 140 ppm of lithium (Flanagan and Gottfried, 1980); a similar sample prepared by the P.P. Shirshov Oceanology Institute of the USSR Ac. Sci. contains 129 ppm. The South Pacific nodules contain less lithium (22 to 77 ppm; 40 ppm on average) with respect to the nodules of the radiolarian zone. The minimum concentration of lithium (2 to 5 ppm) was assessed in ore crusts at the

TABLE 85
Lithium content (ppm) in nodules

Region	Sample	Li Range	Mean	N	Reference
Pacific Ocean	Whole samples	<100- 600	200	19	McKelvey et al., 1983
	- " -	23-1055	160	25	Haynes et al., 1986
northern part	- " -	26- 120	80	8	Dvoretskaya and Boiko, 1979
	Outer shells	27- 91	44	5	- " -
	Nuclei	38- 75	57	3	- " -
radiolarian belt	Outer shells	38- 240	117	9	Author's data
	Nuclei	22- 180	70	7	- " -
	Referens sample	-	140	1	Flanagan and Gottfried, 1980
northern tropical part	Whole samples	16- 110	42	4	Sano et al., 1985
southern part	- " -	22- 77	40	9	Author's data
Mid Pacific mountains	manganese crusts	2- 5	3	7	- " -
Indian Ocean	Whole samples	6- 73	28	42	Levitan and Gordeev, 1981
Atlantic Ocean	- " -	2- 260	70	29	Author's data
southern part	Tops	2- 180	50	5	- " -
	Bottoms	14- 210	140	5	- " -
Blake Plateau	Reference sample	-	76	1	Flanagan and Gottfried, 1980
Sediments			29	-	Lisitsin, 1978
			64	-	Morozov, 1968
			53	-	Dvoretskaya and Boiko, 1979

TABLE 86
Rubidium content (ppm) in nodules of the Pacific Ocean

Region	Rb Range	Mean	N	Reference
Whole ocean	5 - 60	14	18	Calvert and Price, 1977
	5 - 60	15	43	Haynes et al., 1986
radiolarian belt	15 - 20	18	19	Calvert et al., 1978
hights	-	7	17	Calvert and Piper, 1984
valleys with thick sediment cover	-	11	19	- " -
valleys with thin sediment cover	-	14	30	- " -
Northern part	17 - 39	24	8	Dvoretskaya and Boiko, 1979
shells	16 - 30	22	4	- " -
nuclei	14 - 42	26	3	- " -
Southern part	8.1- 30	14	12	Rankin and Glasby, 1979
Sediments	-	139	-	Morozov, 1969
	-	100	-	Calvert and Price, 1977
	-	44	-	Lisitsin, 1978
	-	112	-	Dvoretskaya and Boiko, 1979

Mid Pacific sea mountains in the northern tropical part of the Pacific.

The Indian ocean nodules contain 6-73 ppm of lithium, 28 ppm on the average.

In the South Atlantic, bulk samples (from the Brazil and Cape Basins), the lithium content in nodules varies from 2 to 260 ppm; in their top parts it is 2 to 180 ppm, in the bottom parts it is 14 to 260 ppm. Similar to the Pacific radiolarian zone nodules, they tend to a relative accumulation of lithium in their bottom parts, which contact the sediments. This seems to testify to a prevailing diagenetic mechanism of lithium supply to nodules. The lithium average in nodules is about 70-80 ppm (Table 85). The lithium average in deep oceanic sediments is from 29 to 64 ppm according to different estimates (Morozov, 1968; Dvoretskaya and Boiko, 1979; Lisitsin, 1978). Thus, a concentration coefficient for lithium in the World Ocean nodules is 1.5 to 2.0.

Rubidium was analysed in the Pacific nodules, where its content varies from 5 to 60 ppm (Calvert and Price, 1977; Table 86). In the South Pacific, the rubidium content is 8 to 30 ppm, being 14 ppm on average (Rankin and Glasby, 1979). In the radiolarian zone nodules, rubidium was analysed more thoroughly. From the first estimates the rubidium average was 18 to 20 ppm (Calvert et al., 1978; McKelvey et al., 1979). Later, the rubidium content in nodules was assessed many times at Site A of the DOMES project. The Site location is 8 °27 N, 150 °47'W, its area is 55 x 65 km^2; the average depth being 5100 m; sediments are mainly Quaternary siliceous-clayey muds underlain by Tertiary siliceous oozes.

The rubidium content in nodules from the Site in question was determined within three geomorphological features, namely, submarine hills, valleys with thick and valleys with thin sedimentary cover (Calvert and Piper, 1984). The rubidium average in the nodules of the first type is 7, of the second 11, and of the third 14 ppm, which testifies to the influence of diagenetic processes upon rubidium accumulation in the nodules of the latter two sites. Nodules from submarine hills are mainly hydrogeneous (Mn/Fe ratio is 1.92 on average), nodules from submarine valleys are mainly diagenetic (Mn/Fe ratio is 3.2 to 4.3 on average).

A reference nodule sample from the Pacific contains 22 ppm of rubidium and that from the Atlantic contains 11 ppm (Flanagan and Gottfried, 1980).

Rubidium distribution in various parts of individual nodules from the North Pacific turned out to be regular: in bulk samples it is 17 to 39, in shells it is 16 to 30, in nuclei it varies from 14 to 42 ppm; their averages being 14, 22 and 26 ppm, respectively (Dvoretskaya and Boiko, 1979).

On the basis of the data presented, the rubidium average in the North Pacific nodules is 17 ppm.

The rubidium average in deep oceanic sediments is 44 to 139 ppm according to various authors (Morozov, 1968; Calvert and Price, 1977; Lisitsin, 1978; Dvoreskaya and Boiko, 1979).

Thus, on the whole, the rubidium content in nodules is 2.5 to 8 times lower than in deep oceanic sediments. This ratio may vary widely from region to region. For instance, according to the data of 21 stations in the North and South Pacific, it varies from 0.05 to 1 (Calvert and Price, 1977).

Cesium. Up to now, determinations of cesium content in nodules have been scarce, the Pacific samples being the only source of data. In the North Pacific, it varies from less than 1 to 2 ppm. Samples do not show any evidence of enrichment or depletion of cesium either in nodule shells or in their nuclei (Dvoretskaya and Boiko, 1979) (Table 87).

In various parts of the South Pacific, the cesium content in nodules is less than 0.7 to 0.82 ppm and only in one nodule from the equatorial zone is it 2.6 ppm (Rankin and Glasby, 1979; Table 87). The cesium average in the Pacific nodules is probably about 1 ppm. Its average in deep oceanic sediments was estimated by various authors as being 4.2 to 10 ppm (Morozov, 1968; Lisitsin, 1978; Dvotetskaya and Boiko, 1979; Rankin and Glasby, 1979). So, cesium concentration in nodules is 4 to 10 times lower than that in host sediments.

RARE METALS OF THE SECOND GROUP OF PERIODIC TABLE

In this section we shall consider beryllium, strontium, cadmium and mercury.

TABLE 87
Cesium content (ppm) in nodules of the Pacific

Region	Cs		N	References
	Range	Mean		
Whole ocean	< 0.5-2.6	0.75	7	Haynes et al., 1986
northern part	< 1-2	1	8	Dvoretskaya and Boiko, 1979
shells of nodules	< 1-1	1	4	- " -
nuclei of nodules	< 1-2	1	3	- " -
south-western part	-	0.82	2	Rankin and Glasby, 1979
southern part	< 0.5-0.7	0.6	10	- " -
equatorial zone	-	2.6	1	- " -
Sediments	-	8	-	Morozov, 1960
	-	4.2	-	Lisitsin, 1978
	-	10	-	Dvoretskaya and Boiko, 1979
	-	5.8	-	Rankin and Glasby, 1979

TABLE 88
Berillium content (ppm) in nodules

Region	Be		N	References
	Range	Mean		
Pacific ocean	2-5	3	7	Mero, 1965
	2.2-8.3	4.4	29	Inoue et al., 1983
	2-15	4	29	Haynes et al., 1986
radiolarian belt	-	2	1	McKelvey et al., 1983
	-	2.8	1[1]	Flanagan and Gottfried, 1980
	2.4-4.6	3.0	5	Krishnaswami eet al., 1982
northern part	2.2-5.0	3.2	10	Sharma and Somayajulu, 1982
	2.9-9.8	6.9	15	Kusakabe and Ku, 1984
	2.7-6.4	3.7	18	Segl et al., 1984a
central part[2]	1.3-13.5	5.4	59	Aplin and Cronan, 1985a
South China Sea	4.9-5.8	5.4	2	Mangini et al., 1986
Peru Basin	0.35-1.85	1.3	16	Reyss et al., 1985
Indian Ocean	3.1-7.8	4.8	7	Krishnaswami et al., 1982
	3.2-12.1	6.1	12	Sharma and Somayajulu, 1982
Atlantic Ocean	-	5.57	1[1]	Flanagan and Gottfried, 1980
	5.7-9.3	7.4	3	Sharma and Somayajulu, 1982
Sediments	-	2.6	-	Turekian and Wedepohl, 1961

[1] - reference samples; [2] - manganese crusts.

Beryllium is a lithophilic element which properties are closer to aluminium than to magnesium. Strontium belongs to the rare alkaline-earth elements. Other alkaline earths (magnesium, potassium, barium) are not rare elements and they were considered in the previous chapter. Radium also belongs to the rare alkaline-earth elements. It is one of the intermediate unstable elements of the uranium family of radioactive decay. We shall touch upon its behaviour in the chapter devoted to the dating of nodules. Cadmium and mercury are chalcophylic metals.

Beryllium content was assessed in only a few samples (Table 88).

The first estimates were made by Mero (1965) in the Pacific nodules and they were 2 to 5 ppm.

In one sample from the radiolarian zone, the beryllium content was 2 ppm
(McKelvey et al., 1979). This result was verified by the analysis of a reference
sample from the same zone, 2.8 ppm (Flanagan and Gottfried, 1980).

However, in the latter paper by McKelvey et al. (1983), beryllium content was
assessed as less 10 to 100 ppm in the Pacific and Indian ocean nodules. These
estimates seem to be semi-quantitative.

A reference sample of the Atlantic nodule contains 5.57 ppm of beryllium
(Flanagan and Gottfried, 1980). The beryllium average in the Pacific nodules is
probably 2 to 3 ppm, whereas in the crusts from sea mountains it is somewhat
higher, up to 5.4 ppm. The average concentration of beryllium in deep oceanic sediments is 2.6 ppm (Turekian and Wedepohl, 1961). So a concentration coefficient of
beryllium in nodules with respect to host sediments is about 1.

Strontium. The data on strontium content in the World Ocean manganese
nodules are considerably more representative, in contrast to that on beryllium.
Volkov (1879) in his review determined the range of strontium concentration in the
Pacific nodules as being 0.024 to 0.16% (151 samples), and in the Atlantic nodules
as 0.058 to 0.19% (16 samples). According to our results (32 samples were analysed
from the Atlantic: Brazil and Cape basins) this range is somewhat wider, i.e. 0.024
to 0.228%. For the World Ocean nodules this range is 0.010 to 0.285% (McKelvey et
al., 1983).

From the results of various authors, strontium averages in nodules of the
Pacific, Atlantic and Indian oceans are 0.084 to 0.088, 0.089 to 0.110 and 0.079 to
0.086%, respectively. The World Ocean nodules contain 0.0825 to 0.085% of strontium
(Table 89).

Regional and local differences in strontium concentration in nodules are
considerable. At one Site within the Pacific radiolarian zone (8°20'N, 153°W; average
depth is 5000 m), strontium distribution is rather regular and its relative concentration is rather low, 0.0420% (Calvert et al., 1978). Within the same zone, at the
neighbouring Site (8°27'N, 150°47'W; average depth is 5100 m), strontium showed
a higher concentration, its average being 0.0889% at submarine hills, 0.0682 in
valleys with thick sediments and 0.0588% in valleys with thin sedimentary cover
(Calvert and Piper, 1984).

Strontium content in nodules was analysed with respect to the latitudinal
position of the latter; it was revealed that in each 20° interval, from 60°N to 80 °S,
strontium average concentration does not vary considerably (from 0.071% in the
interval of 0 to 20°S to 0.097% in the interval from 20 to 40 °N). Thus, strontium
does not show any distinct correlation between its concentration and climatic position.
However, its concentration obviously depends upon the depth: in shallow-water
nodules (to 3000 m), strontium average is 0.122%, in deeper nodules, it is 0.075%
(McKelvey et al., 1983).

The selective leaching technique was used to determine the forms of strontium
occurrence in nodules. On the average, 54% of total strontium along with 65% of
iron and only 2% of manganese are oxalate leachable in the Pacific nodules from the
radiolarian zone (Calvert et al., 1978). On the whole, the relative share of mobile
strontium bound to the oxide-hydrooxide fraction of the Pacific nodules is 57 to 100%
(Calvert and Price, 1977). The correlation obtained shows that strontium is associated with iron and titanium. At one Site within the Pacific radiolarian zone,
strontium shows a positive correlation with iron and titanium (Calvert and Piper,
1984); at the other, it shows a strong positive correlation with titanium ($r = 0.068$),
a weak positive correlation with iron and a strong negative correlation with
manganese ($r = -0.78$) (Calvert et al., 1978). However, the strontium content in
manganese minerals of hydrothermal origin may be relatively high. It is 0.41% in
hydrothermal todorokite in the nodules sampled at a sea mountain in the Atlantic
(19°24'N, 17°23.4'W; depth is 1925 m). It is considerably lower, 0.032%, in goethite
in the same sample (Halbach and Manheim, 1984).

The average strontium concentration in deep oceanic sediments is 0.0755%
(Lisitsin, 1978) and 0.0275% in the Pacific radiolarian sediments (Bischoff et al.,
1979). Consequently, an average concentration coefficient for strontium in nodules
is 1.1. For the Pacific, this coefficient seems to be somewhat higher - arising from
the data presented by Calvert and Price (1977) - 1.6 on average, fluctuations
being from 0.8 to 3.4.

TABLE 89
Strontium content (%) in nodules

Region	Sr		N	References
	Range	Mean		
Pacific Ocean	0.06 -0.16	0.100	3	Riley and Sinhaseni, 1958
	0.024-0.16	0.088	151	Volkov, 1979
	-	0.085	-	Cronan, 1980
	-	0.084	-	McKelvey et al., 1983
	⟨0.005-0.30	0.083	493	Haynes et al., 1986
radiolarian belt	⟨0.005-0.16	0.045	78	- " -
Mid Pacific mountains	⟨0.005-0.30	0.13	27	- " -
depths ⟩ 3000 m	⟨0.005-0.18	0.08	320	- " -
depths ⟨ 3000 m	⟨0.005-0.28	0.135	68	- " -
Atlantic Ocean	0.058-0.190	0.110	16	Volkov, 1979
	-	0.093	-	Cronan, 1980
	-	0.094	-	McKelvey et al., 1983
	0.024-0.228	0.089	32	Author's data
Indian Ocean	-	0.086	-	Cronan, 1980
	-	0.079	-	McKelvey et al., 1983
World Ocean	-	0.0825	-	Cronan, 1980
	0.010-0.185	0.085	369	McKelvey et al., 1983
Sediments	-	0.0755	-	Lisitsin, 1978
Radiolarian belt				
(Pacific Ocean)	-	0.0275	-	Bischoff et al., 1979

Cadmium content was assessed in manganese nodules of the Pacific, Atlantic and Indian oceans by Ahrens et al. (1967), the range was 3 to 21.2 ppm. Later, these results were verified by many other researchers (Table 90).

The cadmium content in the bulk samples of nodules of the Pacific radiolarian zone varies from 4 to 30 ppm and is 10 ppm, on average (McKelvey et al., 1979). A reference sample from the region considered is richer in cadmium, 22.3 ppm (Flanagan and Gottfried, 1980).

A layer-by-layer analysis of the Pacific nodules from various parts of the ocean showed that nodule shells contain from 2.5 to 15 ppm of cadmium, whereas ore nuclei contain from 0.15 to 6.8 ppm, the average concentration being 6.0 and 4.2 ppm, respectively. In the bulk samples of these nodules the cadmium concentration is 2.2 to 10.6 ppm, being 5.8 ppm on average (Baturin and Oreshkin, 1984; 1986).

Cadmium content in ore crusts at the Mid Pacific and Line sea mountains varies from 1.4 to 10.0 ppm (Baturin and Oreshkin, 1984; Aplin and Cronan, 1985).

In nodules of the Indian ocean the cadmium content is less than 3 to 13.2 ppm, being 3.5 ppm on average (Ahrens et al., 1967). Later, a higher concentration of cadmium was determined in the Indian ocean nodules; 8 to 13 (10 ppm, on average) in the Madagascar Basin and 11 to 25 ppm (20 ppm, on average) in the Central Basin (Moorby and Cronan, 1981).

In the Atlantic nodules, the cadmium content is 5 to 16.2 ppm, 9 ppm on average, from the results of two series of determinations (Ahrens et al., 1967; Moorby and Cronan, 1981). It is interesting to note that the first group of authors determined the maximum content of cadmium in the Atlantic nodules, whereas according to other data, cadmium content in the Atlantic nodules is a minimum. In particular, in the reference sample from the Blake Plateau, the cadmium content is 6.5 ppm (Flanagan and Gottfried, 1980).

A layer-by-layer analysis of the Atlantic nodules (Brazil and Cape basins), performed in the P.P. Shirshov Oceanology Institute of the USSR Academy of

TABLE 90
Cadmium content (ppm) in nodules and crusts

Region	Sample	Cd		N	Reference
		Range	Mean		
Pacific Ocean	whole sample	5.1 - 8.4	6.7	2	Mullin and Riley, 1956
		7 -11	9	2	Riley and Sinhaseni, 1958
		< 3 -21.2	6	15	Ahrens et al., 1967
		2.2 -18.6	5.8	16	Baturin and Oreshkin, 1984, 1986
	shells	2.5 -15.0	6.0	26	- " -
	nuclei	0.15- 6.8	4.2	9	- " -
radiolarian belt	whole nodules	4 -30	10	17	McKelvey et al., 1979
	- " -	1 -35	12.3	127	Haynes et al., 1986
	reference sample	-	22.3	1	Flanagan and Gottfried, 1980
subsea mountains	nodules	1 -35	8.3	15	Haynes et al., 1986
	crusts	3.3 -10.0	6.0	31	Baturin and Oreshkin, 1984
	- " -	1.4 - 8.1	4.0	59	Aplin and Cronan, 1985a
depth >3000 m	nodules	1 -35	10.7	133	Haynes et al., 1986
depth <3000 m	- " -	1 -35	10.2	23	- " -
Indian Ocean		<3 -13.2	3.5	9	Ahrens et al., 1967
Madagascar Basin		8 -13	10	11	Moorby and Cronan, 1981
Central Basin		11 -25	16	16	- " -
Atlantic Ocean		5 -16.2	9	7	Ahrens et al., 1967
		6 -14	9.5	14	Moorby and Cronan, 1981
Blake Plateau	reference sample	-	6.5	1	Flanagan and Gottfried, 1980
Sediments	-	-	0.3	-	Oreshkin, 1977

Sciences showed the following values of cadmium content in various parts of nodules: 20 ppm on average, 13 ppm in nodule tops, 13.5 ppm in inner parts and 6 ppm in nuclei. Though, on the whole, these results are somewhat overestimated in contrast to other estimates for the Atlantic nodules, they demonstrate a distinct enrichment by cadmium of the bottom parts of the nodules directly affected by diagenetic processes in the underlying sediments.

Forms of cadmium occurrence in the nodules of the Atlantic and Indian oceans were determined by the traditional methods of selective leaching (Moorby and Cronan, 1981). The results obtained demonstrate that the major part of cadmium is leached by hydroxylamine. The share of this form in the total cadmium content is 67 to 100% in the nodules of the Madagascar Basin, 79 to 100% in the Central Indian Basin and 25 to 84% in the Atlantic nodules. The share of an acid soluble form is

TABLE 91
Mercury content (ppm) in nodules

Region	Hg		N	Reference
	Range	Mean		
Nodules				
Pacific Ocean	-	2000	1	Riley and Sinhaseni, 1958
	< 1- 775	-	7	Harriss, 1968
	2- 780	150	68	Haynes et al., 1986
northern part	36- 45	40	3	Toth, 1980
	6- 26	12	5	Baturin et al., 1986d
southern part	3- 7	5	3	- " -
East Pacific Rise	-	92	1	Toth, 1980
Bauer Deep	-	20	1	- " -
Atlantic Ocean	< 1- 810	-	5	Harriss, 1968
Indian Ocean	< 1- 3	1	4	- " -
Hydrothermal crusts				
Pacific Ocean	3- 296	40	5	Toth, 1980
Atlantic Ocean	136-1080	390	7	- " -
Hydrogenous crusts				
Pacific Ocean	13- 300	112	10	Toth, 1980
Atlantic Ocean	640-2520	1720	3	- " -
Sediments of the Pacific	0.7- 58	30	170	Cox and McMurtry, 1981

0 to 33%, 0 to 12% and 0 to 50%, respectively; the share of cadmium in the residue is 0 to 20%, 0 to 16%, 0 to 55%. These estimates show that cadmium enters the predominantly readily dissolved oxides and hydrooxides of manganese.

All researchers note that in nodules cadmium is associated mainly with manganese; that can be seen from the graphs of Cd-Mn and Cd-Fe correlation in the Pacific nodules and ore crusts (Figure 47).

The Cd-Zn correlation is also distinct. From Ahrens et al. (1967), we find that Zn to Cd ratios in the nodules they studied were from 47 to 117, being 77 on average.

Cadmium average in manganese nodules of the World Ocean is about 10 ppm (deduced from the data in Table 90), in deep-sea sediments it is 0.3 ppm (Oreshkin, 1977). This allows us to estimate its average concentration coefficient as about 30.

Mercury content in manganese nodules was first assessed as 2 ppm (Riley and Sinhaseni, 1958). Then mercury behaviour was studied by Harriss (1968). He determined that the mercury content in nodules of the Pacific, Atlantic and Indian oceans varies from less 1 to 775, less 1 to 810, less 1 to 3 ppb, respectively (Table 91).

Later, Toth (1980) estimated the mercury content in five samples of the nodules from the East Pacific Rise region, Bauer depression and North Pacific. Its distribution was considerably more regular in the material analysed than in the previous samples; from 20 to 92 ppb, being 47 ppb on average. The mercury content in hydrogenous ore crusts of the Pacific is 20 to 300 ppb, being 140 ppb on average. In hydrothermal crusts it is 3 to 296, being 40 ppb on average. In the Atlantic, the values are 136 to 1080 ppb and 390 ppb, respectively.

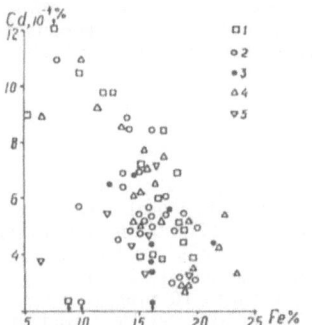 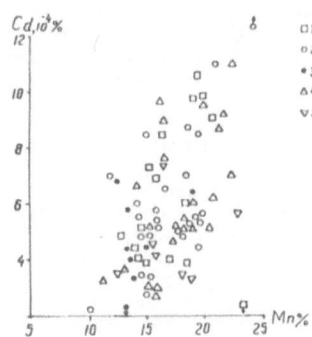

Fig. 47. Correlation between Cd, Mn and Fe concentration in the Pacific nodules and ore crusts (Baturin and Oreshkin, 1986). (1) bulk nodule samples; (2) nodule shells; (3) nodule nuclei; (4) ore crusts, tops; (5) ore crusts, bottoms.

Some 100 samples of the Pacific nodules and ore crusts were analysed by Oreshkin et al. (Baturin, Oreshkin, et al., 1986). The material analysed was subdivided into three categories:
(1) nodules and ore crusts sampled during cruises 43 and 48 of R/V "Vityaz" (1968 and 1970); samples were first processed on board the vessel and then the powdered material was kept in paper envelopes;
(2) hand specimens of nodules (lump) and ore crusts collected in the cruises mentioned; they were powdered directly before the analysis;
(3) samples collected in the cruises by R/V "Dmitry Mendeleev" and "Akademik Mstislav Keldysh".
The mercury content in the samples of the first category turned out to be anomalously high, from 150 to 3320 ppb, the average concentration of mercury in bulk samples, shells and nuclei of nodules and also in ore crusts being almost similar, 600 to 770 ppb. In the samples of the second category, the mercury content was generally lower, the shells of nodules being highly enriched in contrast to the inner parts: 87 and 10 ppb, respectively. In the samples of the third category the mercury content is a minimum, from 3 to 26 ppb, being 10 ppb on average.
The high mercury content in the samples of the first category might result from the ability of powder samples to sorb mercury from the air through paper and polyethylene films. This phenomenon has been observed many times when analysing standard samples of sediments (Oreshkin et al., 1985). The outer parts of hand specimens (lump samples) from the Vityaz seem to be contaminated by mercury, whereas the content of mercury in their inner parts represents its natural background.
The mercury content in the samples of the third category is 1.5 to 2 times higher in nodule shells with respect to their inner parts. It is not yet clear, however, whether this is a superficial sorption effect resulting from long-time storage of the samples, or whether the distribution is natural.
One of the samples from the East Pacific Rise zone, analysed by Harriss, showed a high concentration of mercury (775 ppb). A high mercury content was determined also by Toth in the majority of hydrothermal crusts. The high content of these samples seems to result from hydrothermal activity which affects sedimentation and ore formation (Bostrom and Peterson, 1966; 1969; Metalliferous sediments.... 1979).

From the available data, the average mercury content is 20 ppb in the "pure" material of oceanic manganese nodules. In the Pacific sediments, mercury content is 0.7 to 58 ppb (Cox and McMurthy, 1981; Baturin et al., 1986); it can be concluded that mercury does not accumulate in manganese nodules when no hydrothermal processes occur in their vicinity.

RARE METALS OF THE THIRD GROUP OF THE PERIODIC TABLE

Rare metals of this group are most numerous. The major subgroup includes Sc, Y and REE, namely lanthanum, cerium praseodymium, neodymium, samarium, europium, gadolinium, terbium, dysprosium, holmium, erbium, thulium, ytterbium and lutecium whereas the subordinate subgroup includes Ga, In and Tl namely lanthanum.

Gallium content in manganese nodules was first determined by Mero (1965) who analysed 54 samples; the amount varies from 2 to 30 ppm (10 ppm on average). Later, similar results were obtained for the nodules of the Pacific radiolarian zone, Indian and Atlantic oceans, and for the World Ocean: they vary from 6 (or less 10) to 70 ppm (Volkov, 1979; McKelvey et al., 1979; 1983; Table 92). On the basis of these data, one can accept that the average content of gallium in the World Ocean nodules is 10 ppm; its average in deep-sea sediments is twice as great: 20 ppm (Lukashin and Lisitsin, 1980). That gives us evidence of the passive behaviour of gallium in the process of oceanic nodule formation.

Indium. Data on indium concentration in manganese nodules are scarce. From Glasby (1973), its content is 0.15 ppm in nodules from the Campbell Plateau; 0.41 ppm in the equatorial Indian ocean; 0.09 ppm in the Bay of Aden; and 0.37 ppm in nodules from the Blake Plateau. In nodules from Loch Fyne (Scotland) the indium content is 0.13 ppm. From these data the average content of indium in oceanic nodules can be estimated as 0.25 ppm. The indium content of deep-sea sediments is three times lower: 0.08 ppm on average (Turekian and Wedepohl, 1961). This leads us to suppose some participation of indium in the formation of nodules.

Thallium. The first determinations of thallium in manganese nodules were made by Riley and Sinhaseni (1958). They analysed three samples from the Pacific and found from 80 to 100 ppm of this element.

In the course of further analyses, the range of thallium concentration in nodules of the Pacific was specified by Ahrens et al. (1967); it is from 2 to 614 ppm, average values being from 71 to 173 ppm depending on the region (Table 93).

The thallium content in nodules of the Pacific radiolarian zone varies from 60 to 300 ppm, being 160 ppm on average (McKelvey et al., 1979). The average thallium content in the Pacific nodules is 91 to 170 ppm according to various estimates.

In the Indian ocean, the range of thallium content variation is from 2 to 242 ppm (Ahrens et al., 1967), being 80 to 134 ppm on average (from various estimates). In the Atlantic, these estimates are 50 to 349 and 77 to 180 ppm, respectively (Table 93).

From the total amount of the data considered, the most probable average estimates of thallium content in nodules of the Pacific, Indian and Atlantic oceans are 170, 100 and 100 ppm, respectively, its average being 150 ppm in the World Ocean nodules.

The latitudinal distribution of thallium in nodules shows two distinct tendencies, namely, its minimum content (100 ppm on average) is confined to sub-Arctic latitudes (40 to 60°N), and its maximum (180 ppm on average) to temperate zone of the northern hemisphere (20 to 40°N). Thallium content does not seem to be dependent upon the depth, its average being 180 ppm in the nodules from the depth to 3000 m and 160 ppm from deeper nodules (McKelvey et al., 1983).

Various methods of selective leaching were used to determine forms of thallium occurrence in nodules. Results obtained, however, were ambiguous.

The share of thallium leached from nodules by various solvents varies from 24 to 89% (Description..., 1977). Low-temperature leaching of nodules by SO_2 extracts some 45% of the thallium, whereas 70% of it may be extracted by high-temperature processing; this may result from the difference in the degree of thallium transformation from a trivalent compound to a univalent one, i.e. from Tl_2O_3 to Tl_2SO_4 (Iskowitz et al., 1982).

TABLE 92
Gallium content (ppm) in nodules

Region	Ga Range	Ga Mean	N	Reference
Pacific Ocean	-	4	1	Goldschmidt and Peters, 1931a
	16-22	19	3	Riley and Sinhaseni, 1958
	2-30	10	54	Mero, 1965
	2-72	11	39	Haynes et al., 1986
radiolarian belt	7-30	20	4	McKelvey et al., 1979
Indian Ocean	6-42	14	7	Volkov, 1979
World Ocean	‹10-70	10	57	McKelvey et al., 1983
Sediments	-	20	-	Lukashin and Lisitsin, 1980

TABLE 93
Thallium content (ppm) in nodules

Region	Tl Range	Tl Mean	N	Reference
Pacific Ocean	80-110	90	3	Riley and Sinhaseni, 1958
	1-110	83	6	Willis and Ahrens, 1962
	2-614	173	13	Ahrens et al., 1967
	150-300	170	7	Mero, 1965
	144-160	152	2	Glasby et al., 1978
	23-226	123	9	Iskowitz et al., 1982
	15-250	91	42	Volkov and Sokolova, 1984
	-	170	-	Cronan, 1980
	-	170	-	McKelvey et al., 1983
	2-675	169	141	Haynes et al., 1986
radiolarian belt	60-300	160	17	McKelvey et al., 1979
Indian Ocean	55- 80	74	4	Willis and Ahrens, 1962
	3-242	55	10	Ahrens et al., 1967
	59- 64	62	2	Glasby et al., 1978
	45-128	82	4	Iskowitz et al., 1982
	77-290	134	7	Volkov, 1979
	-	100	-	Cronan, 1980
	-	80	-	McKelvey et al., 1983
Atlantic Ocean	15-135	70	8	Willis and Ahrens, 1962
	109-349	177	7	Ahrens et al., 1967
	50-138	106	8	Iskowitz et al., 1982
	-	77	-	Cronan, 1980
	-	180	-	McKelvey et al., 1983
Average for the World Ocean	-	129	-	Cronan, 1980
	10-610	160	130	McKelvey et al., 1983
Sediments	-	0.28	-	Lisitsin, 1978
	-	1.8	-	Volkov and Sokolova, 1984
	0.002-6	0.21	104	Oreshkin et al., 1986

Volkov and Sokolova (1984) believe that in most nodules, about 40% of thallium is in a sorbed form (Tl^+). The rest of it (Tl^+ and Tl^{3+}) is included in manganese and iron oxides.

The affinity of Tl for Mn is clear from the correlation of these metals determined in the Pacific nodules ($r = 0.54$); the correlation coefficient of thallium content and Mn/Fe ratio is even higher, 0.78 on average. One can suppose that thallium accumulation in nodules results from Tl^+ oxidation in sea water until it reaches Tl_2O_3, which is resistant to solvents. Its accumulation probably goes along with the process of Mn^{2+} oxidation to Mn^{4+} and their coprecipitation (Volkov and Sokolova, 1984). Thallium content is high (370 ppm) in some nodules rich in cobalt (McKelvey, 1983).

Some researchers have noticed that a high content of thallium in technological solutions remaining after the processing of manganese nodules can be toxic for land and water environments since thallium salts are very toxic (Fairhall, 1969; Iskowitz et al., 1982).

According to Lisitsin (1978), the thallium content in deep-sea sediments of the World Ocean is 0.28 ppm whereas its content in the Pacific deep-sea sediments is 1.8 ppm (Volkov and Sokolova, 1984). Thus, the average concentration coefficient of thallium in manganese nodules with respect to host sediments is 80 to 500. However, the former is the more realistic value. So, in the Pacific, this coefficient varies from 24.1 to 200, being 72.8 on average. According to this coefficient, thallium exceeds all other elements accumulating in nodules, namely, molybdenum, nickel, manganese and cobalt, for which this coefficient is only 29 to 40 (Volkov and Sokolova, 1984).

Scandium was mainly analysed in nodules of the Pacific. In most analyses, its content in various oceanic regions varies from 1.9 to 18.1 ppm (Piper and Williamson 1977; Piper et al., 1979; Dymond et al., 1984). In the south-eastern Pacific, where the hydrothermal effect upon sedimentation is strong, the scandium content of nodules is somewhat higher and reaches 66.9 ppm (Girin et al., 1979).

The maximum content of scandium, 2700 ppm, was determined in the sample from the Pacific radiolarian zone (McKelvey et al., 1979); though, from other data, on the whole, scandium content in the zone in question is similar to that of other regions, from 7.3 to 18.1 ppm, being 11.3 ppm on average (Piper et al., 1979).

From McKelvey et al., (1983), the scandium average in nodules of the Pacific is 50 ppm, it is 10 in nodules of the Indian ocean, 20 in the Atlantic nodules, and 50 ppm in the World Ocean. All these estimates are considerably higher by comparison with all other data available (Table 94), so one is forced to accept 10 ppm as its average.

When analysing the latitudinal distribution of scandium in nodules of the World Ocean, it was determined that its minimum content (10 ppm on average) is confined to the zones of 40 to 60 °N and 0 to 20 °S, and its maximum (230 ppm on average) is confined to the interval of 0 to 20 °N. Scandium shows a distinct correlation of its content with depth: 10 ppm of scandium is the average for the depths to 3000 m, and it is 70 ppm for deeper nodules. It was recorded that nodules rich in nickel and copper (over 1.8% in total) are also rich in scandium (to 180 ppm on average) as compared to all other samples (McKelvey et al., 1983). Though absolute values of scandium content from the data of the authors mentioned seem to be overestimated, this regularity is preserved.

The analysis of nodules from various parts of the eastern hemipelagic zone in the Pacific showed that the average scandium content is two times higher in the nodule tops than in their bottoms: 6.0 and 3.1 ppm, respectively, so its supply is mainly a hydrogenous one (Dymond et al., 1984).

When comparing the content of scandium and that of other elements in nodules of the Pacific radiolarian zone, a weak positive correlation with iron and cobalt was determined ($r = 0.24$). It showed a stronger positive correlation with hafnium ($r = 0.55$) and strong negative correlation with manganese ($r = -0.63$) (Piper et al., 1979).

Deep-sea sediments of the ocean contain 14 ppm of scandium on average (Lisitsin, 1978). Its content is twice as high in the Pacific radiolarian zone, 24-28 ppm on average (Bischoff et al., 1979; Piper et al., 1979). Thus, on the whole, scandium accumulation in nodules relative to sediments does not occur; its average concentration coefficient is 0.3 to 1.

TABLE 94
Scandium content (ppm) in nodules

Region	Sc Range	Sc Mean	N	Reference
Pacific Ocean	-	4	1	Goldschmidt and Peters, 1931b
	5 - 10	8	3	Riley and Sinhaseni, 1958
	1.9 - 16.4	8.4	16	Piper and Williamson, 1977
	-	50	-	McKelvey et al., 1983
	1 - 29	10	159	Haynes et al., 1986
radiolarian belt	9 -2700	-	5	McKelvey et al., 1979
	7.3 - 18.1	11.3	41	Piper et al., 1979
northern tropical part	7 - 22	15	4	Sano et al., 1985
south-eastern part	4.1 - 66.9	15.6	78	Girin et al., 1979
eastern tropical part	3.1 - 4.6	4.0	6	Dymond et al., 1984
tops of nodules	3.1 - 7.6	6.0	11	- " -
bottoms of nodules	2.1 - 3.8	3.1	13	- " -
Indian Ocean	-	10	-	McKelvey et al., 1983
Atlantic Ocean	-	20	-	- " -
World Ocean	10 -2700	50	76	- " -
Sediments	-	14	-	Lisitsin, 1978
Radiolarian belt of the Pacific Ocean	-	28	118	Bischoff et al., 1979
	13.01- 37.36	24.46	38	Piper et al., 1979

TABLE 95
Yttrium content (ppm) in manganese nodules

Region	Y Range	Y Mean	N	Reference
Pacific Ocean	-	10	1	Goldschmidt and Peters, 1931a
	30- 30	30	3	Riley and Sinhaseni, 1958
	160-450	330	54	Mero, 1965
	80-170	127	18	Calvert and Price, 1977
	13-161	77	42	Volkov, 1979
	-	310	-	Cronan, 1980
	-	150	-	McKelvey et al., 1983
	17-950	133	132	Haynes et al., 1986
Radiolarian	40-150	100	26	McKelvey et al., 1979
Site A:				
highs	-	146	17	Calvert and Piper, 1984
depressions	-	149	49	Ibid.
Campbell Plateau	-	48	1	Glasby, 1973
Indian Ocean	54-107	72	12	Volkov, 1979
	-	110	-	McKelvey et al., 1983
Atlantic Ocean	-	240	-	Ibid.
World Ocean	20-950	150	134	- " -
Sediments:				
World Ocean	-	53	-	Lisitsin, 1978
Pacific Ocean	24-420	124	109	Gurvich et al., 1980a
Radiolarian	-	131	118	Bischoff et al., 1979

Yttrium content was mainly assessed in nodules of the Pacific where, according to various authors, it varies from 13 to 450 ppm (Mero, 1965; Glasby, 1973; Calvert and Price, 1977; Volkov, 1979; McKelvey et al., 1979; Calvert and Piper, 1984). Estimates of Mero (1965) were 160 to 450, 330 ppm on average. Successive analyses showed that yttrium content is 2 to 3 times lower (Table 95) that allowed average estimation of yttrium content in the Pacific manganese nodules as 100 ppm.

A number of analyses was made to determine the yttrium content in nodules of the Indian and Atlantic oceans; it is 72-110 in the former and 240 ppm in the latter (Volkov, 1979; McKelvey et al., 1983).

In the nodules of the World Ocean, the range of yttrium content variations is from 13 to 950 ppm; its average is 150 ppm (McKelvey et al., 1983).

Two tendencies are observed in the latitudinal distribution of yttrium. Its minimum (130 ppm on average) was recorded in the southern latitudes (20 to 60 °S) and its maximum (280 ppm on average) was in the sub-Arctic latitudes (40 to 60 °N). The vertical pattern of yttrium content is as follows: it is double at depths down to 3000 m what it is in nodules at greater depth (240 ppm contrast to 120 ppm, on average). Comparison of contents of yttrium and of major ore components showed that the nodules rich in cobalt (over 1%) are also rich in yttrium (310 ppm on average). In this aspect yttrium behaves like cobalt. The average concentration of yttrium in deep-sea sediments of the ocean is 53 ppm (Lisitsin, 1978), in the Pacific radiolarian deposits it is 131 ppm (Bischoff et al., 1979). From these data, it is possible to conclude that yttrium does not accumulate in nodules relative to host sediments; the average concentration coefficient is approximately 1, with slight deviations.

Rare-earth elements (REE), or lanthanides, represent the most extensive group of rare elements - 14 elements - that always occur in manganese nodules.

The first data on REE in nodules were published by Goldberg et al. (1963) and Pachadzhanov et al., (1963). Later, many researchers studied the behaviour of REE in nodules and partly in sediments (Mero, 1965; Fomina, 1966; Volkov and Fomina, 1967; 1973; Piper, 1972; 1974; Glasby, 1973; Glasby et al., 1978; Addy, 1979; Girin et al., 1979; McKelvey et al., 1979; Rankin and Glasby, 1979; Courtois and Clauer, 1980; Gurvich et al., 1980a; Toth, 1980; Elderfield et al., 1981a,b; Elderfield and Greaves, 1981; Moore et al., 1981; Tlig, 1982; McKelvey et al., 1983; Calvert and Piper, 1984; Murphy and Dymond, 1984; Aplin, 1985).

By now, considerable analytical data on REE content in nodules (mainly in the Pacific) have been obtained. The most extensive data have been collected on lanthanum and cerium, the least on holmium, thulium and lutecium.

The results of REE estimates in manganese nodules and crusts of the Pacific, Indian and Atlantic oceans and also in deep-sea sediments of the ocean are given in Tables 96-100.

Lanthanum content in the Pacific nodules varies from 16 to 979 ppm (from various authors). Its average content in various parts of the Pacific is 100-150 ppm in the radiolarian zone to 235 ppm in the southeastern region, its average for the entire ocean is 180 ppm.

The average concentration of lanthanum in nodules of the Indian, Pacific and World Ocean are assessed as 110, 177 and 122 ppm (Table 99).

The average concentration of lanthanum in deep-sea sediments of the oceans varies from 12 to 129 ppm, being mainly 42 to 65 ppm; thus the average concentration coefficient of lanthanum in nodules relative to host sediments is 3.8.

Cerium content in the Pacific nodules varies from 12 to 3000 ppm and usually considerably exceeds the lanthanum content. In various parts of the Pacific radiolarian zone, the average content of cerium differs by two times: from 350 ppm (McKelvey et al., 1979) to 843 ppm (Calvert and Piper, 1984). The widest range of cerium variations at its general high averge (880 g/t) is recorded in nodules from the southeastern region of the ocean (Girin et al., 1979)

Data on cerium concentration in nodules of the Indian ocean are controversial. It varies from 249 to 937 ppm (from Volkov, 1979), being 585 ppm on average; however, McKelvey et al. (1983) estimated its average as 1250 ppm - that is twice as high. The cerium content of nodules and micronodules of the Indian ocean is even higher, 2057 and 1360 ppm, on average (Addy, 1979).

The average concentration of cerium in the World Ocean nodules estimated from the data of Tables 96-98 is 665 ppm.

The cerium content in deep-sea sediments of the oceans varies from 11 to 180 ppm and is 90 ppm, on average, which permits evaluation of its concentration coefficient in nodules relative to sediments as 7.6.

Praseodymium content in the Pacific nodules varies from 5 to 232 ppm and in the Indian ocean from 12 to 148 ppm. Its average in the Indian ocean nodules is about twice as high (107 ppm) than in the Pacific nodules (60 ppm) and is about 70 ppm for the entire World Ocean.

Estimates of praseodymium average concentration in oceanic sediments vary considerably and may be assumed to be 11 ppm. Thus, its average concentration coefficient in nodules relative to sediments is 6.4.

Neodymium content in the Pacific nodules is 11 to 700 ppm. In nodules of the radiolarian zone and southwestern part of the ocean, neodymium content variations and its averages are close; they are 102-210 ppm (143 ppm on average) and 120-180 ppm (160 ppm on average), respectively. Its average content in the Pacific nodules is 130 ppm.

According to various authors, the average neodymium content in the Indian ocean nodules varies from 124 to 187 ppm, being 220 ppm in the Atlantic and 160 ppm in the World Ocean nodules. Average concentration of neodymium in oceanic sediments can be assessed as 54 ppm. An average concentration coefficient for neodymium deduced from these estimates is 3.

Samarium content in the Pacific nodules varies from 2.3 to 141 ppm and its average in the radiolarian zone nodules is 34, in the southwestern part of the ocean it is 40 ppm, and for the entire Pacific it is 30 ppm.

In oceanic sediments, the average samarium content is 12.6 ppm. Its average concentration coefficient is thus 2.8.

Europium content in the Pacific nodules varies from 0.5 to 27 ppm. The widest range of its variations was registered in the nodules of the southwestern Pacific. Meantime, its average concentration in nodules from various regions of the Pacific (4 to 10 ppm) testifies to the considerably regular distribution of this element over the ocean. A similar pattern is observed in the nodules of the Indian ocean and Atlantic ocean. Its average content in nodules of the World Ocean is about 9 ppm. In oceanic sediments the europium content is 3.2 ppm. Thus, deduced average concentration coefficient of the element in nodules is about 2.8.

Gadolinium content in the Pacific nodules varies from 5 to 61 ppm, its average in various regions and in the ocean as a whole being 32 ppm. In nodules of the Indian ocean the average gadolinium content is 48 ppm and in nodules of the World Ocean it is 35 ppm.

Thus, one can deduce the average concentration coefficient for gadolinium in nodules of the World Ocean relative to sediments as about 3.5.

Terbium content lies within the range from 0.4 to 19 ppm in the Pacific, 3.6 to 8.0 ppm in the Indian ocean and 4.3 to 7.8 ppm in the Atlantic nodules, with average values 5.0, 5.8 and 6.3 ppm. The average content of terbium in nodules of the World Ocean is 5.3 ppm, in deep-sea sediments of the ocean its average concentration is assessed as 1.5 ppm. Terbium average concentration coefficient seems to be 3.5.

Dysprosium content in the Pacific nodules varies from 2 to 68 ppm. Its average content in nodules from various parts of the Pacific is from 7.5 ppm (Campbell Plateau) to 33 ppm (southwestern Pacific); it is 27.6 ppm in the radiolarian zone and 23 ppm for the Pacific as a whole.

The dysprosium content in nodules of the Indian ocean varies from 5 to 71 ppm (in one sample it was 246 ppm). Its average in nodules of the World Ocean is 30 ppm.

The average concentration of dysprosium in sediments of the World Ocean is about 9 ppm. Thus, the average concentration coefficient of this element in nodules is 3.3.

Holmium content in the Pacific nodules varies from 0.5 to 24 ppm, in the Indian ocean from less than 0.5 to 42 ppm.

The average concentration of holmium in nodules of the Pacific is assessed as 4.8 ppm, whereas in the Indian ocean it is 16 ppm. These data show that nodules of the Indian ocean are about three times richer in holmium than are nodules of the Pacific.

TABLE 96

Rare earth elements content (ppm) in manganese nodules and sediments of the Pacific Ocean

Ele-ment	1	2		3	4	5		6		7	
	Mean	Range	Mean	Mean	Mean	Range	Mean	Range	Mean	Range	Mean
La	115	66 –152	100	152	73	120 – 170	150	56 –211	118	67 – 90	79
Ce	517	184 –823	420	600	451	>200 ->200	>200	71 –783	293	65 –115	96
Pr	–	–	–	–	18	32 – 46	40	–	–	–	–
Nd	138	102 –210	143	–	54	120 – 180	160	56 –162	136	63 – 85	72.7
Sm	64	23.3 – 48.0	34	38.4	7.82	32 – 51	40	12.5- 44	19.5	13.2 – 17.6	15.3
Eu	10.3	5.59- 11.1	8.06	1.0	1.89	7.8- 13	10	3.6- 10.3	6.0	3.62- 4.71	4.09
Gd	46	23.0 – 49.6	34.7	–	7.56	28 – 38	34	–	–	16.4- 21.2	18.6
Tb	6.9	–	7.3*1	3.2	1.45	4.5- 7.4	5.8	2.5 – 9.1	3.8	–	–
Dy	29.9	22.2 – 42.3	27.6	–	7.51	27 – 40	33	–	–	16.6 – 20	18.6
Ho	5.75	–	4.6*2	–	1.59	6.6- 9.4	7.9	–	–	–	–
Er	13.8	12.1 – 22.8	16.4	–	5.94	19 – 27	22	–	–	11.6 – 13.6	12.6
Tm	2	–	–	–	0.55	–	–	–	–	–	–
Yb	4.6	11.0 – 20.7	15.0	17.8	2.88	17 – 25	21	11.1- 29.6	5.8	10.8 – 13.0	12.1
Lu	I	2.9- 3.1	3.0*1	6.4	0.40	–	–	1.9- 4.3	2.8	–	–

1. "Horizon" nodule, 40 °14′N, 155 °52′W, depth 5000 m (Goldberg et al., 1963);
2. Clarion-Clipperton zone, 7 samples (Elderfield et al., 1981a);
3. Reference sample from the same zone (Toth, 1980);
4. Campbell Plateau (Glasby, 1973);
5. South-western part of the Ocean, 12 samples (Rankin and Glasby, 1979);
6. South-eastern part of the Ocean, 8 samples (Courtois and Clauer, 1980);
7. Bauer Deep, 4 samples (Elderfield and Greaves, 1981);
8. East equatorial hemipelagic zone, 6 samples (Murphy and Dymond, 1984);
9. The whole Ocean, 42 samples (Volkov, 1979; Piper, 1974);
10. The same after Haynes et al., 1986 (number of samples for each element: La-151, Ce-131, Pr-8, Nd-96, Sm-115, Eu-115 Gd-57, Tb-104, Dy-18, Ho-66, Er-18, Tm-41, Yb-171, Lu-76);
11. Sediments of the Clarion-Clipperton zone, 7 samples (Elderfield et al., 1981a);
12. Sediments of the south-western part of the Ocean; 12 samples (Rankin and Glasby, 1979).

*1 - Piper, 1972; *2 - Moore et al., 1981; *3 - Volkov and Fomina, 1973.

Table 96 (continued)

Ele-ment	8		9		10		11	12
	Range	Mean	Range	Mean	Range	Mean		
La	17 -35	20.5	16 - 514	184	66- 979	157	65	42
Ce	12 -38	19.7	39 -2126	678	74-3000	530	91	106
Pr	-	-	5 - 232	78	26- 46	36	-	13
Nd	11 -31	19	18 - 596	204	60- 700	158	92	51
Sm	2.3 - 6.9	4.0	4 - 114	44	14- 141	35	23	13
Eu	0.54- 1.70	0.99	4 - 24	9.8	1- 27	9	5.7	3.5
Gd	-	-	5 - 61	32	14- 53	32	-	13
Tb	0.41- 1.27	0.70	3.2- 19	7.2	1- 11	5.4	-	1.9
Dy	-	-	2 - 68	23	22- 42	31	23	12
Ho	-	-	<0.5- 24	4.5	1- 8	4	-	2.8
Er	-	-	1.2- 42	15	11- 27	18	13.4	7.2
Tm	-	-	<0.5- 19	1.7	1- 9	2.3	-	2.3*3
Yb	2.2 - 6.0	3.5	<0.5- 54	17.7	8- 100	20	12.3	7.3
Lu	0.45- 0.92	0.55	<0.5- 9,3	2.2	1- 6	1.8	-	-

TABLE 97

Rare earth elements content (ppm) in manganese nodules and sediments of the Indian Ocean

Element	Nodules								Sediments			
	1	2	3	4	5 Range	5 Mean	6 Range	6 Mean	7	8	9	10
La	31	131	86	180	87 -292	163	116.7- 180.1	155	10	16	18	24
Ce	549	906	>200	>200	249 -937	585	866 -1409	1150	27	61	19	69
Pr	12	117	26	47	75 -148	114	-	-	2.8	11	5	30
Nd	75	110	110	190	69 -192	124	141 -241	187	13	19	27	-
Sm	24	29	28	51	20 -109	56	28.7- 50.9	38.6	4,8	5,6	6.7	8.0
Eu	8	7.4	7.2	12	6.5- 18,1	8.5	3.5- 11.3	7.8	-	-	-	1.9
Gd	20	7.4	24	37	29 -100	56	4.7- 7.4	5.8	5,4	5,8	7.5	-
Tb	4	4	3.6	5.8	5.6- 8,0	7.0	-	-	-	-	-	1,2
Dy	31	246(?)	17	34	5 - 71	46	-	-	0,7	5,3	2,2	-
Ho	8	22	3.9	7.7	<0,5- 42	18	-	-	0,6	8,4	2,5	-
Er	20	22	10	21	5 - 41	26	-	-	1,5	5,9	1,4	-
Tm	4	7.5	-	-	<0,5- 14	3.2	-	-	-	-	-	-
Yb	40	52	9.6	19	4 - 81	25.6	13.8- 17.8	15.6	1.5	10,1	3.0	4.1
Lu	4	7.5	-	-	<0,5- 9	1.7	2.1- 8.83	3.9	-	-	-	0.5

1,2 – St. 4555 and St. 4575, "Vityaz" (Pachadjanov et al., 1963);
3,4 – St. Z2140 and 6269 (Glasby et al., 1978);
5 – Volkov, 1979, 12 samples;
6 – Tlig, 1982, 5 samples;
7,8 – red clays, and
9 – foraminiferal oozes, 4 samples (Volkov and Fomina, 1967);
10 – clayey-calcareous sediments, 4 samples (Tlig, 1982).

TABLE 98

Rare earth elements content (ppm) in manganese nodules and sediments of the Atlantic Ocean (Addy, 1979)

Ele-ment	Nodules (9 samples)		Micronodules (7 samples)		Red clays (3 samples)		Terrigeneous muds (3 samples)	
	Range	Mean	Range	Mean	Range	Mean	Range	Mean
La	125 - 256	177	140 - 511	263	43 - 50	46	36 -41	39
Ce	1698 -3272	2057	835 -2180	1360	122 -152	133	78 -84	81
Nd	210 - 230	220	393 - 640	517	-	-	-	-
Sm	39 - 61	49	20 - 192	77	8.3 - 9.6	9.0	6.4 - 6.8	6.6
Eu	6.5- 13.0	10.9	5.6- 45.5	17.5	2.0 - 2.7	2.3	1.4 - 1.7	1.6
Tb	4.3- 7.8	6.3	5.7- 28.5	11.0	1.2 - 1.4	1.3	0.8 - 1.1	1.0
Yb	9.6- 18.7	13.1	8.0- 43.2	19.4	3.6 - 4.7	4.2	7.7 - 3.5	3.1
Lu	1.4- 2.6	2.1	1.0- 6.4	2.8	0.64- 0.73	0.68	0.53- 0.59	0.55
Ce/Ce*	3.5- 10.5	5.5	1.2- 8.1	3.3	1.35- 1.52	1.4	0.96- 1.13	1.05

Ce/Ce* - ratio of analized to normalized content

TABLE 99

Average contents of REE (ppm) in nodules and their concentration coefficients relative to sediments*

Element	Pacific Ocean Range	Mean	N	Indian Ocean Range	Mean	N	Atlantic Ocean Range	Mean	N	Average Nodules	Sediments	Concentration coefficient
La	16 – 979	160	151	91 – 291	110	21	125 – 256	177	9	160	42	3.8
Ce	12 –3000	530	131	249 –1409	820	19	1698 –3272	2037	9	660	90	7.6
Pr	5 – 232	60	58	12 – 148	107	16	-	-	-	70	11	6.4
Nd	11 – 700	160	96	69 – 241	144	21	210 – 230	220	2	160	54	3.0
Sm	2.3– 141	35	115	20 – 109	46	21	35 – 56	49	9	35	12.6	2.8
Eu	0.5– 27	9	115	3.5– 12	8.3	21	6.5– 13	-	9	9.0	3.2	2.8
Gd	5 – 61	32	67	7.4– 100	48	16	-	10.9	-	35	10	3.5
Tb	0.4– 19	5.4	104	3.6– 8	5.8	21	4.3– 7.8	-	9	5.3	1.5	3.5
Dy	2 – 68	23	67	2 – 246	50	16	-	6.3	-	30	9	3.3
Ho	<0.5– 24	4.2	66	<5 – 42	16	16	-	-	-	6.4	2.4	2.7
Er	1.2– 42	15	67	5 – 41	24	16	-	-	-	18	7.3	2.5
Tm	<0.5– 19	2.0	44	<5 – 14	3.5	14	-	-	-	2.2	1.4	1.6
Yb	<0.5– 100	18	171	4 – 81	23	21	9.6– 18.7	13.1	9	18	6.5	2.8
Lu	0.4– 9.3	2.0	76	<5 – 9	2.7	19	1.4– 2.6	2.1	9	2.2	1.9	1.2

* Data for nodules are taken from Tables 99-101, the average content of REE in sediments is calculated on the basis of data from Wildeman and Haskin, 1965; Volkov and Fomina, 1965; Addy, 1979; Migdisov et al., 1979; Rankin and Glasby, 1979; Courtois and Clauer, 1980; Gurvich et al., 1980; Elderfield et al., 1981; Tlig, 1982.

The average content of holmium in manganese nodules of the World Ocean is assumed to be 6.4 ppm, and in oceanic sediments as 2.4 ppm. Its average concentration coefficient, thus, seems to be about 2.7.

Erbium content in the Pacific nodules is from 1.2 to 42 ppm; for various regions of the ocean it is: 5.9 within the Campbell Plateau, 16.4 in the radiolarian zone, 22 in the southwestern part of the ocean, 12.6 ppm in the Bauer Depression.

In the Indian ocean, nodules contain from 5 to 41 ppm of the element, its average being 24 ppm.

Erbium average content in nodules of the World Ocean is 18 ppm, in the World Ocean sediments it is about 7.3 ppm. Consequently, erbium concentration coefficient in nodules of the World Ocean relative to sediments is 2.5.

Thulium content in the Pacific nodules varies from < 0.5 to 19 ppm in the Indian ocean, it is from less than 0.5 to 14 ppm. Averages for both oceans are 2 and 3.5 ppm, respectively.

The average content of thulium in nodules of the World Ocean is estimated as 1.8 ppm and in oceanic sediments as 1.4 ppm. Its average coefficient of concentration is 1.3.

Ytterbium content in the Pacific nodules varies from < 0.5 to 66 ppm. Its average content in nodules from the radiolarian zone is 15 ppm; it is 21 ppm in the southwestern part of the ocean, 5.8 ppm in the southeastern part, 12.1 ppm in the Bauer Depression and 3.5 ppm i nthe eastern hemipelagic zone.

Ytterbium average in nodules of the Pacific is 18 ppm.

Ytterbium content in nodules of the Indian ocean varies from 4 to 81 ppm, its average being 23 ppm. The former estimate by McKelvey et al.. based on semi-quantitative analyses, was lower, namely 10 ppm. That might partly result from the composition of the samples analysed.

Nodules of the Atlantic contain 9.6 to 18.7 ppm of ytterbium, 13.1 ppm, on average.

The average ytterbium content in nodules of the World Ocean is 18 ppm, in oceanic sediments it is 6.5 ppm, so its average concentration coefficient is 2.8.

Lutetium content in the Pacific nodules varies from 0.40 to 9.3 ppm. Its average concentration in nodules of the Pacific radiolarian zone is 3.0, in south-eastern pelagic zone it is 2.8, in eastern hemipelagic zone it is 0.55 ppm.

Nodules of the Indian ocean contain from less than 0.5 to 9 ppm of lutetium, nodules of the Atlantic contain 1.4 to 2.6 ppm, the average being 2.7 and 2.1 ppm, respectively.

The average concentration of lutetium in nodules of the World Ocean is 2.2 ppm and it is 1.9 ppm in deep-sea sediments, thus the average concentration coefficient is 1.2.

The above data demonstrate that nodules of the Indian ocean are enriched by all REE except lanthanum, as compared to the Pacific nodules. Nodules of the Atlantic, in their turn, are enriched in light and intermediate REE as compared to other oceans.

The content of the majority of REE is higher in hydrogeneous crusts and considerably lower in hydrothermal crusts as compared to nodules (Table 100).

Distribution of REE in manganese nodules with respect to latitudinal occurrence was demonstrated by McKelvey et al. (1983). They determined that lanthanum minimum (170 ppm) is confined to 40-60°S and its maximum (270 ppm) to 0-20°N. Cerium minimum (450 ppm) coincides in latitude with lanthanum minimum, but its maximum (1260 ppm) is confined to 20-60 °S. Neodymium minimum (180 ppm) is located within 40-60 °N and its maximum (340 ppm) coincides with lanthanum maximum (0-20 °N). Samarium maximum (60 ppm) falls within the same interval. No distinct trends in distribution were observed for the rest of REE.

When analysing REE distribution in nodules with respect to depth, the authors mentioned above discovered that the content of lanthanum, europium and terbium is similar both at depths to 3000 m and deeper; cerium and ytterbium show a tendency to relative accumulation in shallow nodules, whereas neodymium and samarium are concentrated in deep-sea nodules.

Analysing the composition of REE in nodules with respect to the depth of their occurrence, Piper (1972) determined that shallow nodules are relatively enriched with heavy REE (ytterbium and lutetium); a similar distribution of REE was detected

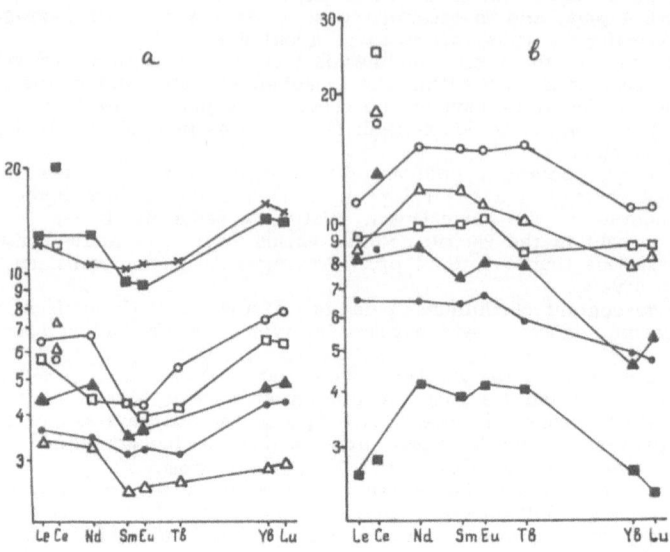

Fig. 48. Normalized REE concentrations in the Pacific nodules (Piper, 1972):
(a) depth to 3000 m; (b) deeper 3000 m.

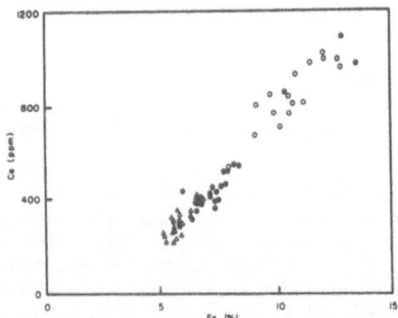

Fig. 49. Ce - Fe correlation in nodules from the northeastern zone of the tropical
Pacific (Calvert and Piper, 1984).

TABLE 100

REE content (ppm) in ferromanganese crusts

Element	1	2	3	4 Range	4 Mean	5 Range	5 Mean	6	7 Range	7 Mean
						Hydrogeneous crusts				
La	121	246	236	146 -174	156	209 - 462	335	134	139 - 387	307
Ce	123	181	82.1	148 -230	185	807 -1730	1268	203	210 -2916	1100
Pr	-	-	-	-	-	-	-	-	21.6- 77.6	56
Nd	98.2	-	-	-	-	-	-	112	109 - 309	261
Sm	20	43.6	12.2	29.9- 37.5	34	44.3- 82.0	62.1	24	18.8- 68.0	54
Eu	5.44	10.4	4.0	1.0- 10.9	8.0	2.3- 12.3	7.3	6.15	5.2- 16.5	14
Gd	24.9	-	-	-	-	-	-	-	30.6- 73.2	55
Tb	-	7.4	2.4	4.5- 5.5	5.0	4.6- 9.0	6.8	3.6	-	-
Dy	25.2	-	-	-	-	-	-	-	24.4- 67.8	52
Er	16.4	-	-	-	-	-	-	-	16.3- 37.9	31
Yb	15.6	22.9	12.9	20.3- 28.0	20.6	18.1- 27.4	22.7	18	14.2- 33.8	28
Lu	-	4.3	2.7	3.2- 3.9	3.6	1.8- 3.6	2.7	2.75	2.9- 5.5	4.2

1. Bauer Deep, Pacific Ocean (Elderfield and Greaves, 1981); 2. The same (Toth, 1980); 3. East Pacific Rise (Toth, 1980); 4. Juan de Fuka, 5 samples (Toth, 1980); 5. Subsea mountains in the north-eastern Pacific (Toth, 1980); 6. Eastern equatorial Pacific, hemipelagic zone (Murphy and Dymond, 1984); 7. Subsea mountains of the Central Pacific, 19 samples (Aplin, 1985); 8. Subsea mountains north of Tuamotu, 13°09'S, 2 samples (Lalou et al., 1979a); 9. Northern Mid Pacific Ridge, 3 samples (Toth, 1980); 10. East Pacific Rise, 19 °S (Toth, 1980); 11. North-Eastern Pacific (Toth, 1980); 12. Galapagos region, 3 samples (Toth, 1980); 13. Mid-Pacific Ridge, 37°N (Toth, 1980); 14. The same, 26°N, 3 samples (Toth, 1980).

Table 100 (continued)

| | | | | | | | | | | | Average for crusts | |
| | | | Hydrothermal crusts | | | | | | | | | |
8 Range	8 Mean	9 Range	9 Mean	10	11	12 Range	12 Mean	13	14 Range	14 Mean	Hydro-geneous	Hydro-thermal
274 -309	292	235 -295	266	94	1.3	1.0 -4.5	1.3	3.7	3.1 -10.1	5.6	223	21
876 -701	693	960 -970	963	50.9	0.9	0.5 -2.1	1.3	3.8	0.5 - 5.8	2.2	388	12
-	-	-	-	-	-	-	-	-	-	-	56	-
-	-	-	-	-	-	-	-	-	-	-	157	-
282 -361	321	50.5- 62.5	57	10.9	0.27	-	0.13	0.7	0.55- 2.0	1.1	38	2.6
56 - 59	57	1.3- 12.8	5.2	2.3	0.09	-	0.16	0.17	0.17- 6.2	2.2	7.6	1.0
-	-	-	-	-	-	-	-	-	-	-	40	-
9.3- 9.9	9.6	0.8- 6.2	4.0	2.2	0.06	0.04-0.18	0.12	0.16	0.12- 0.37	0.2	4.9	0.5
-	-	-	-	-	-	-	-	-	-	-	38	-
-	-	-	-	-	-	-	-	-	-	-	23	-
21.8- 25.5	23.6	24.6- 29.0	26.5	9.2	0.46	0.17-0.60	0.4	0.9	0.49- 1.8	1.2	21	2.4
3.5- 3.6	3.55	1.4- 3.3	2.6	1.7	0.11	0.09-0.39	0.2	0.16	0.08- 0.28	0.2	3.3	0.4

TABLE 101

REE content in top and bottom parts of manganese nodules of the Pacific Ocean*

NN of samples	La		Ce		Nd		Sm		Eu		Gd		Tb		Dy		Ho		Er		Yb		Lu	
	Top	Bottom	Top	Bottom	Top	Bottom	Top	Bottom	Top	Bottom	Top	Bottom	Top	Bottom	Top	Bottom	Top	Bottom	Top	Bottom	Top	Bottom	Top	Bottom
1	117	101	625	297	210	134	54.7	32.7	10.63	6.67	-	-	6.33	4.47	-	-	4.7	2.76	-	-	17.4	11.1	2.71	1.67
2	117	89.3	624	441	170	125	52.5	31.1	9.25	7.10	41.4	27.9	-	-	33.4	23.7	-	-	16.3	11.0	-	10.7	-	-
3	97.2	94.1	403	394	136	133	32.3	32.9	7.89	7.78	-	29.7	-	-	26.2	25.5	-	-	13.4	12.7	11.6	11.4	-	-
4	83.8	63.2	259	215	110	109	28.1	27.4	6.80	6.76	26.3	25.6	-	-	24.0	22.4	-	-	12.6	11.5	11.6	10.3	-	-

* Sampe (1) radiolarian belt, site S, MANOP (Moore et al., 1981), (2) and (3) radiolarian belt, Clarion-Clipperton field; (4) Southern equatorial part of the Pacific (Elderfield et al., 1981a).

TABLE 102

Rare earth element content in manganese nodules and associated sediments of the Pacific ocean

(Elderfield et al., 1981a)*

NN Station	La		Ce		Nd		SM		Eu		Gd		Dy		Er		Yb	
	Nod.	SEd.	Ned.	Sed.	Ned.	Sed.	Nod.	Sed.	Nod.	Sed.	Nod.	Sed.	Nod.	Sed.	Nod.	Sed.	Nod.	Sed.
1 WAH13F8	66.3	87.6	184	90.3	102	121	23.3	21.7	5.97	8.34	23	33	22.6	31.2	12.1	17.8	11.0	17.1
2 WaH18F1	112	64.5	450	103	165	94	41.3	23.9	9.52	5.77	36.1	25.8	35.1	24.0	18.6	14.4	17.1	13.6
3 WAH18F3	78.3	77.9	218	90.2	107	101	23.9	25.3	5.59	6.14	24.3	24.9	22.2	25.3	12.1	14.9	11.0	12.3
4 WA24F2	85.6	61.8	301	91.3	115	88.4	28.0	21.9	6.84	5.34	27.3	22.7	24.5	21.8	12.8	12.8	11.5	12.0
5 WAH24F6	95.5	64.8	313	90.4	135	103	32.7	25.0	8.02	6.23	31.9	26.0	30.2	26.0	15.7	14.7	14.3	14.2
6 WAH2Pg	152	52.3	823	93	210	72.4	48.6	16.7	11.1	4.14	49.6	18.7	42.3	16.7	22.8	9.89	20.7	9.13
7 DODO20C	119	47	660	78.5	170	67.8	39.8	16.0	9.41	3.89	38.9	–	36.5	16.0	20.6	9.49	19.4	8.10
8 TRIP9G	108	41.9	529	150	156	65.8	37.6	15.4	8.85	3.54	35.1	14.8	29.1	13.1	14.4	7.23	13.2	6.54
9 JYN118G	104	18.7	521	35	175	22.2	43.0	4.85	10.0	1.13	40.8	4.35	32.2	4.09	15.1	2.19	13.0	1.92
10 AMPH85PG	161	105	1098	465	189	161	40.6	35.9	9.03	8.05	42.3	39.3	40.1	36.7	23.5	22.4	21.6	19.7

* Sediments associated with nodules: (1-7) siliceous oozes; (8) pelitic clayey mud; (9) silt; (10) clay (the slope of the sea hill).

TABLE 103
Neodymium isotopes in manganese nodules and sediments
(Goldstein and O'Nions, 1981)

Sample	Location	Depth, m	$^{143}Nd/^{144}Nd$
Pacific Ocean			
Nodule	0°N,179°W	3594-2587	0.512465±28
Hydrogenous crust	3°02'S, 95°09'W	2420 87	0.512486±32
Hydrothermal crust	3°02'S, 95°09'W	2520	0.512468±30
Sediment	18°23'S, 165°44'W	4839	0.512227±30
-"-	17°11'N, 133°16' O	5997	0.512251±30
-"-	19°04'N, 161°33' O	4819	0.512210±24
Indian Ocean			
Nodule	34°45'S, 44°46' O	2479-2050	0.512142±30
-"-	18°11'S, 99°24' O	5759	0.512214±28
-"-	28°06'S, 51°19' O	5037	0.511996±32
Sediments (6 samples)	28°06'S, 51°19' O	5037	÷0.512011±26
Sediment	38°38'S, 70°00' O	4050	0.512210±24
Atlantic Ocean			
Nodule	13°04'S, 24°41'W	4415	0.512078±28
Inscl.residue of nodule	13°04'S, 24°41'W	4415	0.512444±30
Nodule	56°N, 21°W	1641-1509	0.512109±24
-"-	49°35'S, 48°05'W	2725	0.512206±32
Sediment	20°46'S, 24°08'W	5297	0.512150±20
-"-	25°58'N, 60°19'W	4682	0.512056±24
-"-	35°05'N, 7°56'W	5868	0.5119825±26
-"-	30°56'S, 39°37'W	4665	0.512241±24

in sea water (Figure 48a). That was a proof of the earlier supposition that sea water is the source of REE supply to nodules (Goldberg, 1954; 1963). However, in deep-sea nodules, the composition of REE turned out to be different (Figure 48b) that supposes another source of their accumulation in nodules, i.e. underlying sediments.

In most cases, REE concentration in various parts of individual manganese nodules is not homogeneous: their tops are richer with REE relative to their bottoms (Table 101).

To shed light on possible influence of sediment composition upon REE content in manganese nodules, REE were analysed in 10 samples of nodules and host sediments from the Pacific, mainly from the radiolarian zone (Elderfield et al., 1981a; Table 102). It was deduced from the analysis in question that, along with some meaningful average coefficients of REE concentration in nodules with respect to host sediments (Table 96), there are individual concentration coefficients that may vary within a wide range under various geochemical situations. Table 102 demonstrates that REE concentration coefficients in nodules with respect to host sediments may vary for individual elements, namely, 0.75 to 5.56, lanthanum; 2.0 to 14.9, cerium; 0.84 to 7.95, neodumium; 0.73 to 8.86, samarium; 0.71 to 8.85, europium; 0.7 to 9.4, gadolinium; 0.72 to 7.9, dysprosium; 0.68 to 6.9, erbium; 0.64 to 6.8 ytterbium.

Comparing REE and major element concentrations in manganese nodules, it is clear that behaviour of the former mainly controlled by iron (Figure 49), and to less degree by phosphorus. This inference is verified by similarity in the behaviour of REE and cobalt. Nodules rich in cobalt are also rich in REE but to various degrees. So, cerium and ytterbium maxima are associated with over 1% of cobalt in nodules, whereas lanthanum, neodynium and samarium maxima are recorded in nodules with 0.5 to 1% of cobalt (McKelvey et al., 1983).

In siliceous sediments, the significant part of iron hydrooxides is supposed to react with amorphous silica thus forming smectite and holding iron away from nodule formation (Lyle et al., 1977; Calvert et al., 1978). Meantime, REE and other metals are released from iron hydrooxide and get associated with sediment particles or are transferred into pore water. However, in contrast to manganese, nickel, copper, zinc and molybdenum which are included in nodule composition, REE are fixed by biogenic phosphate and remain in sediments, which may explain the lower concentration of REE in bottom parts of nodules which have maximum Mn/Fe ratios (Elderfield et al., 1981a).

The behaviour of cerium differs from that of three-valent REE. Nodules with maximum Mn/Fe ratio are characterized by minimum cerium, an anomaly which Elderfield et al. (1981a) proposed to define as

$$\log \left(Ce / \frac{2}{3} \ La + \frac{1}{3} \ Nd \right).$$

In contrast to other REE elements, cerium has the most pronounced correlation with iron (Elderfield et al., 1981b) that was supposed to result from its fall out of sea water due to oxidation of three-valent form and its subsequent sorption by iron hydrooxides (Goldberg, 1954).

Isotope analysis of neodymium in manganese nodules, ore crusts and deep oceanic sediments (Goldstein and O'Nions, 1981) showed that $^{143}Nd/^{144}Nd$ ratio differ from ocean to ocean, but are very close within each ocean (Table 103). Besides, $^{143}Nd/^{144}Nd$ ratios in the ore matter of nodules and in their insoluble residue also turned out to be almost identical, though they differ from the ratios typical of sea water (Figure 50).

So, sediments seem to play an important and sometimes even decisive role in the REE supply to manganese nodules.

The Ce/La ratios in nodules and sediments observed in some regions also indicate that nodules inherit the REE composition of sediments; however, the cerium share is always greater in the REE of nodules than in sediments (Girin et al., 1979).

It may car no relation to dissolved cerium deposition from sea water but may be induced by its diffusion out of sediment and its scavenging by nodules (Murphy and Dymond, 1984).

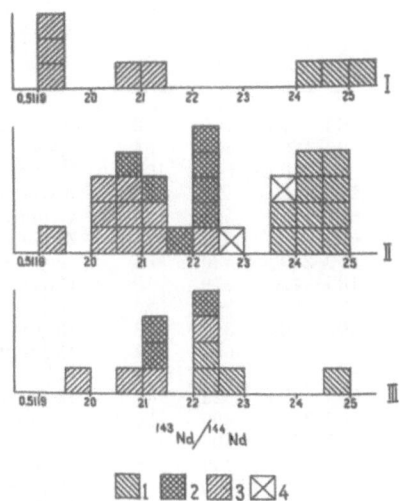

Fig. 50. Nd isotopes in sea water (I), nodules (II) and sediments (III) of the Pacific,
Indian and Atlantic oceans (Goldstein and O'Nions, 1981). (1) the Pacific;
(2) the Indian Ocean; (3) the Atlantic; (4) Antarctic and Scotia Sea.

Few determinations of REE in pore water showed no enrichment by these
elements as compared to sea water (Ridout, 1984), but their relatively high
concentration in bottom water defined in some oceanic regions might be a result of
such diffusion (Elderfield and Greaves, 1982; De Baar et al., 1985).

The study of REE fluxes by means of sedimentary traps showed that these
elements reach the bottom in the form of suspended matter. A certain cerium dificit
in hemipelagic nodules seems to be inherited from biogenic suspended matter
scavenging REE from sea water without their fractioning (Murphy and Dymond,
1984).

RARE METALS OF THE FOURTH GROUP OF PERIODIC TABLE

Rare metals of Group IV of the Periodic Table are zirconium and hafnium (main
subgroup), germanium and tin (subordinate subgroup). Lead and titanium, which
are also in this group, were considered in the chapters devoted to the major elements of nodules.

Zirconium. The first representative data on zirconium concentration in nodules
were obtained by Mero (1965), who analysed 54 samples from the Pacific and determined from 90 to 1200 ppm of zirconium. Later analyses verified these estimates in
general.

Zirconium content in nodules from the Pacific radiolarian zone varies from 185
to 800 ppm (Table 104).

From various authors, its average in the nodules from the zone in question
is 319 to 666 ppm with maximum at submarine hills and minimum in
valleys (Calvert et al., 1978; McKelvey et al., 1983; Calvert and Piper, 1984).

In nodules from the South Pacific, the zirconium content is considerably higher,
being 629 ppm on average (Landmesser et al., 1976).

The average zirconium concentrations in the Pacific nodules range from 417 to 630
ppm, according to various estimates (Table 104). In the Indian ocean nodules the
zirconium content varies from 20 to 1490 ppm and is 350 ppm on average (Volkov,
1979).

TABLE 104
Zirconium content (ppm) in nodules

Region	Zr		N	Reference
	Range	Mean		
Pacific ocean	90- 110	100	3	Riley and Sinhaseni, 1958
	90-1200	630	54	Mero, 1965
	90-1200	500	102	Volkov, 1979
	290- 550	417	18	Calvert and Price, 1977
	-	520	-	Cronan, 1980
	-	610	-	McKelvey et al., 1983
	ʿ50-2000	580	304	Haynes et al., 1986
radiolarian belt	185- 420	319	19	Calvert et al., 1978
	190- 800	360	26	McKelvey et al., 1979
	100- 900	350	33	Haynes et al., 1986
highs	-	666	17	Calvert and Piper, 1984
valleys	-	400	19	- " -
southern part	540- 760	629	12	Landmesser et al., 1976
depth ʾ3000 m	ʿ50-2000	600	226	Haynes et al., 1986
depth ʿ3000 m	ʿ50-2000	500	27	- " -
Indian ocean	10-1490	350	38	Volkov, 1979
	-	340	-	McKelvey et al., 1983
Atlantic ocean	210- 450	350	3	Volkov, 1979
	-	560	-	McKelvey et al., 1983
World Ocean	20-3600	570	289	- " -
Sediments	112- 400	165	189	Lukashin et al., 1980

Two average estimates of zirconium concentration in the Atlantic nodules are 350 and 560 ppm (Volkov, 1979; McKelvey et al., 1983).

Its average in the World Ocean nodules is 560 ppm deduced from 289 samples (McKelvey et al., 1983), its average in the Pacific deep oceanic red clays is 165 ppm from 189 samples (Lukashin et al., 1980).

The average concentration coefficient of zirconium in nodules with respect to deep oceanic red clays is some 3. However, when comparing individual samples of nodules and host sediments, it varies over a wide range, from 1.3 to 19.7 (Calvert and Price, 1977).

Zirconium concentration in nodules was considered in relation to the latitude and depth. Its minimum, 360 ppm on average, is confined to the interval of 20-40°S and its maximum, 730 ppm on-average, falls within the interval 60-80°S, i.e. to the Antarctic zone. On a global scale, zirconium is indifferent to depth (McKelvey et al., 1983).

In general, zirconium distribution in nodules seems to be controlled by iron. In the nodules of the South Pacific, the correlation coefficient for these two elements is 0.88 (Landmesser et al., 1976). On the whole, nodules of the World Ocean rich in cobalt are also rich in zirconium (McKelvey et al., 1983). In nodules from the Pacific radiolarian zone, zirconium shows a strong positive correlation with titanium, cerium, lead and zinc, the correlation coefficients being 0.82, 0.79, 0.73 and 0.73, respectively (Calvert et al., 1978).

From Calvert and Price (1977), 71 to 94% of the total zirconium in nodules is in a mobile form in the oxide and hydrooxide phases.

Hafnium content in manganese nodules was first determined by Glasby (1973) in a sample from the submarine Campbell Plateau in the South Pacific. It was 0.01 to 0.2 ppm. A recent analysis specified that hafnium content in nodules from various parts of the ocean varies from 0.3 to 24 ppm (Table 105). Minimum hafnium content, 0.5 to 1.9 ppm, and 0.9 ppm on average, is typical of nodules from the hemipelagic eastern part of the Pacific (Dymond et al., 1984). Nodules from the Pacific radiolarian zone contain from 2.6 to 14.2 ppm of hafnium; its average is 6.2 ppm (Piper et al., 1979). Hafnium maximum, 2.2 to 24.0 ppm, was recorded in nodules of the southeastern Pacific, in the vicinity of the East Pacific Rise, its average is 12 ppm (Girin et al., 1979).

As far as can be determined from scanty analyses, ore crusts are also rich in hafnium (to 11 ppm, from Dymond et al., 1984).

Analysing the distribution of elements in various parts of nodules, it was found that in the Pacific hemipelagic zone, tops of nodules are on average 5.5 times richer in hafnium than their bottoms; the average hafnium content being 2.2 and 0.4 ppm, respectively (Dymond et al., 1984). These estimates testify to the importance of the hydrogenous mechanism in its supply to nodules.

In nodules, hafnium is predominantly associated with iron. Hf to Fe ratio is stable along the entire geochemical cross-section through the southeastern Pacific, fluctuations recorded are from 0.8 to 1.3 ppm (Girin et al., 1979).

The hafnium average in nodules is 6 to 9 ppm and its average in deep oceanic sediments is some 4 ppm, so, its average concentration coefficient is 1.5 to 2.

Germanium was first determined in two nodule samples from the Pacific by Riley and Sinhaseni (1958). The result was 5 and 6 ppm.

Later analyses of samples from the Pacific and Indian oceans showed considerably lower concentrations of germanium, 0.5 to 1.3 ppm, being 0.8 ppm on average (Volkov and Sokolov, 1970; Volkov, 1979).

TABLE 105

Hafnium content (ppm) in nodules

Region	Hf		N	Reference
	Range	Mean		
Pacific ocean	‹10 -10	10	8	McKelvey et al., 1983
	3 -14	6	96	Haynes et al., 1986
	4.93- 8.76	6.85	2	White et al., 1986
radiolarian belt	2.6 -14.2	6.2	40	Piper et al., 1979
	-	4.3	1*	Flanagan and Gottfried, 1980
northern tropical part	5.1 - 8.5	6.8	4	Sano et al., 1985
Campbell Plateau	-	0.2	1	Glasby, 1973
South-eastern part	2.2 -24.0	12	78	Girin et al., 1979
South-western part	8.4 - 9.8	9.2	2	Glasby et al., 1978
eastern equatorial hemipelagic part	0.5 - 1.9	0.9	6	Dymond et al., 1984
tops of nodules	0.4 - 3.7	2.2	11	- " -
bottoms of nodules	0.3 - 0.6	0.4	13	- " -
Mn-crusts	11 -11	11	2	- " -
Indian ocean	6.1 - 9.3	7.7	2	Glasby et al., 1978
	-	5.78	1	White et al., 1986
Atlantic ocean	-	7.54	1	- " -
Blake Plateau	-	6.2	1*	Flanagan and Gottfried, 1980
Sediments	-	4.2	-	Lisitsin, 1978
Red clays (The Pacific)	3.3 - 7.1	4.9	36	Gurvich et al., 1980b
Radiolarian oozes (The Pacific)	3.0 - 4.8	4.0	38	Piper et al., 1979

1* Reference sample

TABLE 106
Tin content (ppm) in nodules

Region	Sn		N	Reference
	Range	Mean		
Pacific ocean	240 -320	267	3	Riley and Sinhaseni, 1958
	1.5- 4.5	2	15	Ahrens et al., 1967
	0.2- 5.8	3	6	Smith and Burton, 1972
	‹10 -400	-	-	McKelvey et al., 1983
	2 -450	108	87	Haynes et al., 1986
Campbell Plateau	-	1.68	1	Glasby et al., 1973
south-western part	5.3- 6.0	5.6	2	Glasby et al., 1978
	1.6- 5.7	5	13	Rankin and Glasby, 1979
Indian ocean	‹1.5- 3.0	1.5	12	Ahrens et al., 1967
	2.7- 2.7	2.7	2	Glasby et al., 1978
Atlantic ocean	1.5- 6.2	2.7	5	Ahrens et al., 1967
Sediments	-	11	-	Lisitsin, 1978
Sediments of the south-western Pacific	‹1 - 3.5	3.4	15	Glasby et al., 1978; Rankin and Glasby, 1979

According to the semi-quantitative analyses of four nodule samples from the Pacific, the germanium content is less than 10 to 90 ppm (McKelvey et al., 1983). The average germanium content in deep oceanic sediments is 1.6 ppm (Lisitsin, 1978), which is approximately twice as high, in contrast to nodules.

Tin was first determined in nodules of the three Pacific samples, and its content turned out to be considerably high: 240 to 320 ppm (Riley and Sinnaseni, 1958). Later, when more correct analyses were made, the tin content in nodules appeared to be two orders lower (0.2 to 6.2 ppm), its average is about 2 in the Pacific nodules, 1.7 in the Indian ocean, 2.7 ppm in the Atlantic (Ahrens et al., 1967; Smith and Burton, 1972; Glasby, 1973; Rankin and Glasby, 1979; Table 106).

According to the semi-quantitative analyses, the tin content in nodules is from less than 10 to 400 ppm (McKelvey et al., 1983); these estimates do not seem to be reliable.

Tin average in deep oceanic sediments of the World Ocean is 11 ppm (Lisitsin, 1978). It is three times lower in sediments of the southeastern Pacific, being 3.4 ppm on average (Glasby, 1973; Rankin and Glasby, 1979).

The average tin content in nodules of the World Ocean seems to be 2 ppm, whereas in sediments, it is 4 to 11 ppm. Thus, no accumulation of tin occurs in nodules.

RARE METALS OF THE FIFTH GROUP OF PERIODIC TABLE

Rare metals of Group V are vanadium, niobium, tantalum (main subgroup), antimony and bismuth (subordinate subgroup).

Vanadium content in manganese nodules from the Pacific radiolarian zone varies from 20 to 1300 ppm (McKelvey et al., 1979). It is somewhat higher (456 ppm on average) at submarine hills, within the zone in question, in contrast to valleys with thick and thin sedimentary cover, 443 and 406 ppm, respectively (Calvert and Piper, 1984).

In nodules from the southeastern Pacific, vanadium content is 234 (Girin et al., 1979) to 310 ppm (Cronan, 1980) on average.

Vanadium maxima were recorded in nodules and ore crusts at the Mid Pacific sea mountains and in the southern borderland of the Pacific, on average 540 and 650 ppm, respectively (Cronan, 1980).

The vanadium average in nodules for the entire Pacific is from 510 to 560 ppm (Table 107).

Vanadium contents in nodules of the Indian ocean vary from 170 to 910 ppm. its average being 440 to 540 ppm, according to various estimates; the corresponding values for the Atlantic nodules are 200 to 1100 ppm and 530 to 700 ppm (Mero, 1965; Volkov, 1979; Cronan, 1980; McKelvey et al., 1983).

In nodules of the World Ocean, the vanadium content varies over a range of 10 to 5000 ppm, being 520 ppm on average (McKelvey et al., 1983).

The average concentration of vanadium in deep oceanic sediments is 96 ppm (Lisitsin, 1978; Bischoff et al., 1979); so, the vanadium concentration coefficient in nodules with respect to host sediments is about 5.

When considering vanadium concentration in nodules with respect to the latitude, it shows its minimum, 410 ppm on average, in the Antarctic zone, and its maximum, 710 ppm on average, within the interval of 40-60°S. Distribution of vanadium with depth is very distinctive. Shallow-water nodules are rich in vanadium, whereas deep-sea nodules are poor in it; its average being 710 and 490 ppm, respectively.

Comparison of concentrations of vanadium and major elements in nodules showed that nodules with manganese maximum (over 35%) are poor in vanadium (to 410 ppm on average); a vanadium maximum (690 ppm, on average) is associated with nodules rich in cobalt (to 1% and more) (McKelvey et al., 1983).

Vanadium behaviour in nodules is mainly controlled by iron, a relatively constant V/Fe ratio being preserved in the southeastern Pacific in various lithological-facial zones (Girin et al., 1979)

Niobium content was first determined by Rancama (1948) in one nodule sample (24 ppm). Then Pachadzhanov et al. (1963) found 29 ppm of niobium in the nodule shell and 8 ppm in the nodule inner parts. Later, Mero(1965) analysed 8 nodules from the Pacific and determined 30 to 150 ppm of niobium, 85 ppm on average.

Further analyses of niobium content in nodules in individual samples of the Pacific, Indian and Atlantic oceans showed considerably lower values, from 2 to 60 ppm (Glasby, 1973; Volkov, 1979; McKelvey et al., 1979).

From the semi-quantitative analyses made by McKelvey et al. (1983), the niobium content in manganese nodules of the World Ocean varies from 10 to 300 ppm, being 70 ppm on average. That is consistent with the results obtained by Mero (1965) but is overestimated contrast to the results of other investigators (Table 108).

The niobium average in nodules is probably from 30 to 70 ppm.

The niobium concentration in deep oceanic sediments varies from less 1.5 to 15 ppm and is 9.5 ppm on average (Lisitsin, 1978; Rankin and Glasby, 1979). Therefore, an average concentration coefficient of niobium in nodules with respect to host sediments may be 5.

Tantalum content was first determined in the sample from the Indian ocean in which niobium was analysed (Pachadzhanov et al., 1963). The outer shell of the nodule analysed contained 0.17 ppm and its inner parts 2.2 ppm of tantalum.

Manganese nodules from the South Pacific turned out to contain less than 20 ppm in all 13 samples analysed (Landmesser et al., 1976). Another attempt to determine tantalum in nodules showed from less than 10 to 20 ppm in 9 samples (McKelvey et al., 1983).

The tantalum average in deep oceanic clayey sediments is 0.8 ppm (Turekian and Wedepohl, 1861), in the Pacific radiolarian zone (38 samples) it is 0.87 ppm (Piper et al., 1979), in continental clays it is 3.5 ppm (Vinogradov, 1962).

On this basis it is possible to deduce that the average concentration coefficient of tantalum in nodules with respect to host sediments is 1.

Antimony content in the Pacific nodules varies from 1.4 to 118 ppm (Table 109).

Nodules of the Pacific radiolarian zone contain from 14.5 to 47.4 ppm of antimony, its average being 31.2 ppm (Piper et al., 1979). Nodules of the south-

TABLE 107
Vanadium content (ppm) in nodules

Region	V		N	Reference
	Range	Mean		
Pacific ocean	380- 700	490	3	Riley and Sinhaseni, 1958
	390- 750	530	6	Willis and Ahrens, 1967
	210-1100	540	54	Mero, 1965
	90-1500	560	153	Volkov, 1979
	-	530	-	Cronan, 1980
	-	510	-	McKelvey et al., 1983
	‹50-3000	500	507	Haynes et al., 1986
radiolarian belt	20-1300	300	43	McKelvey et al., 1979
	50- 800	470	70	Haynes et al., 1986
highs	-	456	17	Calvert and Price, 1984
valleys	-	424	49	- '' -
south-eastern part	70- 580	234	78	Girin et al., 1979
	-	310	8	Cronan, 1980
central part	-	360	9	- '' -
western part	-	440	23	- '' -
Mid Pacific mountains	-	540	5	- '' -
	‹50-3000	860	29	Haynes et al., 1986
subsea mountains of southern borderland	-	650	5	Cronan, 1980
depth ›3000 m	‹100-3000	480	370	Haynes et al., 1986
depth ‹3000 m	‹50-3000	670	38	- '' -
Indian ocean	580- 640	600	4	Sano et al., 1985
	170- 910	490	56	Volkov, 1979
	-	440	-	Cronan, 1980
	-	540	-	McKelvey et al., 1983
Atlantic ocean	320- 840	570	8	Willis and Ahrens, 1967
	200-1100	700	4	Mero, 1965
	-	530	-	Cronan, 1980
	-	600	-	McKelvey et al., 1983
World Ocean	-	558	-	Cronan, 1980
	10-5000	520	437	McKelvey et al., 1983
Sediments	-	96	-	Lisitsin, 1978
Sediments of the Pacific radiolarian belt	-	97	118	Bischoff et al., 1979

TABLE 108
Niobium content (ppm) in nodules

Region	Nb		N	Reference
	Range	Mean		
Pacific ocean	-	24	1	Rancama, 1948
	30-150	85	8	Mero, 1965
	-	70	-	McKelvey et al., 1983
	6-150	74	42	Haynes et al., 1986
radiolarian belt	-	30	1	McKelvey et al., 1979
Campbell Plateau	-	24	1	Glasby, 1973
Indian ocean	-	18	1	Pachadjanov et al., 1963
	2- 60	31	14	Volkov, 1979
	-	70	-	McKelvey et al., 1983
Atlantic ocean	9- 28	21	3	Volkov, 1979
	-	40	-	McKelvey et al., 1983
World Ocean	10-300	70	68	- '' -
Sediments	-	9.5	-	Lisitsin, 1978
south-western Pacific	1.5- 15	9.5	13	Rankin and Glasby, 1979

eastern Pacific contain from 1.4 to 46 ppm of antimony, 27.4 ppm on average (Girin et al., 1979). The antimony maximum was recorded in nodules from the eastern equatorial hemipelagic zone of the Pacific, 43 to 119 ppm; 83 ppm on average (Calvert and Piper, 1984).

From the results of the 10 semi-quantitative analyses, the antimony content in the Pacific and Atlantic nodules is 50 ppm on average (McKelvey et al., 1983).

The antimony average in the World Ocean nodules can be deduced as being 30 to 50 ppm from the data of Table 109.

The antimony average in the deep oceanic sediments is from 1.0 to 2.5 ppm, according to various authors (Turekian and Wedepohl, 1961; Lisitsin, 1978; Piper et al., 1979), and seems to be about 2 ppm. The average concentration coefficient of antimony in nodules with respect to their host sediments is 20.

In nodules from the eastern equatorial hemipelagic zone of the Pacific, differentiation of antimony concentration is pronounced: in the bottom parts of nodules' antimony concentration is double that in their tops, on average, 111 and 54 ppm, respectively (Dymond et al., 1984).

In nodules of the southeastern Pacific where an oxiding environment of sediments prevails, antimony shows a distinct correlation with iron at a relatively constant Sb/Fe ratio (Girin et al., 1970); whereas in the nodules of the eastern equatorial hemipelagic zone of the Pacific, antimony behaviour is controlled by a suboxic diagenesis. In this case antimony follows manganese, not iron. According to the normative model suggested by Dymond (1981), 72% of the total antimony was supplied to these nodules as a result of diagenetic processes (Dymond et al., 1984).

Bismuth. According to the first determinations made by Mero (1965), the bismuth content of the Pacific nodules varies from 25 to 45 ppm, being 30 ppm on average.

From other estimates (Table 110), the bismuth content in nodules of the Pacific, Indian and Atlantic oceans is considerably lower, from less than 0.5 to 24.4 ppm, its average being from 5.3 ppm in the Pacific to 8.2 ppm in the Indian ocean (Ahrens et al., 1967; Glasby, 1973).

From the semi-quantitative analyses of 41 samples from various regions of the World Ocean, the bismuth content in nodules is from less 10 to 90 ppm, being 20 ppm on average (McKelvey et al., 1983). According to these data, the bismuth average in nodules of the ocean is 5 to 10 ppm.

The bismuth average in continental clays is 0.01 ppm from Vinogradov (1962).

If we suppose that oceanic sediments contain one order more of bismuth, on the average, i.e. 0.1 ppm, its concentration coefficient in nodules with respect to sediments is 50 to 100. According to this estimate bismuth may be probably regarded as the element most actively accumulating in nodules.

RARE ELEMENTS OF THE SIXTH GROUP OF PERIODIC TABLE

Three rare metals, namely, chromium, molybdenum and tungsten belong to the Group VI of the Periodic Table. Their behaviour in sedimentary processes differs considerably.

Chromium content in manganese nodules of the Pacific varies from 1 to 700 ppm, the widest range of its content being in the radiolarian zone.

According to McKelvey et al. (1979), the chromium content of the nodules of the zone in question is 200 ppm; however, from other estimates it is considerably lower both in the radiolarian zone and in other zones of the Pacific, being from 5 to 51 ppm on average (Table 111).

Chromium average in the Pacific nodules was determined by some researchers as 13 to 90 ppm. The upper limit seems to be overestimated (as follows from Table 111) and, therefore, it is possible to estimate its average as 20 to 25 ppm.

Chromium content in the Indian ocean nodules varies from 1.4 to 155 ppm; in the Atlantic nodules, it varies from 11 to 252 ppm; its corresponding averages are 20 and 60 ppm.

Average chromium content in nodules of the World Ocean is 35 ppm (Cronan, 1980) and 70 ppm (McKelvey et al., 1983). The data presented in Table 111 speak in favour of the former.

TABLE 109
Antimony content (ppm) in nodules and crusts

Region	Sb		N	Reference
	Range	Mean		
Nodules				
Pacific ocean	-	50	-	McKelvey et al., 1983
	14 - 72	37	103	Haynes et al., 1986
radiolarian belt	14.5- 47.4	31.2	41	Piper et al., 1979
northern tropic part	9.1- 48	26.6	4	Sano et al., 1985
Campbell Plateau	-	18.63	1	Glasby, 1973
south-eastern part	1.4- 46.7	27.4	78	Girin et al., 1979
eastern equatorial				
hemipelagic part	43 -118	83	6	Dymond et al., 1984
topps of nodules	32 - 74	54	11	- " -
bottoms of nodules	72 -163	111	13	- " -
Atlantic ocean	-	40	-	McKelvey et al., 1983
World Ocean	40 - 50	50	10	- " -
Hydrothermal crusts				
Pacific ocean	6 - 67	23	5	Toth, 1980
Atlantic ocean	2 -150	60	6	- " -
Hydrogenic crusts				
Pacific ocean	19 -151	48	10	Toth, 1980
Atlantic ocean	35 - 49	42	2	- " -
Sediments	-	1.0	-	Turekian and Wedepohl, 1961
	-	1.6	-	Lisitsin, 1978
The Pacific radiolarian belt	1.1- 5.6	2.5	38	Piper et al., 1979

TABLE 110
Bismuth content (ppm) in nodules

Region	Bi		N	Reference
	Range	Mean		
Pacific ocean	25 - 45	30	12	Mero, 1965
	<0.5- 18.3	5.3	14	Ahrens et al., 1967
	-	30	-	McKelvey et al., 1983
	6 - 31	21	13	Haynes et al., 1986
	1.2- 4.6	2.8	23	Author's data
radiolarian belt	-	6.2	1	Ahrens et al., 1967
Campbell Plateau	-	9.88	1	Glasby, 1973
Indian ocean	<0.5- 24.4	8.2	10	Ahrens et al., 1967
Gulf of Aden	-	5.09	1	Glasby, 1973
	-	10	-	McKelvey et al., 1983
Atlantic ocean	<0.5- 12.1	6.6	7	Ahrens et al., 1967
Blake Plateau	-	10.81	1	Glasby, 1973
	-	10	-	McKelvey et al., 1983
World Ocean	10 - 90	20	41	- " -
	5 - 14	8	-	Cronan, 1980
Sediments	< 0.005- 0.5	0.03	104	Oreshkin et al., 1986

Chromium average in deep oceanic sediments is close to 60 ppm (Lisitsin, 1978; Bischoff et al., 1979; Piper et al., 1979). On this basis, one may conclude that chromium does not accumulate in nodules and its average concentration coefficient in nodules is 0.5 to 0.7.

When chromium concentration was considered with respect to its latitudinal occurrence, its minimum was recognized in the interval of $0 - 20^{\circ}S$ and its maximum in the interval of $20 - 40^{\circ}N$. Shallow-water nodules contain 150 ppm of chromium, on average, deep oceanic nodules contain 60 ppm of this element. The richest in chromium nodules are those with a maximum content of copper and nickel (over 1.8% in total, from McKelvey et al., 1983).

Nodules from the eastern equatorial hemipelagic zone of the Pacific show a distinct differentiation of chromium in their tops and bottoms: 20.6 and 11.8 ppm, respectively. This supposes a preferential chromium supply to the nodules from the sea water. However, according to the normative model by Dymond et al. (1984), the share of hydrogenous supply to nodules turned to be subordinate (25%) to the predominant suboxic diagenesis (60%).

The method of selective leaching, applied to determine the forms of chromium occurrence in nodules of the Indian and Atlantic oceans, showed that, on the whole, the range of relative concentration of chromium in each form is extremely wide. The share of a hydroxylamine-hydrochloride leachate is less than 1 to 80%, the share of hydrochloric leachate is less than 5 to 100% and that of the insoluble residue is 20 to 75%. The most typical chromium concentrations in these fractions are as follows: in nodules of the Madagascar Basin (the Indian ocean), less 20; 40 to 80; 40 to 60%; in nodules of the Central Basin (the Indian ocean), less 20; 30 to 50; 50 to 60%; in the Atlantic nodules, 30 to 50; 10 to 40; 20 to 40, respectively (Moorby and Cronan, 1981).

Molybdenum content in the Pacific nodules was first assessed by Mero (1965) as 100 to 1500 ppm. Later, this range became even wider (from 7 to 1500 ppm, according to Volkov, 1979).

Manganese nodules from the Pacific radiolarian zone contain from 40 to 1300 ppm of molybdenum; 480 ppm on average (McKelvey et al., 1979). According to another series of analyses, nodules from submarine hills in the zone considered contain 388 ppm of molybdenum on average; nodules from depressions with thick sedimentary cover contain 489 ppm of the element; and its content in nodules from depressions with thin sedimentary cover is 520 ppm (Calvert and Piper, 1984). After Cronan (1980), nodules from the Mid Pacific sea mountains contain 420 ppm of molybdenum on average; nodules at sea mountains of the southern borderland contain 400 ppm; it is 370 ppm in the Western Pacific; 410 ppm in the central part of the ocean; 470 ppm in the northeastern part; 370 ppm in the southeastern and 350 ppm in the southern parts.

According to more generalized data (Skornyakova et al., 1986), deep oceanic nodules of the Pacific, except for its equatorial zone, contain from 110 to 800 ppm of molybdenum, 347 ppm on average; nodules from the equatorial zone contain from 250 to 890 ppm of the element, 490 ppm on average. Molybdenum concentration in nodules from sea mountains is from 170 to 710 ppm, being 400 on average; it is from 17 to 780, being 300 ppm in nodules from hemipelagic regions.

Estimates of molybdenum average in the Pacific nodules are within the range from 330 to 520 ppm (Table 112). The most probable average seems to be 420 ppm.

Molybdenum concentration in the Indian ocean nodules varies from 16 to 910 ppm, being about 300 ppm on average; its content in the Atlantic nodules varies from 110 to 780 ppm, being 370 ppm on average; the World Ocean nodules contain from 7 to 2200 ppm of the element, its average being 400 ppm (Volkov, 1979; Cronan, 1980; McKelvey et al., 1983).

From various authors, the molybdenum average in oceanic sediments varies from 4.2 ppm (the entire World Ocean; Lisitsin, 1978) to 27 ppm (deep-sea red clays; Turekian and Wedepohl, 1961). According to the data presented in other works for various oceanic regions (Cronan and Tooms, 1969; Calvert and Price, 1977; Bischoff et al., 1979; Geochemistry of diagenesis..., 1980; Skornyakova et al. 1986), molybdenum average in sediments within the zones where manganese nodules occur is about 10 ppm, which allows estimation of its average concentration coefficient in nodules with respect to host sediments as 40. Direct determination of molybdenum in nodules and host sediments at 18 stations in the Central Pacific showed that this coefficient varies from 4.5 to 106, being 26 on average (Calvert and Price, 1977).

TABLE 111
Chromium content (ppm) in nodules

Region	Cr		N	Reference
	Range	Mean		
Pacific ocean	9.1- 55.6	27.5	12	Piper and Williamson, 1977
	-	13	-	Cronan, 1980
	-	90	-	McKelvey et al., 1983
	1 -150	31	394	Haynes et al., 1986
radiolarian belt	2 -700	200	49	McKelvey et al., 1979
	2 - 58	25	40	Author's data
north-eastern part	-	7	10	Cronan, 1980
western part	-	7	23	- " -
central part	-	12	9	- " -
southern part	-	7	11	- " -
south-eastern part	-	5	8	- " -
Mid-Pacific mountains	-	11	5	- " -
	1 -150	58	22	Haynes et al., 1986
subsea mountains of				
southern borderland	-	51	5	Cronan, 1980
depth ˃3000 m	1 -150	25	277	Haynes et al., 1986
depth ˂3000 m	1 -130	60	38	- " -
eastern equatorial				
hemipelagic zone	6.3- 16.4	10.6	6	Dymond et al., 1984
tops of nodules	6.4- 40.0	30.6	11	- " -
bottoms of nodules	3.7- 20.6	11.8	13	- " -
Indian ocean	1.4-110	15	69	Volkov, 1979
	-	29	-	Cronan, 1980
	-	20	-	McKelvey et al., 1983
Central Basin	15 -155	36	22	Cronan, 1980
Madagascar Basin	20 - 65	28	15	- " -
Atlantic ocean	10 - 30	20	4	Mero, 1965
	38 -150	62	14	Moorby and Cronan, 1981
	11 -252	38	32	Author's data
	-	70	-	Cronan, 1980
	-	60	-	McKelvey et al., 1983
World Ocean	-	35	-	Cronan, 1980
	10 -2310	10	274	McKelvey et al., 1983
Sediments	-	60	-	Lisitsin, 1978
sediments of the				
Pacific radiolarian belt	19.4- 73.1	58.7	38	Piper et al., 1979

As for the latitudinal distribution, the molybdenum maximum is distinct in the interval of 0-20°N. Its concentration also varies with depth. At depths down to 3000 m its average is 450 ppm, it is 370 ppm in the deeper nodules. The molybdenum maximum, 600 ppm on average, is associated with nodules where nickel concentration is over 1% and with nodules where manganese concentration is over 35% (molybdenum content in the latter being 560 ppm, on average) (McKelvey et al., 1983). That supposes several factors as determining molybdenum behaviour during nodule formation.

The pattern of molybdenum distribution in nodules of the Pacific (Figure 51) shows that its low concentration (less 300 ppm) in nodules is confined to the western, northwestern and southern periphery of the ocean, whereas nodules from the eastern periphery (near California) are rich in molybdenum (500 ppm and over). A high molybdenum content is observed in nodules from the near equatorial zone of the ocean. On the whole, this pattern is similar to the distribution pattern of manganese, nickel and copper (Figures 36, 38, 39).

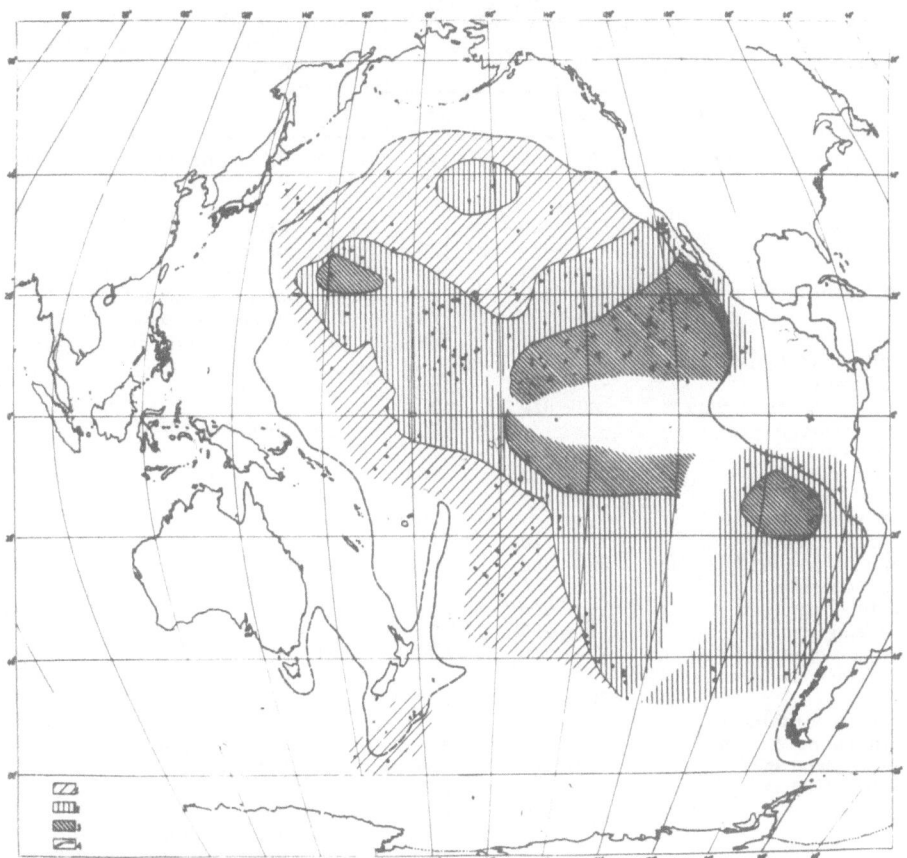

Fig. 51. Mo distribution in the Pacific nodules (Skornyakova et al., 1986). Mo concentration, ppm: (1) less 300; (2) 300 to 500; (3) over 500; (4) boundary of the oxidized sediment layer over 1 m.

High correlation coefficients of molybdenum and manganese testify to a close relationship of the two elements. This coefficient is 0.694 in deep-sea nodules; 0.512 in deep-sea nodules in the near equatorial zone; 0.460 in nodules at sea mountains; 0.658 in nodules from pelagic zones; 0.641 in nodules of hemipelagic regions (Skornyakova et al., 1986).

The bulk of molybdenum (87 to 100%) in nodules is associated with their oxic phases; both todorokite and vernadite nodules are rich in molybdenum (Calvert and Price, 1977). Molybdenum differentiation in various parts of the nodule is not so distinct. At variable ratios, tops of 20 nodules from the Pacific contain, on average, 500 ppm and their bottoms contain 640 ppm of the element (Raab, 1972). Correlation between molybdenum concentration in nodules and host sediments is weak, from 0.078 to 0.453 (Skornyakova et al., 1986). On the whole, that supposes a two-fold mechanism of molybdenum accumulation in nodules: diagenetic and hydrogenous. Molybdenum occurs mainly in its sorbed form which results from its association with manganese hydrooxides and, to a lesser extent, with iron hydrooxides.

TABLE 112
Molibdenum content (ppm) in nodules

Region	Mo		N	Reference
	Range	Mean		
Pacific ocean	400- 500	430	3	Riley and Sinhaseni, 1958
	100-1500	520	54	Mero, 1965
	7-1500	330	192	Volkov, 1979
	-	440	-	Cronan, 1980
	-	410	-	McKelvey et al., 1983
	‘50-1500	400	1157	Haynes et al., 1986
radiolarian belt	40-1300	480	108	McKelvey et al., 1979
	‘50-1200	520	265	Haynes et al., 1986
highs	-	388	17	Calvert and Price, 1984
valleys with thick sediment cover	-	489	19	- " -
northern tropical part	40- 820	420	4	Sano et al., 1985
north-western part	-	470	10	Cronan, 1980
western part	-	370	23	- " -
central part	-	410	9	- " -
southern part	-	350	11	- " -
south-eastern part	-	370	8	- " -
Antarctic zone	200- 220	210	2	Author's data
deeps of equatorial zone	250- 890	490	96	Scornyakova et al., 1986
deeps outside equatorial zone	110- 800	347	122	- " -
hemipelagic zone	17- 780	300	29	- " -
pelagic mountains	170- 710	400	31	- " -
Mid Pacific mountains	-	420	-	Cronan, 1980
	‘50-1100	520	56	Haynes et al., 1986
subsea mountains of southern borderland	-	400	-	Cronan, 1980
depth ’3000 m	‘50-1300	360	746	Haynes et al., 1986
depth ‘3000 m	‘50-1500	500	80	- " -
Indian ocean	16- 910	313	95	Volkov, 1979
	-	290	-	Cronan, 1980
	-	290	-	McKelvey et al., 1983
Atlantic ocean	130- 560	350	4	Mero, 1965
	140- 550	369	25	Volkov, 1979
	-	490	-	Cronan, 1980
	-	310	-	McKelvey et al., 1983
	110- 780	350	29	Author's data
World ocean	-	412	-	Cronan, 1980
	20-2200	380	836	McKelvey et al., 1983
Sediments	-	4.2	-	Lisitsin, 1978
red clays	-	27	-	Turekian and Wedepohl, 1961
radiolarian oozes (Pacific Ocean)	5- 40	19	17	Calvert and Price, 1977
pelagic clays (Pacific Ocean)	2.1- 17	6.3	31	Skornyakova et al., 1986

Tungsten was first determined in two nodule samples from the Pacific. Its
content was 47 and 95 ppm (Riley and Sinhaseni, 1958); later this was verified by
more representative data (Table 113).

Tungsten content in nodules from the Pacific radiolarian zone is 26 to 130 ppm,
its average being 60 ppm; it is 55 to 164 ppm in the southern ocean, 108 ppm on
average (Baturin and Isaeva, 1986); 60 to 110 ppm in the southwestern part, 84
ppm on average (Landmesser et al., 1976). In ore crusts at the Mid Pacific sea
mountains, the tungsten is considerably higher, 124 ppm on average (Baturin and
Isaeva, 1986).

On the whole, the tungsten content in the Pacific nodules varies from 23 to
580 ppm, being 163 ppm on average (Volkov, 1979). McKelvey et al. (1983) found
that the tungsten average in the Pacific nodules is half of the first estimate, 80
ppm.

Tungsten contents in nodules of the Indian ocean vary from 10 to 127 ppm,
being from 67 to 120 ppm on average according to various estimates. So, the
average content of tungsten in nodules of the World Ocean seems to be close to
100 ppm (McKelvey et al., 1983).

The tungsten average in oceanic sediments is about 10 ppm as deduced from
the data presented in several works (Isaeva, 1960; 1977; Amiruddin and Ehmann,
1962; Gramm-Osipov and Kiseleva, 1970; Levashov et al., 1975; Volkov et al., 1976;
Landmesser et al., 1976). That allows estimation of its average concentration
coefficient in nodules with respect to sediments as 10. For nodules in the Pacific
radiolarian zone it is 4. Diagenetic nodules, classified so by Skornyakova, of the
Pacific radiolarian zone contain 50 ppm of tungsten on average, whereas hydrogenous
nodules contain 120 ppm of this element. So, one can suppose that it behaves similar
to iron. The correlation of tungsten to iron in the Pacific and Indian ocean nodules
and lack of such correlation with manganese (Figure 52) also testify to this
similarity.

When comparing tungsten and iron concentration in sediments of various basins
it becomes clear that the W/Fe · 1000 ratio decreases from 0.67-1.11 in seas (the
White and Baltic Seas and in oxygenated areas of the Black Sea) to 0.23 in the
Pacific and Indian oceans, probably due to the greater migration ability of iron.
However, in nodules this ratio is inverted: in seas, the average W/Fe · 1000 ratio
is 0.15-0.33; in oceans it is 0.66-1.00. That shows a more active relative
accumulation of tungsten in nodules of the oceanic environment contrast, those of
marine origin (Baturin and Isaeva, 1986).

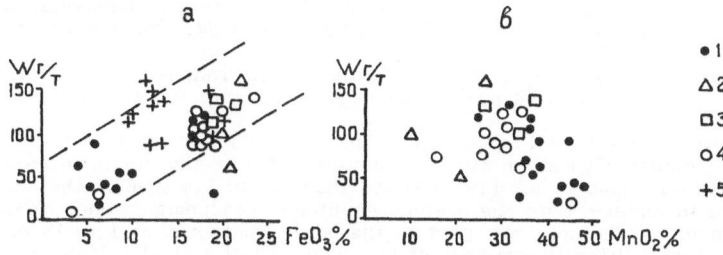

Fig. 52. Ratios between W–Fe (a) and W–Mn (b) concentrations in nodules (Baturin
and Isaeva, 1986). (1) nodules from the Pacific radiolarian zone; (2) nodules from the South Pacific; (3) ore crusts at the Mid Pacific sea mountains; (4) nodules from the Indian ocean (after Isaeva, 1967); (5) nodules
along the Hawaiian profile of the Pacific (after Volkov et al., 1976).

RADIOACTIVE METALS

Radioactivity of manganese nodules is determined primarily by concentrations of
uranium and thorium and of their decay products, namely uranium 234, protactinium
ionium, radium, etc.

TABLE 113
Tungsten content (ppm) in nodules

Region	W		N	Reference
	Range	Mean		
Pacific ocean	47- 95	66	2	Riley and Sinhaseni,
	23-580	163	32	Volkov, 1979
	-	80	-	McKelvey et al., 1983
	26-120	76	7	Haynes et al., 1986
radiolarian belt	26-130	60	13	Baturin and Isaeva, 1986
northern tropical part	20-250	115	4	Sano et al., 1985
south-western part	60-110	84	13	Landmesser et al., 1976
southern part	55-164	108	3	Baturin and Isaeva, 1986
Mid Pacific mountains				
(crusts)	129-139	124	3	- " -
Indian Ocean	26-126	88	-	Isaeva, 1967
	10-127	67	20	Volkov, 1979
	-	120	-	McKelvey et al., 1983
World Ocean	30-600	100	22	- " -
Sediments	-	10	-	*
Sediments of the				
south-western Pacific	10- 20	10	16	Landmesser et al., 1976

* Average calculated on the basis of data taken from: Isaeva, 1960, 1977;
Amiruddin and Ehmann, 1962; Gramm-Osipov and Kiseleva, 1970; Levashov et al.,
1975; Volkov et al., 1976.

Here we shall dwell upon the parent elements of these families, namely,
uranium and thorium. The data on the behaviour of their decay products are given
in the next chapter devoted to the dating of nodules.

Uranium. A systematic analysis of uranium in manganese nodules started after
the work by Tatsumoto and Goldberg (1959) who discovered from 3.6 to 5 ppm of
the element in five samples of the Pacific nodules.

Later, uranium in nodules was analysed mainly for the purpose of their dating;
however some works appeared that concerned the geochemical behaviour proper
(Baturin, 1975; Kunzendorf and Friedrich, 1976a,b; 1977; Kunzendorf et al., 1982;
1983).

The uranium content in nodules from the Pacific radiolarian zone varies from
1 to 35 ppm, being 4.5 to 13 ppm on average, according to various estimates
(Table 114).

Japanese researchers (Yabuki and Shima, 1973) determined 10 to 11 ppm of
uranium in nodules associated with foraminiferal ooze of the northern tropical zone
of the Pacific; Dymond et al. (1984) found that 2.8 to 5.4 ppm of the element is
accumulated in nodules from the eastern equatorial hemipelagic zone. Nodules from
various regions of the southern part of the ocean contain from 1 to 18 ppm of
uranium (Glasby, 1973; Landmesser et al., 1976; Glasby et al., 1978; Rankin and
Glasby, 1979; Kunzendorf et al., 1983; Baturin et al., 1986). The greatest number
of determinations was made by Kunzendorf and Friedrich (1976a; 1977) and
Kunzendorf et al. (1982; 1983) in the nodules from the North Pacific (over 600).
In the material they analysed, the uranium content varies from 3 to 10 ppm, being
4.6 ppm on average. The highest content of uranium was recorded in nodules and
ore crusts at the Mid Pacific sea mountains (14 to 22 ppm, and 17 ppm on the
average, after Baturin et al., 1986).

In the Indian ocean nodules the uranium content is 1.6 to 12.2 ppm, its
average being 3.7 to 9.8 ppm (Nikolaev and Efimova, 1963; 1965; Ku and Broecker,
1969).

TABLE 114
Uranium content (ppm) in nodules

Region	U		N	Reference
	Range	Mean		
Pacific Ocean	3.6 - 5.0	4.2	5	Tatsumoto and Goldberg, 1959
	3.9 - 9.3	6.5	11	Nikolaev and Efimova, 1963, 1965
	3.7 -13.3	8.1	28	Ku and Broecker, 1969
	5 -47	17	69	Cherdintsev et al., 1971
	3 - 8.3	6	15	Baturin, 1975
	4.2 -10.1	4.9	129	Kunzendorf and Friedrich, 1978
	1 -68	6.8	255	Haynes et al., 1986
radiolarian belt	2 -35	13	6	McKelvey et al., 1979
	1.0 - 8.6	4.5	41	Piper et al., 1979
	-	4.6	458	Kunzendorf et al., 1983
	6 -14	11	19	Baturin et al., 1986c
northern tropical part	10 -11	10.5	2	Yabuki and Shima, 1973
eastern equatorial hemipelagic zone	2.8 - 5.4	4.7	6	Dymond et al., 1984
tops of nodules	2.5 - 4.2	7.1	11	- " -
bottoms of nodules	3.1 -10.0	8.6	13	- " -
southern tropical part	3.52± 0.62		213	Kunzendorf et al., 1983
southern part	5.8 - 6.4	6.1	2	Glasby et al., 1978
	6.1 ± 0.5		31	Kunzendorf et al., 1983
	4 -12	7	6	Baturin et al., 1986c
south-western part	8 -12	12	13	Landmesser et al., 1976
	6.1 - 9.4	7.9	13	Rankin and Glasby, 1979
	0.68- 5.27	2.3	17	Immel and Osmond, 1976
	7.71± 0.84		31	Kunzendorf et al., 1983
Peru Basin	3.53± 0.50		230	Kunzendorf et al., 1983
Mid Pacific mountains	14 -22	17	8	Baturin et al., 1986c
Indian Ocean	1.6 - 5.5	3.5	9	Nikolaev and Efimova, 1963, 1965
	8.0 -12.2	9.8	5	Ku and Broecker, 1969
	3.4 - 7.2	5.3	2	Glasby et al., 1978
Atlantic Ocean	8.8 -11.1	9.8	15	Ku and Broecker, 1969
	3 -20	5	13	Author's data
Sediments:				
red clays	-	2.2	-	Baturin, 1975
calcareous oozes	-	1.0	-	- " -
siliceous oozes	-	0.9	-	- " -
sediments of the Pacific radiolarian belt	1.2 - 2.6	2.0	38	Piper et al., 1979
	-	1.1	18	Kunzendorf et al., 1983

In the Atlantic nodules, the uranium content was determined by Ku and Broecker (1969) as 8.8 to 11.1 ppm; according to our data, it is from 3 to 20 ppm.

One may expect less discrepancy in the data on nodules from certain oceanic regions since their macroelement composition is relatively homogeneous. In fact, the discrepancy in the data given by various authors is much greater. It may result from variations in the analytical methods applied, namely, chemical, luminescence, radiometric, X-ray fluorescent, neutron-activation, and also delayed - neutron counting techniques. On this basis, uranium average in nodules of the World Ocean can be presumtively assessed as 5-10 ppm.

The uranium content in nodules of the Kara, Barentz and Baltic Seas is 2.7 to 7.8 ppm with rare exceptions (Nicolaev and Efimova, 1963; 1965); in nodules from Loch Fyne (Scotland) and Jervis Inlet (Vancouver Is.), it is 10.1 to 21.7 ppm (Ku and Glasby, 1972). The uranium average in the main types of deep sea sediments is from 0.9 (siliceous ooze) to 2.2 ppm (red clays) (Baturin, 1975). Therefore, the uranium average concentration coefficient in oceanic nodules is 2.5 to 4.5 with respect to red clays and 5 to 10 with respect to biogenic siliceous and calcareous sediments.

From the data by Ku and Broecker (1969), Cherdyntsev et al., (1971) and Kunzendorf and Friedrich (1976b), the tops of nodules from the pelagic zone are richer in uranium in contrast to their bottoms which is typical of iron and associated elements. The layer-by-layer analysis of deep-sea nodules from the North Pacific showed that their tops contain 6.3 to 8.3 ppm of uranium, the intermediate layers contain some 5, the lower parts about 3, and the lowermost layer adjacent to sediments contain only 2.7 ppm of the element (Kunzendorf and Friedrich, 1976b). In the eastern equatorial hemipelagic zone of the Pacific, uranium has an inverse pattern of concentration in nodule layers: 2.5 to 4.2 ppm (3.1 ppm on the average) in the upper layers, 3.1 to 10 ppm (8.6 ppm, on the average) in the lower layers (Dymond et al., 1984).

Uranium in manganese nodules is associated with the oxidic phase and is extracted completely by hydrochloric acid (Efimova and Nikolaev, 1965). Its form in nodules seems to be a sorbed one.

In nodules from some regions of the ocean, one can observe U-Fe correlation with correlation coefficient of 0.31 to 0.82 (Friedrich et al., 1983). In other regions no U-Fe correlation was observed (Landmesser et al., 1976; Baturin et al., 1986). In pelagic nodules, no U-Mn correlation has been discovered but it was revealed in hemipelagic nodules (Dymond et al., 1984).

Uranium accumulation in nodules can result from its co-precipitation out of sea water with iron hydrooxides, its sorption by iron and manganese oxide phases, and also from its diagenetic flux out of underlying sediments (Boulegue et al., 1978; Kunzendorf et al., 1983; Dymond et al., 1984).

Thorium was first determined in several samples of the Pacific nodules. Its concentration was from 1.3 to 143 ppm (Riley and Sinhaseni, 1958). Further analyses widened that range due to greater precision in determining its lower concentration limit (0.2 to 1 ppm) (Table 115).

Thorium content in nodules from the Pacific radiolarian zone does not vary considerably and its average is 20 ppm according to various estimates (Calvert et al., 1978; McKelvey et al., 1979; Piper et al., 1979; Baturin et al., 1986).

Thorium content is higher in nodules from the southern and southwestern part of the Pacific. Its average is 4.0 ppm (Landmesser et al., 1976; Rankin and Glasby, 1979; Baturin et al., 1986), however, in nodules from the southeastern Pacific in the vicinity of the East Pacific Rise, thorium content is extremely irregular (0.2 to 106 ppm) and lowers to 19 ppm, on average (Girin et al., 1979).

Thorium content is a minimum (1.3 ppm on average) in the nodules from the eastern equatorial hemipelagic zone (Dymond et al., 1984). A low thorium content (14 ppm on average) is also typical of nodules and ore crusts at the Mid Pacific sea mountains (Baturin et al., 1986), however, in three samples of ore crusts at one of the sea mountains south of the Hawaii Is., the thorium content was higher: from 26 to 64 ppm (Kunzendorf and Friedrich, 1976b).

Thorium content in nodules of the Indian ocean varies from 19 to 75 ppm, it is 6 to 152 ppm in the Atlantic nodules. Thorium average in nodules of the World Ocean was assessed as 30 ppm (McKelvey et al., 1983).

TABLE 115
Thorium content (ppm) in nodules

Region	Th		N	Reference
	Range	Mean		
Pacific Ocean	-	1.3	1	Koszy, 1949
	38 -143	50	3	Matthews, 1954
	24 -124	40	4	Goldberg and Picciotto, 1955
	24 -143	50	3	Riley and Sinhaseni, 1958
	13 - 34	20	9	Nikolaev and Egimova, 1963, 1965
	2.8-154	59	26	Ku and Broecker, 1969
	12 - 75	33	18	Calvert and Price, 1977
	1.2-138	44	14	Piper and Williamson, 1977
	-	30	-	McKelvey et al., 1983
	5 -154	28	283	Haynes et al., 1986
radiolarian belt	10 - 25	18	19	Calvert et al., 1978
	10 - 30	20	25	McKelvey et al`., 1979
	9.0- 38.4	22	41	Piper et al., 1979
	12 - 32	20	10	Baturin et al., 1986c
northern tropical part	8.9- 39	25.7	4	Sano et al., 1985
south-western part	28 - 60	42	13	Landmesser et al., 1976
	25 - 65	43	12	Rankin and Glasby, 1979
south-eastern part	0.2-106	19	76	Girin et al., 1979
southern part	27 - 51	40	6	Baturin et al., 1986c
eastern equatorial hemipelagic zone	0.6- 2.7	1.3	6	Dymond et al., 1984
tops of nodules	0.8- 4.5	3.0	11	- " -
bottoms of nodules	0.4- 1.0	0.7	13	- " -
Mid Pacific mountains	3 - 21	14	8	Baturin et al., 1986c
Indian Ocean	19 - 42	31	8	Nikolaev and Efimova, 1963, 1965
	54.9- 74.8	70	5	Ku and Broecker, 1969
	30 - 35	32	2	Glasby et al., 1978
	-	30	-	McKelvey et al., 1983
Atlantic Ocean	10.2-152	59	13	Ku and Broecker, 1969
	6.6-138	51	12	Author's data
World Ocean	10 -130	30	121	McKelvey et al., 1983
Sediments	-	7.7	-	Lisitsin, 1978
Sediments of the Pacific:				
red clays	3.5- 22	11.8	132	Gurvich et al., 1980b
radiolarian oozes	6.8- 15.8	13.1	38	Piper et al., 1979
muds of south-western part	10 - 19	15	11	Rankin and Glasby, 1979

The average thorium content in deep-sea sediments of the World Ocean seems to be 12 ppm (Table 115); its average concentration coefficient in nodules with respect to sediments is 2.5.

Analysing 18 samples of nodules and host sediments from the Central Pacific, it was determined that the thorium concentration coefficient varies from 0.7 to 5, being thus 2.1 on average (Calvert and Price, 1977).

The distribution of thorium in nodules with respect to climatic zones is as follows: thorium minimum (20 ppm on average) falls within the interval of 0-40°N, its maximum (80 ppm on average) is in the interval of 40-60°N. Thorium content in deep-sea nodules is somewhat higher in contrast to shallow-water nodules, on

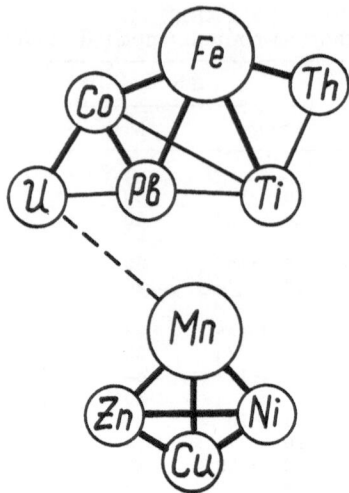

Fig. 53. Correlation between U, Th and ore metals of the Pacific nodules (Baturin
et al., 1986). Solid lines show strong correlation, thin lines show weak correlation.

average, 30 and 20 ppm, respectively (McKelvey et al., 1983). A thorium minimum
(2 to 6 ppm) is typical of nodules from seas and bays (Nikolaev and Efimova, 1963;
1965; Ku and Glasby, 1972).

Thorium distribution within individual nodule samples is irregular. A detailed
analysis of micro-layers of nodules from the Pacific pelagic zone revealed that their
upper layers contain 40 to 130 ppm of thorium, whereas their lower parts only 20
ppm; a similar pattern is typical of ore crusts, 64 ppm and 26 ppm, correspondingly
(Kunzendorg and Friedrich, 1976b). In nodules from the eastern equatorial Pacific
which are poor in thorium, this pattern is also valid, the upper parts of nodules
contain 3 ppm, on average, and their lower parts contain only 0.7 ppm of the
element (Dymond et al., 1984).

The bulk of thorium in nodules is associated with their ore oxide and hydrooxide phases; in nodules from the Central Pacific the share of mobile thorium is
42 to 93%, 78% on average (Calvert and Price, 1977).

Thorium shows a direct correlation with iron (r=0.64 to 0.85) and an inverse
correlation with manganese (r = -0.51 to -070) (Figure 53) which supposes its
association with the vernadite phase of nodules (Landmesser et al., 1976; Baturin
et al., 1986)

To compare thorium and uranium behaviour in natural processes one uses the
Th/U ratio. It is from 5 to 25 in nodules from pelagic zones of the ocean; about
1 in nodules of the Kara, Barentz and Baltic Seas; from 0.1 to 0.5 in nodules from
bays and fjords (Gulf of Finland, Loch Fyne, Jervis Inlet) (Nikolaev and Efimov,
1963; 1965; Ku and Broecker, 1969; Cherdyntsev et al., 1971; Ku and Glasby,
1972). In the upper parts of nodules from the eastern hemipelagic zone of the
Pacific, this ratio is 0.3 to 1.4; in their lower parts, it varies from 0.04 to 0.13.
That supposes that uranium is supplied to nodules out of sediments under suboxic
conditions of hemipelagic diagenesis (Dymond et al., 1984; Figure 54).

Comparing U/Fe, U/Mn, Th/Fe, Th/Mn and Th/U ratios in oceanic water,
suspended matter, sediments and nodules (Table 116) one can see that their
averages in the last three are considerably close and they sharply differ from that
in oceanic water. So, one can conclude that suspended matter and bottom sediments
not oceanic water, are the predominant sources of thorium and uranium supply to
nodules.

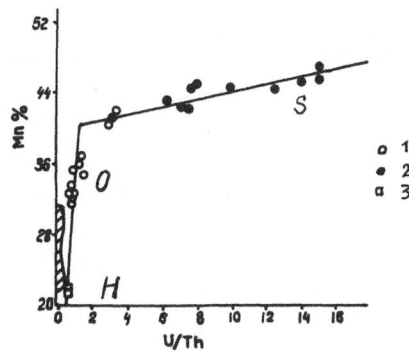

Fig. 54. Correlation between U/Th ratio and Mn concentration in nodules from the eastern equatorial hemipelagic zone of the Pacific (Dymond et al., 1984). (1) nodule tops; (2) nodule bottoms; (3) ore crusts. Hatching shows nodule fields in radiolarian zone. Conditions of ore matter accumulation: S-suboxic diagenesis; O-oxic diagenesis; H-hydrogenous deposition.

NOBLE METALS AND RHENIUM

Silver, gold and the metals of the platinum group, namely, platinum, ruthenium, rhodium, palladium, iridium and osmium comprise the group of noble metals. Rhenium, which has some similar geochemical properties, will be considered with metals of this group.

Silver was first determined by Mero (1965) in five samples of the Pacific nodules. Its content was assessed as being less than 1 to 5 ppm. The same estimate, 5 ppm, was obtained in a reference nodule sample from the Pacific radiolarian zone (Flanagan and Gottfried, 1980). Later estimates of silver content in nodules of seven samples from the Pacific were 10 to 20 ppm (McKelvey et al., 1983).

Silver was also analysed in oceanic sediments. Its average in deep-sea red clays, 0.11 ppm (Turekian and Wedepohl, 1961), coincides with its average, 0.1 ppm (Vinogradov, 1962), in continental clays. The silver content in calcareous sediments of the South Atlantic varies from 0.022 to 0.097 ppm, being 0.06 on average and 0.1 ppm on a carbonate-free basis (Turekian, 1968). From other data, the silver content in oceanic sediments varies from 0.036 to 13.3 ppm, its average being 3.74 ppm in the equatorial and southern Pacific, 0.75 ppm in the Indian ocean and 0.088 ppm in the North Atlantic (Horowitz, 1970).

An original technique (elaborated by Oreshkin et al., 1985b) based on a flameless atomic-absorption analysis of nodules and sediments with precision of 5 ppb (for silver) permitted us to obtain new data on silver concentration.

An analysis of 80 samples of nodules, ore crusts and their separate parts showed (Table 117) that the silver content is 0.02 to 0.26 (average 0.09 ppm) in bulk nodules, 0.03 to 0.45 (average 0.11 ppm) in ore shells, and 0.015 to 0.22 (average 0.045 ppm) in nodule nuclei. The same values are typical of ore crusts at the Mid Pacific sea mountains (0.015 to 0.22; 0.09 ppm on average). Thus, the silver average in nodules and ore crusts of the Pacific is 0.09 ppm, that is, one or two orders lower than in the previous estimates.

The silver average in deep-sea sediments of the Pacific is 0.037 ppm (Baturin and Oreshkin, 1986). Thus, its average concentration coefficient in nodules with respect to sediments is around 3.

No distinct correlation of silver with iron and manganese was detected in nodules; however, in ore crusts at sea mountains it tends to correlate with iron (Figure 55).

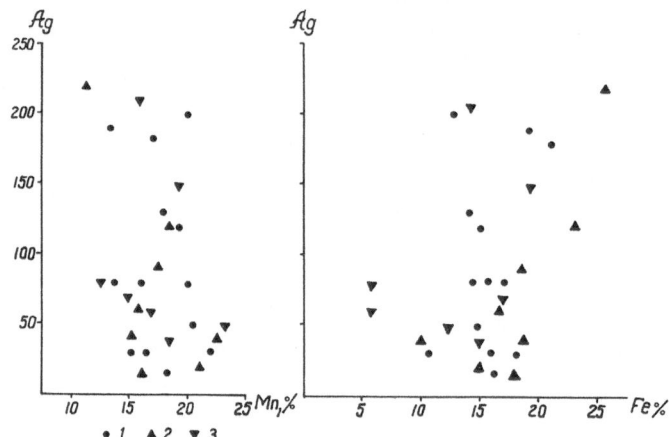

Fig. 55. Correlation between silver, manganese (a) and iron (b) in ore crusts at the Mid Pacific sea mountains (Baturin and Oreshkin, 1986). (1) bulk samples; (2) tops of crusts; (3) bottoms of crusts.

Gold. From the first determination made by Harriss et al. (1968) the gold concentration in nodules varies from 0.21 to 8.28 ppb. Later analysis of nodules from the Pacific, Atlantic and Indian oceans verified these results (Table 118) which indicates that its average content in nodules of the World Ocean is 2 - 3 ppb.

According to recent data, the gold concentration in major types of oceanic sediments varies within a narrower range, from 0.54 to 3.96 ppb, its average being 1.35 ppb in deep-sea red clay, 2.14 ppb in calcareous sediments and 0.85 ppb in siliceous ooze (Crocket and Kuo, 1979).

Thus, the average gold concentration coefficient in nodules with respect to host sediments varies from 1 to 3. The same values result from the comparison of gold averages in nodules and sediments of various lithological types. The average content of gold in nodules resting on red clays and foraminiferal ooze is 2 ppb and 1.5 ppb in nodules lying on diatomaceous ooze (Baturin et al., 1986); that indicates that the concentration coefficient of gold in nodules is 1.5, 1.0 and 1.8, respectively.

Gold distribution in nodules over the Pacific bottom is variable but tends to relative accumulation (1-2 ppm and over) along the northern periphery of the ocean (Figure 56).

A comparison of concentration of gold and major nodule metals shows no correlation. The study of nodules under the electron microscope revealed isometric particles of native gold, their size being 1-2 μm (Figure 57). That may be the reason for its irregular distribution, the absence of correlations, and sporadic extrema in nodules and sediments.

TABLE 116

Average ratios between U, Th, Fe and Mn in oceanic environment

Ratio	Oceanic water*	Oceanic suspension**	Deep-sea sediments***	Manganese nodules and ore crusts		
				Radiolarian belt	Southern part of the ocean	Subsea mountains
U/Fe	50	0.3×10^{-4}	0.5×10^{-4}	1×10^{-4}	0.5×10^{-4}	1.2×10^{-4}
U/Mn	100	10×10^{-4}	4×10^{-4}	0.3×10^{-4}	0.4×10^{-4}	0.8×10^{-4}
Th/Fe	20×10^{-4}	0.1×10^{-4}	1.7×10^{-4}	2.2×10^{-4}	2×10^{-4}	0.7×10^{-4}
Th/Mn	40×10^{-4}	3×10^{-4}	15×10^{-4}	0.7×10^{-4}	2×10^{-4}	0.5×10^{-4}
Th/U	$0.3/10^{-4}$	0.3	3.5	2	5	0.8

TABLE 117

Silver, iron and manganese in nodules and ore crusts of the Pacific ocean (Baturin and Oreshkin, 1986).

Sample	Silver, ppm		Iron, %		Manganese, %		Number of samples
	Range	Mean	Range	Mean	Range	Mean	
Total nodules	0.02-0.26	0.09	5.0-18.8	14.8	12.6-22.5	16.6	18
Shells of nodules	0.03-0.45	0.11	7.0-20.4	14.5	9.9-24.0	16.5	25
Nuclei of nodules	0.015-0.09	0.045	12.8-21.4	16.2	12.5-19.0	13.2	8
Ore crusts on basalts and hyaloclastites	0.015-0.22	0.09	5.6-23.3	15.7	11.3-22.9	16.1	29

TABLE 118
Gold content in manganese nodules

Au. ppb		N	Method	Reference
Range	Mean			
			Pacific Ocean	
1.52-3.98	2.66	7	Neutron activation	Harriss et al., 1968
5-5	5	2	Spectral emission	Landmesser et al., 1976
0.13-0.16	0.15	2	Neutron activation	Glasby et al., 1978
0.1-11.0	2.0	56	- " -	Baturin et al., 1984c, 1986b
2.0-1⊦.0	5.0	8	Atomic absorbtion	- " -
2.0-8.0	4.0	50	Extraction-spectral	- " -
< 2-13	5.0	20	Sorbtion-spectral	- " -
< 2-7 (ore crusts)	2.4*	7	- " -	- " -
			Atlantic Ocean	
1.00-6.62	3.02	7	Neutron activaiton	Harris et al., 1968
0.11-0.36	0.28	3	- " -	Keays, Scott, 1976
50-50	50	5	Spectral emission	Melo et al., 1978
5.0-7.0	6.0	23	Extraction-spectral	Baturin et al., 1986b
			Indian Ocean	
0.21-1.19	0.81	3	Neutron activaiton	Harris et al., 1968
-	10	14	Atomic absorption	Frakes et al., 1976
0.28-0.84	0.56	2	Neutron activation	Glasby et al., 1978
2-5	3	7	Extraction-spectral	Baturin et al., 1986b
			Antarctic Ocean	
4.78-8.28	6.52	2	Neutron activation	Harriss et al., 1968
< 2-13	7	2	Sorbtion-spectral	Baturin et al., 1986b
			Oceanic Sediments	
Red clay 0.98-1.85	1.35	4	Neutron activation	Croket, Kuo, 1979
Calcareous ooze 0.94-3.96	2.14	15	Neutron activation	Crocket, Kuo, 1979
Siliceous ooze 0.94-3.96	2.14	15	- " -	- " -
Metalliferous sediments EPR) 0.36-2.82	1.4	19	- " -	Piper, Graef, 1974

*   in one sample - 67 ppb

TABLE 119
Concentration ratios between silver and gold in the Earth's crust and ocean

Natural object	Average content (ppb)		Concentration
	silver	gold	ratio of Ag to Au
The Earth's crust	70*	4.3*	16.3
Sedimentary rocks	100*	1*	100
Oceanic sediments	37**	1***	37
Manganese nodules	90**	2 - 3	30 - 45
Sea water	0.0025****	0.005****	0.5

From: * Vinogradov, 1962;   **Baturin, Oreshkin, 1986;
      ***Crocket, Kuo, 1979; ****Bruland, 1983.

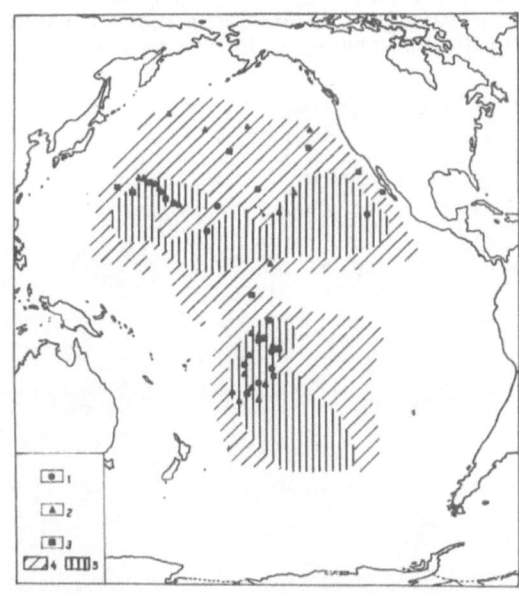

Fig. 56. Distribution of gold in the Pacific nodules (Baturin et al., 1986a). Content
of gold, ppb: (1) less 1; (2) 1 to 2; (3) over 2; (4) zones of occurrence
of nodules; (5) zones of maximum distribution of nodules.

Comparison of gold and silver averages in the Earth's crust, sedimentary
rocks, oceanic sediments, manganese nodules and oceanic water (Table 119) shows
that the closest ratio of the metals in question are typical of sediments and nodules,
30-45. So, one may suppose that host sediments are the main source of gold and
silver supply to nodules, or that both metals are supplied to sediments and nodules
independently from one and the same source, which might be lithogenic matter.

Platinum and platinoids. Platinum was first determined in one sample of a
Pacific nodule by Goldschmidt and Peters (1932). The result was 500 ppb. Analyses
performed almost 50 years later verified that estimate, though the range of
concentrations turned out to be very wide, from 5 to 1000 ppb (Landmesser et al.,
1976; Agiorgitis and Gundlach, 1978; Melo et al., 1978; Flanagan and Gottfried,
1980; Manheim et al., 1980, 1982; Choudhuri and Pal, 1983; Baturin and Fisher,
1984; Baturin et al., 1985; Hodge et al., 1985; Table 120). Platinum concentration
in ore crusts at the Pacific sea mountains also varies over a wide range; being
twice as high in the older parts of the crusts (in their bottoms) in contrast to the
upper younger layers, being 600 and 300 ppb, on average, respectively. Shallow-
water ore crusts are richer in platinum with respect to the deep-sea formations
(Halbach et al., 1984). The platinum average in nodules is 230 ppb (Table 120);
it is 250 ppb if ore crusts at sea mountains are also considered. In nodules from the
Blake Plateau (the North Atlantic) and in ore crusts from the Pacific sea mountains,
a correlation between platinum and nickel was detected (Manheim et al., 1980;
Halbach et al., 1984). In nodules from the Pernambuco Plateau (the South Atlantic),
platinum shows a correlation with lithogenic material (Melo et al., 1978); in the
Pacific nodules it has a correlation with cobalt and lead (Baturin et al., 1985). The
majority of the data available demonstrate that platinum is associated with the ore matter
of nodules, not with its lithogenic part. Results of selective leaching also testify to

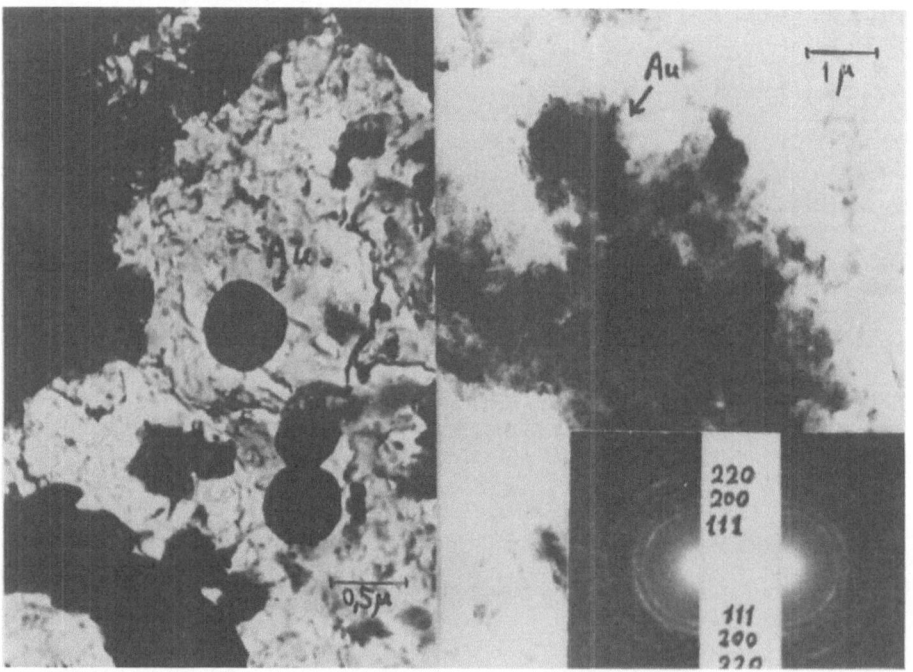

Fig. 57. Ultramicroscopic isometric particles of native gold in the South Pacific
nodules, site 5965 (Baturin et al., 1984), magnification by 15000.

this inference, since platinum was completely extracted by both hydroxylamine and hydrochloric acid (Fisher and Fisher, 1983).

Platinum is believed to be represented in nodules either as a chloridic complex of bivalent platinum (Fisher and Fisher, 1983), or as PtO_2 (Manheim et al., 1980; Halbach et al., 1984). When analysing nodules under electron and scanning microscopes using microdiffraction and microprobe techniques, platinum selenide (Figure 58) was recognized, which was earlier obtained experimentally under high temperature conditions (Grönvold et al., 1960).

Palladium in one sample of a manganese nodule was first analysed by Goldschmidt and Peters (1932); its content was determined as 200 ppb. Subsequently, later analyses showed that the palladium content of nodules varies from 0.24 to 20.9 ppb, its average being 6 ppb (Table 120).

Iridium in nodules was first analysed by Harriss et al., (1968). The value of 0.90 to 23.1 obtained was then verified by later analyses. Its average is about 7 ppb.

Ruthenium content was first determined in reference nodule samples from the Pacific radiolarian zone and from the Blake Plateau (the Atlantic). The corresponding results are 4.7 and 19 ppb (Flanagan and Gottfried, 1980). Special analysis of ruthenium by laser photoionization of atoms verified these results (Bekov et al., 1984). The average content of ruthenium in nodules is 8 ppb.

Rhodium content in one sample of the Pacific nodule was determined by Goldschmidt and Peters (1932) as 200 ppb. Recent analyses showed that it was overestimated by at least one order. The rhodium content in nodules varies from 2 to 30 ppb, its average being 13 ppb (Table 120).

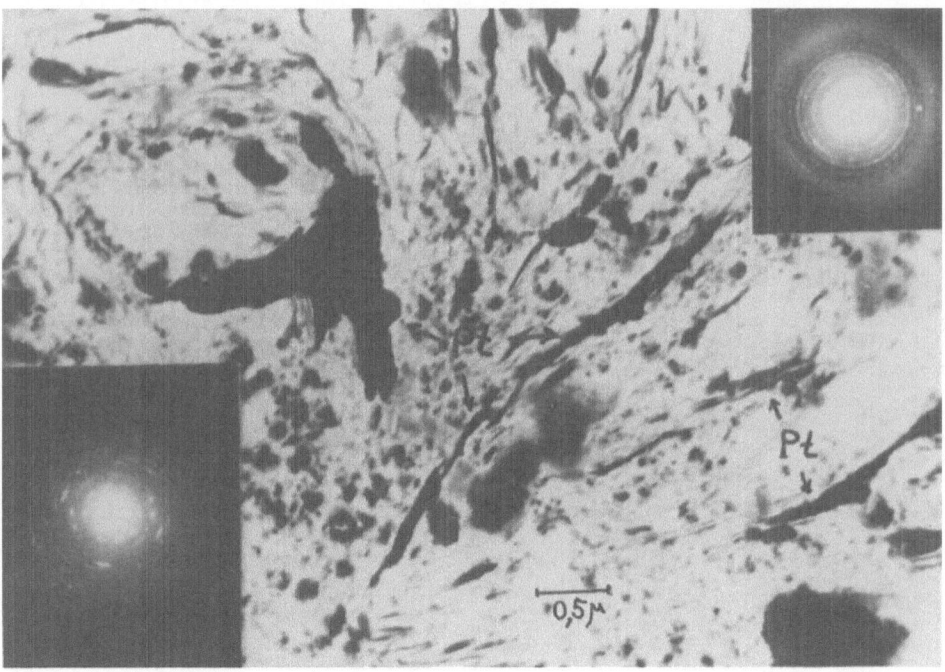

Fig. 58. Lamelli of platinum selenide in nodules from the Pacific radiolarian zone (Baturin et al., 1984), magnification by 10000. Microdiffraction patterns of the mineral are on the left and right hand.

TABLE 120
Concentration of platinum and metals of platinum group in manganese nodules

Ocean	Content of metals (ppb)		average	Number of samples	Method	References
	from-to					
1	2		3	4	5	6

Platinum

Ocean	from	to	average	samples	Method	References
The Pacific	-		500	1	I	Goldschmidt and Peters, 1932
	150	-250	200	2	II	Landmesser et al., 1976
	5	-145	65	4	III	Agiorgitis and Gundlach, 1978
	-		125	1	III	Flanagan and Gottfried, 1980
	144	-752	358	9	IV	Baturin et al., 1985
	10	-200	80	7	III, V	Ibid.
Same, ore crusts	140	-880	420	11	VI	Halbach et al., 1984
The Atlantic	5	-536	158	5	VII	Melo et al., 1978
	-		453	1	III	Flanagan and Gottfried, 1980
	-		370	12	VI	Manheim et al., 1982
Indian Ocean	-		410	1	IV	Baturin et al., 1985
Antarctic	-		238	1	IV	Ibid.
Same	-		30	1	V	Ibid.
Average	5	-880	250	44		

Palladium

Ocean	from	to	average	samples	Method	References
The Pacific	-		200	1	I	Goldschmidt and Peters, 1932
	2.88-	9.25	6.02	7	VI	Harriss et al., 1968
	6.5 -	7.2	6.7	2	VI	Glasby et al., 1978
	6 -	20	11	6	III, V	Baturin et al., 1985
The Atlantic	4.89-	6.62	5.74	7	VI	Harriss et al., 1968
	1 -	7	4	5	VII	Melo et al., 1978
	-		5.6	1	III	Flanagan and Gottfried, 1980
Indian Ocean	0.24-	6.91	3.92	3	VI	Harriss et al., 1968
	6.4 -	20.9	13.6	2	VI	Glasby et al., 1978
Antarctic	2.98-	3.30	3.14	2	VI	Harriss et al., 1968
Average	0.24-	20.9	6	36		

Iridium

Ocean	from	to	average	samples	Method	References
The Pacific	0.90-	23.1	9.39	7	VI	Harriss et al., 1968
	3.62-	4.3	3.9	2	VI	Glasby et al., 1978
The Atlantic	2.65-	16.2	9.32	7	VI	Harriss et al., 1968
Indian Ocean	1.38-	2.57	1.97	2	VI	Glasby et al., 1978
	-		5	1	IV	Baturin et al., 1985
Antarctic	2.29-	3.02	2.65	2	VI	Hariss et al., 1968
Average	0.90-	23.1	7	21		

Table 120 (continued)

Ocean	Content of metals (ppb)		Number of samples	Method	References
	from-to	average			
1	2	3	4	5	6

		Rhuthenium			
The Pacific	-	4.7	1	III	Flanagan and Gottfried, 1980
	1.6 - 3.8	3.0	4	VII	Bekov et al., 1984
The Atlantic	-	18.0	1	III	Flanagan and Gottfried, 1980
Indian Ocean	-	20.0	1	IV	Baturin et al., 1985
Antarctic	-	1.8	1	VIII	Bekov et al., 1984
Average	1.6 - 20	8	8	-	
		Rhodium			
The Pacific	-	200	1	I	Goldschmidt and Peters, 1932
	10 - 23	13	3	III	Baturin et al., 1985
The Atlantic	2 - 30	16	12	VI	Manheim et al., 1980
	6 - 10	8	3	IV	Baturin et al., 1985
Indian ocean	6 - 20	14	4	IV, V	- " -
Average	2 - 30	13	23	-	

* Methods (techniques); (I) fire assay; (II) spectral emission; (III) flameless atomic absorption spectroscopy; (IV) neutron-activation; (V) fire assay-spectral; (VI) spectral chemical; (VII) microadditions; (VIII) laser photoionization of atoms.

TABLE 121

Concentration coefficients of platinum and platinoids in manganese nodules

Element	Pt	Pd	Ru	Ir	Rh	Os
Content in nodules (ppb)	250	5	8	7	13	-
Content in sediments (ppb)	3	2.5	0.2	0.3	0.4	0.2
Concentration coefficients	°0	2	40	23	32	-

Osmium content in nodules was determined by Glasby (1973) in one sample from the Campbell Plateau (the Pacific) and in one sample from the Blake Plateau (the Atlantic). The estimates are 0.1-1 and 0.5-5 ppm, respectively; they seem to be overestimated.

The data considered show that the discrepancy in the estimates of platinum and platinoids in nodules can be large, reaching one or even two orders due to the different methods used. This can be partly explained by the various ways of preconcentration used at all stages of which metal loss is possible (Fisher and Fisher, 1983).

On the other hand, the spectrochemical method systematically gives higher platinum contents in contrast to fire assay and atomic absorption ones. So, until now, the cause of such discrepancies has not been understood, despite numerous experimental tests (Baturin et al., 1985).

TABLE 122

Rhenium concentration in manganese nodules and sediments of the Pacific; composition of samples, %

(Baturin et al., 1985)

Station	Coordinates	Depth (m)	Samples	Re ppb	Mn	Fe	Ni	Cu	SiO$_2$	Al$_2$O$_3$	Mn/Fe
666	14°23'N,117°18'W	4100	Ellipsoidal nodule	0.2	21.6	6.0	1.22	0.45	-	4.6	3.6
2483	10°02'N,146°28'W	5200	Same	0.7	22.3	4.0	1.36	0.98	13.2	4.1	5.6
1936	9°50'N,146°26'W	5050	Polynodules	2.0	28.5	5.2	1.28	0.98	12.9	4.5	5.5
5965	22°41'S,160°50'W	4850	Spheroidal nodule	0.2	16.7	15.5	0.40	0.17	6.0	1.1	1.1
6359	19°02'N,171°09'W	1300	Ore crust	1.2	23.0	13.7	0.51	0.57	7.1	1.4	1.2
666	14°23'N,117°18'W	4100	Clayey-radiolarian ooze	0.5	0.31	2.3	0.02	0.004	-	11.1	0.13
982	32°16'S,159°59'W	5030	Red clay	0.2	1.3	6.3	-	-	-	13.6	0.2

To estimate the average concentration coefficients of platinum and platinoids in nodules one may use the data on the content of platinum (Landmesser et al., 1976; Hodge et al., 1985), palladium, iridium (Crocket and Kuo, 1979; Hodge et al., 1985), ruthenium (Bekov et al., 1984) and osmium (Barker and Anders, 1968) in sediments. The rhodium average was determined only in the Earth's crust (0.4 ppb after Goldschmidt and Peters, 1932). From these data, one can deduce that the average concentration coefficient of platinum in nodules with respect to sediments is 80, it is 40 for ruthenium, 32 for rhodium, 23 for iridium, 2 for palladium (Table 121).

Thus, nodules are poor in palladium as compared with other platinoids. This was also found earlier by Manheim et al. (1980) in the Atlantic nodules.

From Table 121 one may suppose that the concentration coefficient of osmium in nodules is 20 to 40 and, therefore, its content in nodules may be 4-8 ppb. This supposition is verified by the osmium correlation with iridium and by the close absolute content of both metals in deep oceanic sediments (Barker and Anders, 1968).

Rhenium content in nodules was determined by a high precision kinetic technique in three samples from the Pacific radiolarian zone. It varies from 0.2 to 2 ppb (Table 122). In one nodule from the South Pacific it was 0.2 ppb, in the ore crust sampled at the Mid Pacific sea mountains it was 1.2 ppb.

Radiolarian oozes from the Pacific radiolarian zone contain 0.5 ppb of rhenium; red clays from the South Pacific contain less 0.2 ppb of the element.

On the basis of these scanty data, the rhenium average in nodules is 1 ppb, and 0.3 ppb in deep sea sediments.

If we ignore the first sample in Table 122, the rest of the samples show some correlation of rhenium and manganese. So, one may suppose that the behaviour of rhenium in nodules is mainly controlled by manganese (Baturin et al., 1985; Boiko et al., 1985).

HALOGENS

Data on the concentration of halogens in nodules are rather scarce (Table 123).

Fluorine was analysed in nodules of the Pacific and Atlantic oceans. In a sample from the South Pacific (the Campbell Plateau), fluorine concentration is 0.012% (Glasby, 1973); in a sample from the Atlantic (the Blake Plateau) it is 0.09% (Neil, 1980). The concentration values of fluorine, determined in nodules within the Clarion-Clipperton field, differ by a factor of 100 according to various estimates, from 0.035% (Neil, 1980) to 3.00% (McKelvey et al., 1983). The former values seem to be more reliable.

Chlorine concentration in nodules of the Pacific, Indian and Atlantic oceans varies from 0.01 to 1.10%. Its average in nodules of the Pacific and Indian oceans is about 0.8% (Glasby et al., 1978; McKelvey et al., 1983). In a reference nodule sample of the Pacific, the chlorine concentration was assessed as 0.11-0.14% and in the Atlantic reference sample it was 0.32-0.54% (Neil, 1980).

Bromine concentration was determined in several samples from the Pacific Ocean. It was from < 0.002 to 0.080%.

Iodine was analysed in nodules from the Pacific, Indian and Atlantic oceans. In the North Pacific, within the Clarion-Clipperton field, the iodine concentration in one sample was 0.25% (McKelvey et al., 1979), whereas in one sample from the South Pacific it was 0.0546% (Glasby, 1973). In two samples from the Indian ocean the iodine concentration was 0.0125 and 0.022%; in the sample from the Atlantic (the Blake Plateau) it was 0.898% (Glasby, 1973).

At present, according to the data considered, the average concentration of halogens in nodules can be assessed as: 0.1-1% Cl; 0.03-0.09% F; 0.01-0.08% I; ≈0.005% Br. Their approximate ratios are 100 : 10 : 10 : 1 which differ from that in sea water, where the chlorine concentration is 5 to 8 orders higher in contrast to other halogens (Bruland, 1983). The concentration of halogens in sediments is generally higher than in nodules, iodine being an exception: about 2.5% Cl, 0.055% F, 0.008% Br, 0.003% I (Shishkina, 1972); however, this discrepancy needs further investigation.

TABLE 123
Halogens contents (%) in nodules

Region	Element content	N	Reference
	Fluorine		
Pacific Ocean	0.007 -0.072	15	Carpenter, 1967
	<0.01 -0.05	6	Haynes et al., 1986
radiolarian belt	<0.02 -0.04	6	Author's data
	0.035	1	Neil, 1980
	3.00	2	McKelvey et al., 1983
Campbell Plateau	0.0127	1	Glasby, 1973
Atlantic Ocean	<0.02 -0.08	4	Author's data
Blake Plateau	0.09	1	Neil, 1980
Sediments	0.055	-	Shishkina, 1972
	Chlorine		
Pacific Ocean	0.28 -1.01	13	McKelvey et al., 1983
	<0.01 -1.10	10	Haynes et al., 1986
radiolarian belt	0.11	1	Neil, 1980
	0.14	1	Flanagan and Gottfried, 1980
south-western part	0.92 -1.10	2	Glasby et al., 1978
Indian Ocean	0.65 -1.06	2	- " -
Atlantic Ocean (Blake Plateau)	0.37	1	Neil, 1980
Sediments	2 -3	-	Shishkina, 1972
	Bromine		
Pacific Ocean	<0.002 -0.080	7	Haynes et al., 1986
Campbell Plateau	0.0052	1	Glasby, 1973
Sea of Japan (crust)	0.0028	1	Author's data
Black Sea			
(ferruginised shells of			
mollusks)	0.0034	1	- " -
Sediments	0.008	-	Shishkina, 1972
	Iodine		
Pacific Ocean	< 0.01 -0.25	7	Haynes et al., 1986
radiolarian belt	0.25	1	McKelvey et al., 1983
Campbell Plateau	0.0546	1	Glasby, 1973
Indian Ocean:			
equatorial part	0.0125	1	- " -
Bay of Aden	0.0220	1	- " -
Atlantic Ocean (Blake Plateau)	0.0898	1	- " -
Sediments	0.003	-	Shishkina, 1972

TABLE 124
Boron and arsenic content (ppm) in nodules

Region	B		N	Reference
	Range	Mean		
Pacific Ocean	-	500	1	Goldschmidt and Peters, 1932b
	30- 40	35	2	Riley and Sinhaseni, 1958
	70- 600	290	54	Mero, 1965
	-	300	-	McKelvey et al., 1983
	17-1655	273	94	Haynes et al., 1986
radiolarian belt	100- 270	160	7	McKelvey et al., 1979
Atlantic Ocean	90- 500	300	4	Mero, 1965
	-	250	-	McKelvey et al., 1983
Indian Ocean	-	70	-	- " -
World Ocean	20- 900	290	86	- " -
Sediments	-	200	-	Lisitsin, 1978

	As		N	
	Range	Mean		
Manganese nodules				
Pacific Ocean	2-289	114	53	Volkov, 1979
	-	110	-	McKelvey et al., 1983
	20-540	152	122	Haynes et al., 1986
radiolarian belt	45-105	73	9	Calvert et al., 1978
	50-100	80	20	McKelvey et al., 1979
	6- 35	20	18	Sevastyanova et al., 1984
highs	-	82	17	Calvert and Piper, 1984
valleys with thick sedimentary cover	-	41	19	- " -
valleys with thin sedimentary cover	-	20	30	- " -
northern tropical part	120-350	187	4	Sano et al., 1985
Atlantic Ocean	90-210	140	3	Volkov, 1979
	-	200	-	McKelvey et al., 1983
Indian Ocean	39-490	174	15	Volkov, 1979
	-	180	-	McKelvey et al., 1983
World Ocean	20-480	140	63	- " -
Hydrothermal crusts				
Pacific Ocean	23-334	190	5	Toth, 1980
Atlantic Ocean	10- 86	45	5	- " -
Hydrogeneous crusts				
Pacific Ocean	177-710	337	10	Toth, 1980
Atlantic Ocean	329-463	396	2	- " -
Sediments	-	13	-	Turekian and Wedepohl, 1961
	-	17	-	Lisitsin, 1978
Sediments of the Pacific Ocean	5- 40	23	18	Calvert and Piper, 1977

TABLE 124-a

Noble gases (mm^3·g-1) and their ratios in nodules of the Pacific

(Sano et al., 1985)

Samples	Location N	E	Depth (m)	^{4}He 10^{-8}	^{20}Ne 10^{-9}	^{36}Ar 10^{-9}	^{40}Ar 10^{-6}	^{3}He/^{4}He 10^{-6}	^{40}Ar/^{36}Ar	^{38}Ar/^{36}Ar
					Mariana	Trough				
84-1-16-001	20°14'	143°53'	3400-3580	13	1.3	2.2	1.2	4.4±1.0	835± 6	0.186 ± 0.001
84-1-19-A	20°13'	144°04'	2730-2960	13	0.82	2.2	1.2	15.1±1.6	807±16	0.187 ± 0.002
84-1-19-B	20°14'	144°05'	2730-2960	9.4	30	50	1.0	4.3±1.8	316± 4	0.189 ± 0.001
84-1-19-C	20°14'	143°05'	2730-2960	-	-	40	1.5	-	333± 4	0.188 ± 0.002
84-1-25	20°19'	143°43'	2810-1910	4.9	3.8	43	1.2	22.5±1.7	324± 6	0.188 ± 0.001
					Ogasawara	region				
84-1- 2-124	26°16'	142°58'	1430-1570	2.6	10	70	1.1	1.5±0.6	311± 3	0.184 ± 0.003
84-1- 3-606	26°04'	144°19'	1050-1290	0.98	0.25	2.9	0.26	59.3±9.4	384± 2	0.185 ± 0.001
84-1- 3-607	26°04'	144°19'	1050-1290	5.8	4.0	12	-	5.9±1.5	290± 4	0.188 ± 0.001
84-1- 5- 33	26°13'	144°05'	2300-2430	12	10	27	0.47	10.2±1.0	313± 6	0.188 ± 0.001
84-1-27-017	26°12'	144°11'	840- 890	1.2	0.1	1.7	0.23	6.0±5.0	432± 1	0.185 ± 0.002

RARE METALLOIDS

Boron, arsenic, selenium and tellurium are rare metalloids of manganese nodules. The data on their concentration in nodules are given in Table 124.

Boron content in nodules varies from 17 to 1655 ppm, being 300 ppm on average. The Pacific nodules are enriched in boron relative to nodules of the Indian ocean. Its minimum falls within the latitudinal interval of 0-20°N, its maximum is confined to the interval of 40-60°S. The boron content in shallow-water nodules is twice as high, 400 ppm on average, as in deep-sea nodules. The boron maximum is observed in nodules rich in cobalt, which allows us to consider it as an element which is associated with iron in nodules. The boron average in oceanic sediments is 200 ppm (Lisitsin, 1978); therefore the boron concentration coefficient in nodules is only 1.5 with respect to sediments.

Arsenic content in nodules of the World Ocean varies from 2 to 540 ppm; its average being about 110 ppm in the Pacific nodules; 180 ppm in the Indian ocean; 140 to 200 ppm in the Atlantic nodules (according to various estimates); and 140 ppm in the World Ocean (Volkov, 1979; McKelvey et al., 1983). The arsenic minimum is confined to the latitudinal interval of 0-20°N, and it has two maxima, namely within 20-40°N and 20-40°S. Nodules of the radiolarian zone are depleted of arsenic in contrast to other regions, its content being higher at submarine hills than in valleys (Calvert and Piper, 1984). The arsenic content in shallow-water nodules is twice as high as that of deep-sea nodules, which coincides with the behaviour of cobalt (McKelvey et al., 1983). In nodules of the Central Pacific arsenic correlates with iron ($r = 0.77$) and with cobalt ($r = 0.68$) (Calvert et al., 1978). The share of arsenic bound to hydrooxides in nodules is 74 to 98% (Calvert and Price, 1977). In ore crusts, the arsenic content varies from 10 to 710 ppm, the hydrogenous crusts being richer in arsenic in contrast to hydrothermal ones (Toth, 1980).

The arsenic average in deep-sea sediments is 13 to 23 ppm according to various authors (Turekian and Wedepohl, 1961; Lisitsin, 1978; Calvert and Price, 1977); the arsenic concentration coefficient in nodules is about 7.

Selenium content determination in nodules is a great problem because of difficulties in analytical techniques, so that data on the concentration of the element are scarce. From data by Sokolova and Pilipchuk (1973), in the Pacific nodules it is rather uniform, being 0.4 - 0.8 ppm, 0.64 ppm on average. Its average in deep oceanic sediments is 0.2 ppm; so the selenium concentration coefficient in nodules is about 3. But Hagnes et al. (1986) report much higher concentrations of Se in nodules: 30-77 ppm.

Tellurium content in the Pacific nodules is 5 to 125 ppm according to one series of estimates (Lakin et al., 1963), and 170 to 270 ppm from the other (McKelvey et al., 1983); its average being 40 and 220 ppm, respectively. However, the analyses performed in P.P. Shirshov Institute of Oceanology, Academy of Sciences of the USSR, by Ostroumov and Rybakov showed that the tellurium content in nodules is much lower, 0.8 to 10 ppm, with an average of 5 ppm.

No data on tellurium content in oceanic sediments seem to be available. Its average in the Earth's crust is 0.01 ppm (Vinogradow, 1962). Thus, one can suppose that its coefficient of concentration in nodules may be about 10 or even greater.

GASES

The gaseous phase of manganese nodules has not been studied systematically but a few determinations of noble and other gases have been made on samples collected by usual samplers without application of special conservation technique. This implies possibility of drastic change of original gas composition due to rise of temperature and fall of pressure when lifting samples from the depth of several thousands of meters.

Noble gases. Helium isotopes, argon and neon have been determined in nodules from Mariana Trough and Ogasawara region in the Northern Pacific (Sano et al., 1985).

The ^{4}He content in these samples varies in limits $(0.98-13)$ x 10^{-8}mm$^3 \cdot$g^{-1}, whereas ^{3}He/^{4}He ratio varies within the range of $(4.3 - 22.5)$ x 10^{-6} in the first and $(1.5 - 59.3)$ x 10^{-6} in the second area (Table 124a).

The highest value is comparable to those established in the Hawaiian hot spot basalts (Kaneoka and Teraoka, 1978) and in diamonds (Ozima and Zashu, 1983).

The second high value (22.5×10^{-6}) is twice higher as compared to the basaltic glass from the region in question: 11.6×10^{-6} (Poreda and Craig, 1979).

As the mantle sources for helium in this area are evidently absent, Sano et al. (1985) conclude that high $^3He/^4He$ ratio in nodules originates from interplanetary dust where it is 140×10^{-6} on average (Mamyrin and Tolstikhin, 1984).

In deep-sea sediments of the Pacific ocean $^3He/^4He$ varies within the range of $(1.3 - 119) \times 10^{-6}$ (Krylov et al., 1973; Ozima et al., 1984), whereas in magnetic fractions of sediments it is considerably higher: $(138 - 237) \times 10^{-6}$ (Merrihue, 1964; Amari and Ozima, 1985).

In the case the interplanetary matter similar in composition to chondrites of CI type is also thought to be the source of an anomalous helium.

But its manganese nodules are enriched in cosmogenic particles as compared to sediments (Finkelman, 1970). So, the question arises why nodules are not enriched in ^3He as compared to sediments.

It is well known that "anomalous" helium with high $^3He/^4He$ ratio enters the ocean from the mantle and volcanogenic sources and migrates for great distances (Lupton et al., 1977; Jenkins et al., 1978; Lupton and Craig, 1981). It implies that this could be the source of helium in nodules though the mechanism of its incorporation is unknown.

The ^{20}Ne and ^{36}Ar in nodules from the northern part of the Pacific fluctuates within the range of (0.1 ' $30) \times 10^{-9}$ and $(1.7 - 70) \times 10^{-9}$ cm$^3 \cdot$g^{-1} respectively without considerable difference between Mariana and Ogasaware region. But the content of radiogenic ^{40}Ar in nodules of the first region is essentially higher than in the second (averaging 1.2×10^{-6} and 0.5×10^{-6} mm$^3 \cdot$g^{-1}). The Mariana nodules are hydrogenetic and enriched in radioelements as compared to Ogasawara nodules which are diagenetic and lower in this elements; this may explain the observed ^{40}Ar difference.

Other gases. The investigation of gaseous phase of manganese nodules from the North-Western part of the Pacific was performed by one-stage degassing technique with mechanical crushing of samples in vacuum and subsequent mass-spectrometric analysis of gases (Marushkin et al., 1985).

The nodules contain CO_2, N_2, H_2 and CH_4 which volumes are 54.8; 24.6; 19.0 and 1.6%. In some nodules traces of NH_4 have been found.

The nodules could be divided in two types: in the first, CO_2, N_2 and H_2 are contained in comparable proportions, whereas in the second N_2 is predominant.

Special experiments have shown that N_2, H_2 and CH_4 are dissolved in interstitial water of nodules, whereas CO_2 is contained in their voids and pores. The manganese minerals of nodules are evidently free of gases.

The loss of gases during the lift of the samples has not been evaluated.

AVERAGE CONTENTS AND CONCENTRATION COEFFICIENTS OF ELEMENTS IN NODULES

The above review permits a subdivision of microelements into several associations according to their concentration coefficients ("CC") in nodules with respect to host sediments: (1) CC is over 50: thallium, bismuth and probably tellurium; (2) CC is 10 to 50: molybdenum, cadmium, antimony, lead, platinum and metals of platinum group, but for palladium; (3) CC is 5 to 10: zinc, arsenic, tantalum, tungsten; (4) CC is 3 to 5: vanadium, zirconium, niobium, indium; (5) CC is 1 to 3: lithium, sulphur, titanium, selenium, palladium, silver, yttrium, lanthanum, hafnium, rhenium, uranium, thorium, probably radium and protactinium; (6) CC is about 1: beryllium, boron, sodium, magnesium, strontium, barium, gold; (7) CC is less 1: carbon, nitrogen, scandium, chromium, gallium, germanium, rubidium, tin, cesium, mercury, and also all or the majority of halogens.

As the data on many microelements are not sufficiently representative, their position in the classification considered may be changed.

To obtain a general pattern of concentration of elements in nodules, the data presented in this and in the two previous chapters are summarized in table 125 and drawn to a scheme based on Mendeleev's Periodic Table (Table 126). The scheme demonstrates that all elements, the concentration coefficient of which exceeds 10, belong to periods of 4-6 of the Periodic Table. It should be noted that within

TABLE 125
Average concentration of elements (%) in manganese nodules
and deep-oceanic sediments

Element	Average concentration		CC	Element	Average concentration		CC
	Nodule	Sediment			Nodule	Sediment	
Li	0.008	0.0047	1.7	Sr	0.083	0.075	1.1
Be	2.5	2.6	1	Y	0.015	0.01	1.5
B	0.03	0.02	1.5	Zr	0.056	0.017	3
C	0.1	0.3	0.3	Nb	0.005	0.001	5
N	0.02	0.06	0.3	Mo	0.04	0.01	40
O	33	43	0.8	Ru	$8 \cdot 10^{-7}$	0.2×10^{-7}	40
F	0.02	0.05	0.4	Rh	$13 \cdot 10^{-7}$	0.4×10^{-7}	32
Na	2	1.82	1.1	Pd	$6 \cdot 10^{-7}$	3.5×10^{-7}	
Mg	1.6	1.42	1.1	Ag	$90 \cdot 10^{-7}$	37×10^{-7}	2.5
Al	2.7	5.35	0.5	Cd	0.001	0.00003	30
Si	7.7	19.65	0.4	In	$0.25 \cdot 10^{-4}$	0.08×10^{-4}	3
P	0.25	0.11	2.2	Sn	0.0002	0.001	0.2
S	0.5	0.3	1.7	Sb	0.004	0.0002	20
Cl	0.5	2.5	0.2	Te	0.001(?)	‹0.0001(?)	›10(?)
K	0.7	1.33	0.5	I	0.04(?)	0.003	?
Ca	2.3	13.6	0.16	Cs	0.0001	0.0007	0.14
Sc	0.0010	0.0014	0.7	Ba	0.23	0.26	0.9
Ti	0.67	0.26	2.6	La	0.018	0.006	3
V	0.050	0.01	5	Hf	0.0008	0.0004	2
Cr	0.0035	0.006	0.5	Ta	0.001(?)	0.0001	10(?)
Mn	18.6	0.3	60	W	0.01	0.001	10
Fe	12.5	3.8	3.3	Re	$1 \cdot 10^{-7}$	0.3×10^{-7}	3
Co	0.27	0.0065	41	Os	-	0.2×10^{-7}	30(?)
Ni	0.66	0.01	66	Ir	$7 \cdot 10^{-7}$	0.3×10^{-7}	23
Cu	0.45	0.024	19	Pt	$230 \cdot 10^{-7}$	5×10^{-7}	46
Zn	0.12	0.013	9	Au	$2 \cdot 10^{-7}$	2×10^{-7}	1
Ga	0.001	0.002	0.5	Hg	$20 \cdot 10^{-7}$	45×10^{-7}	0.5
Ge	0.8×10^{-4}	1.6×10^{-4}	0.5	Tl	0.0150	0.00018	80
As	0.014	0.002	7	Pb	0.09	0.004	22
Se	6×10^{-5}	2×10^{-5}	3	Bi	$7 \cdot 10^{-4}(?)$	$0.1 \times 10^{-4}(?)$	70(?)
Br	0.004	0.007	0.6	Th	0.003	0.0012	2.5
Rb	0.0017	0.01	0.17	U	0.0005	0.0002	2.5

Groups I-V, all of these individual elements belong to subordinate subgroups, namely, copper (Group I), cadmium (Group II), thallium (Group III), lead (Group IV), antimony and bismuth (Group V). Molybdenum and manganese which belong to Groups VI and VII, respectively, are elements of the main subgroups. The most productive in this aspect is Group VIII: seven of the nine elements of this group accumulate considerably in nodules.

Elements which accumulate in nodules 3-10 times more as compared to sediments also belong to Periods 4-6 of the Periodic Table, except for phosphorus, whose behaviour resembles that of iron. Each Group (II, III, IV and VI) has only one such corresponding element, namely, zinc, indium, zirconium, tungsten. Group V includes five elements: phosphorus, vanadium, arsenic, niobium and tantalum.

The total number of elements, the concentration coefficients of which exceed 3, is 26. Eight of them belong to Group VIII, seven to Group V, and three to Group VI, thus 18 elements altogether, which is the bulk of all elements considered. Such a distribution is probably determined by their chemical properties, individual features of their marine geochemistry and the manner of their association with major ore elements in nodules, i.e. with iron and manganese.

TABLE 126

Concentration coefficients of elements in manganese nodules relative to sediments as a function of elements position in the Periodic Table

CHAPTER XI

AGE AND ACCRETION RATES OF MANGANESE NODULES

Many experimental and theoretical works are devoted to the age and accreation rates
of manganese nodules, since the intensity of metal concentration in time is one of
the major parameters of the ore processes as a whole.

The present chapter considers the data obtained over the last 20-30 years in
the Pacific, Indian and Atlantic oceans.

BRIEF HISTORICAL NOTE AND METHODS

The problem of the age and accretion rate of nodules dates back to the cruise of
the Challenger (1872-1876) when nodules were first discovered on the oceanic
bottom. When describing nodules, in particular those recovered from the Pacific,
Murray recognized Tertiary sharks' teeth in them and magnetic spherules of a cosmic
nature, which made them date those nodules back to ancient geological times (Murray,
1876; Murray and Renard, 1884, 1891).

Early in the XX century, Joly (1908) detected that nodules are radioactive
and so the foundation was laid for a quantitative assessment of their age and for
the estimation of their rate of accretion by radiometric methods. Later, the first
steps were initiated towards their dating from the decrease of radium towards their
center. It was assumed that the excess radium is being supplied to nodules from
sea water and decays over 8000 years (Iimory, 1927; Kurbatov, 1936; Petterson,
1943; etc.). Subsequently Goldberg and Arrhenius (1958) supposed that the
distribution of radium in nodules is not an independent geochemical parameter, but
depends upon the distribution of its parent isotope, ^{230}Th. That supposition was
soon verified by direct measurements of radioisotopes of the uranium series (^{230}Th;
^{231}Pa; ^{234}U); subsequently, the corresponding techniques based on the decrease
of these isotopes towards the nodule centre became of widespread use in the dating
of manganese nodules from various regions of the World Ocean (Nikolaev and Efimova
1963; Bender et al., 1966; Barnes and Dymond, 1967; Ku and Broecker, 1967; 1969;
Bhat et al., 1973; etc.).

Along with the technique just mentioned, such cosmogenic isotopes were used
for dating of nodules as ^{10}Be (Somayajulu, 1967; Bhat et al., 1973; Guichard et
al., 1977; 1978; Krishnaswami et al., 1972; 1982; Ku and Omura, 1979; Ku et al.,
1979; Sharma and Somayajulu, 1982; Inoue et al., 1983), or ^{26}Al (Guichard et al.,
1977; 1978; Reyss and Yokoyama, 1976). The fission track method, which is an
analog of the ionium* method, is also widely used (Shima and Okada, 1968; Heye,
1975; Andersen and MacDougall, 1977; McDougall, 1979).

General characteristics of the methods which were used for dating of nodules
are given in Table 127**.

Besides direct radiometric dating of nodules, a number of indirect methods are
used, namely, biostratigraphic, paleomagnetic (dating of ore matter) and also dating
of non-metalliferous nuclei in nodules, host sediments and rocks.

---

* ionium = ^{230}Th
** The lead isotopes (^{204}Pb, ^{206}Pb, ^{207}Pb, and ^{208}Pb) whose sources are oceanic
   basalts, terrigenous matter and partly seawater proved to be not valid for manganese
   nodules dating (Chow and Pattersson, 1959, 1962; Reinolds and Dash, 1971; O'Nions
   et al., 1978).

TABLE 127

Characteristics of radiometric techniques of nodule dating

Isotope	Half-life, (yr)	Age limit detected (yr)	First datings
^{226}Ra	1622	8000	Iimory, 1927
^{231}Pa	34300	2×10^5	Ku and Broecker, 1969
^{230}Th	75200	4×10^5	Nikolaev and Efimova, 1963
^{234}U	248000	1×10^6	Barnes and Dymond, 1967
^{26}Al	715000	3×10^6	Reyss and Yokoyama, 1976
^{10}Be	1.5×10^6	8×10^6	Somayajulu, 1967

The biostratigraphic method is appropriate for detection of the relative geological age of organic fossils trapped in nodules, mainly of foraminifera and nannofossils (coccoliths and discoasters), and permits one to determine age intervals between layers in nodules (Harada and Nishida, 1976; 1979). By using the paleomagnetic method one can determine direct and inverse residual magnetism in nodules and estimate the thickness of the layer accumulated since the last inversion of the geomagnetic field. This method permitted reliable data, but its application is restricted since it is difficult to distinguish in one nodule distinct and reliable stratigraphic layers without contaminations (Crecelius et al., 1973).

The potassium-argon method can be applied to the dating of basaltic nuclei of nodules. In this case, it is possible to find out the age of nodules since the beginning of their growth. This method can be also applied to manganese nodule crusts on basalts (Barnes and Dymond, 1967; Lalou et al., 1979a).

The age of basalts which form the nuclei of nodules or on which the nodule crust rests can also be dated by the thickness of the palagonitized layer, knowing the rate of basalt submarine palagonitization (2 to 4 mm/m yr) (Moore, 1966; Hekinian and Hoffert, 1975; Morgenstein and Riley, 1975).

The age of bone remnants which surve as nodule nuclei can be determined by both paleontological methods and by the racemization rate of amino acids, i.e. from the ratio between their optical isomers, which varies regularly as the sample is aging, the reaction constant in various amino acids being 6.5×10^{-7} to 6.5×10^{-8} (Bada, 1972; Bada et al., 1973; Kvenvolden and Blunt, 1979).

A general description of the methods reviewed is given in Kuznetzov (1976) and Ku (1977). New methods of dating of nodules are also being worked on, in particular, thermoluminescent and gamma-spectral ones (Schvoerer et al., 1979; Yokoyama and Nguyen, 1979). The growth of nodules is also being modelled under controlled laboratory conditions (Krylov et al., 1981; Tikhomirov, 1982).

Dating of nodules from the age of the host sediments is of particular concern. When nodules rest on ancient sediments within the zones where recent sedimentation has not occured, the age of ancient sediments can be assumed as being the lower age limit of nodules (Bezrukov, 1976). However, in many of the oceanic regions, nodules rest on late Quaternary sediments. In such cases, a sharp disagreement arises between the ancient age of the nodules and the young age of the superficial layer of sediments determined by the same methods of absolute geochronology.

ACCRETION RATES OF NODULES

The accretion rates of nodules were first obtained from the gradients of radium concentration. The results were 0.7 to 65 mm/m yr (Iimory, 1927; Kurbatov, 1936; 1937; Petterson, 1943; 1955; Von Buttlar and Houtermans, 1950). The accretion rate of the pelagic nodules from the Pacific, determined by other radiometric techniques, turned out to be lower by two to three orders; several millimeters per million years (Table 128). As the age limits determined by these techniques are from 200-400 thousand years (^{230}Th, ^{231}Pa) to 3-8 million years (^{10}Be, ^{26}Al) they allow the dating of only those outer nodule layers that contain an excess of isotopes.

Samples of nodules from the northwestern Pacific were analysed by the biostratigraphic method; the accretion rates of various layers turned out to be from 1 to 10 mm/m yr, and 39 mm/m yr for the inner layer of one sample (Harada and Nishida, 1976; 1979).

An attempt was made to estimate the growth rates of nodules by gradients of the radial changes of their chemical composition. In the samples from the Suiko sea mountain in the northwestern Pacific, four layers were detected with manganese maxima that were believed to correspond to four periods of glaciation. On this basis, the accretion rates of individual nodule layers were determined to be 2 to 5 mm per thousand years (Nohara and Nasu, 1977).

The accretion rates of pelagic nodules determined by the paleomagnetic method from the reversal at the Brunhes-Matuyama boundary are several millimeters per million years. This boundary was not recorded in shallow-water nodules, which grow considerably faster (Crecelius et al., 1973).

The growth rates of nodules determined by potassium-argon dating of basalt nuclei are a minimum since no manganese crust usually occurs at a young basalt surface. Given the age of nuclei and underlying basalts as 1.7 to 29 m yr and the thickness of ore matter from 5 to 100 mm, the average rate of its accumulation is 0.1 to 23 mm/m yr (Barnes and Dymond, 1967; Lalou et al., 1979a).

The data on the growth rates of nodule and ore crusts determined by the palagonitization of basalt inclusions or underlying basalts are scarce. Since these analyses have no independent value, they are usually accompanied by radiometric dating, and the results obtained are not always consistent. The rates obtained by this method for ore crusts are 2 to 4 mm/m yr in the rift zone of the North Atlantic (Hekinian and Hoffert, 1975), 10 to 20 mm/m yr within the Hawaiian shelf (Ku, 1977) and 333 to 816 mm/m yr within the East Pacific Rise (Burnett and Morgenstein, 1976; Burnett and Piper, 1977; Lalou et al., 1979a).

Sharks' teeth are widespread over the sea bottom, especially in the zones where recent sedimentation has not occurred, and nodule nuclei are often built of sharks' teeth. The paleontological method was applied and tooth age was dated back to the Tertiary-Recent time interval (Belyaev, 1959; Belyaev and Glikman, 1965; 1970 a,b). This dating method has its difficulties, as teeth in nuclei are usually ruined and it is difficult to recognize them. Better preserved teeth, which occur over the bottom, are filled with ore matter inside, whereas from the outside, they are merely covered by a thin film of hydrooxides.

It seems more pertinent to determine the age of bone remnants (sharks' teeth and ear bones of whales) trapped in nodules by the extent of racemization of amino acids (Bada, 1972). In the North Pacific, the growth rates of nodule matter estimated from such a dating method were 0.4 to 2.0 mm/m yr by isoleucine and 1.3 to 8.1 mm/m yr by asparagine (decay constant being 6.47×10^{-7} per year) and somewhat higher, 1.6 to 10.3 mm/m yr (decay constant being 8.24×10^{-7} per year) (Kvenvolden and Blunt, 1979). It was assumed that nodules started to grow on the bones immediately after they had sunk to the bottom. The results of age-dating of nodules by various techniques are given in Table 129.

In most cases, the discrepancies in the dating results are within one order (Table 128, 129). However, nodules and ore crusts from the area adjacent to the East Pacific Rise are characterized by considerably higher growth rates (Figure 59). Thus, for two samples of ore crusts on basalts in the region considered, the accretion rates determined by ionium-thorium ratio and potassium-argon dating of basalts were 23-48mm/m yr; 333-816 mm/m yr by palagonitization of basalts; and up to 1000 mm/m yr by ionium excess (Burnett and Morgenstein, 1976; Burnett and Piper, 1977).

On the whole, according to the data of absolute geochronology, accretion rates of 1 to 12 mm/m yr are typical of the nodules from both South and North Pacific,

TABLE 128

Accretion rates of manganese nodules and ore crusts in the oceans from
radiometric data

Sample (station)	Coordinates	Depth	Accretion rate, $(mm/10^6 yr)$	Method	References
1	2	3	4	5	6
			The North Pacific		
Fan Bd-20	40°15'N,128°27'W	4500	3	K-Ar	Barnes and Dymond 1961
Horizon	40°14'N,155°05'W	5500	2.5	K-Ar,^{230}Th	Barnes and Dymond 1967; Goldberg, 1963
V21-D2	35°54'N160°19'W	5400	4	^{230}Th ^{231}Pa,^{234}U	Ku and Broecker, 1969
V21-71a	27°54'N,162°31'E	5870	2.5	^{230}Th	Ku and Broecker, 1969, Bender et al., 1966
V21-D46	14°25'N,145°52'W	4618	3	^{230}Th, ^{231}Pa	Ku and Broecker, 1969
6A	19°39'N,113°44'W	4000	4	same	Ibid.
Carr 5	9°26,5'N,113°16.5W	3700	17-24	^{230}Th, ^{234}U	Barnes and Dymond 1967
MP 26	19°N, 171°W	1464	10	^{230}Th	Ibid.
Zetes-3D	40°16'N,171°20'W	3000	0.8-2.3	^{10}Be^{230}Th track	Krishnaswamy et al., 1972; Bhat et al., 1973
Tripod-2D	20°45'N,112°47'W	3000	4	^{10}Be,^{230}Th	Ibid.
Dodo-9D	18°16'N,161°50W	5500	2.1	^{230}Th, track	Ibid.
Dodo-15-1	19°23'N,162°20'W	4160	1.8	^{10}Be	Somayajulu, 1967
Wah 24F-8	8°18'N,153°03'W	5143	10(3 samp)	^{230}Th	Somayajulu et al., 1971
V21-116	19°34'N,134°30'E	5826	28	track	Shima and Okada, 1968
2P-52	9°57'N,137°47'W	4930	7.3	^{230}Th	Krishnaswamy, Lal, 1972
3996	4°56'N,135°29'E	4580	33	same	Nikolaev and Efimova, 1963
3782	23°55'N,173°40'E	4970	40	same	Ibid
TE 8-1	near Oahu Is;	1230	3.6-4.4	^{230}Th, ^{230}Th, ^{232}Th	Burnett and Morgenstein, 1976; Burnett and Piper, 1977
KK73	40°N,147°W	5163	3.3-4.1	same	- " -
1052	2°17'8N,101°1.5W	3150	47.9-'100	- " -	- " -
AH 69D1	36°N,157°W	5800	22.6-23.3	- " -	- " -
A47-16-1.2	9°23'N,151°11.4'W	5040	1.5-4.3	^{230}Th ^{230}Th ^{232}Th,^{231}Pa	Krishnaswamy and Cochran, 1978
C57-58-1	15°19.5'N,125°54.4'W	4638	1.8-2.4	same	Ibid.
VA 17863	29°58'N,125°57°W	4324	4.1	track	Macdougall, 1979
MRA 17875	22°30'N,113°00'W	3603	2.9		
Ris 4G	22°18'N,117°24'W	3890	7.6	"	"
Styxx, DWD					
BDI	21°27'N,126°43'W	4300	6.6-8.9	"	"

Table 128 (continued)

1	2	3	4	5	6
MR 55	20°45'N,114°15'W	3740	10.9	track	Macdougall, 1979
Exp.A17865	19°20'N,114°12'W	3400	7.8	"	"
DSV,Mn7402	14°58'N,125°00 W	4460-5000	5.0-9.2 (4 samp)	"	"
Expo 60-4	14°28'N,107°51'W	3678	5.9	"	"
Mid.Pac.5-1	14°22'N,133°71'W	4990	6.9	"	"
Mid.Pac.16-1	13°45'N,149°15'W	4564	2.8	"	"
Mn 7402 77G37	14°61'N,140° W	4800	4.6	"	"
Mn 7601 20-B2	11°4'N,140°03'W	4722	6.7	"	"
MSN 150-G	10°59'N,143°22'W	4978	3.8	"	"
DOMES	14°15'N,124°59'W	1560	5.9-8.2 (8 samp)	"	"
PUMICE Mn 68	40°18'N,175°38'W	5307	6.8; 10.7	"	"
PUMICE Mn67	40°18'N,173°46'E	4399	6.7; 10.7	"	"
PUMICE Mn61	31°18'N,174°02'E	5362	6.7	"	"
PUMICE 70-1	38°56'N,178°36'E	5398	11.2	"	"
ARIES 13D	20°45'N,173°40'E	3816-2955	2.4	^{10}Be	Ku et al., 1979
Mn 139	20°01'N,136°36'W	3916	1.3	^{10}Be	Ibid.
DPSN L2	9°08.65'N,105°W 12.94'W	3214-3265	13-114	^{230}Th,K-Ar track	Lalou et al., 1979-1
ARIES 12D	20°45'N,173°26'E	1824	3.2-3.8	^{10}Be,^{10}Be, ^{9}Be	Sharma and Somayajulu, 1979, 1982
ARIES 15D	20°47'N,173°20'E	1285	2.3-3.0	same	Ibid.
ARIES 39D	34°15'N,143°51'E	1819	2.0	"	"
ANTR 50D	32°42'N,158°15'E	2607	8.0	"	"
A47-16-2	9°2.3'N,151°11.4W	5049	2.8-3.3	^{230}Th, ^{231}Pa	Krishnaswamy et al., 1979
C57-58-1	15°19.5'N,125° 54.43'W	4338	2.4-2.7	^{230}Th ^{231}Pa	Ibid.
A47-16-4	9°2.3'N,151°11.4'W	5040	2.0-2.4	^{10}Be,^{10}Be ^{9}Be	Krishnaswamy et al., 1982
-	30°N,140°W	3840	1.4	^{10}Be	Ku et al., 1982
Mn7601-20	11°07'N,140°05'W	4722	4.7-29	^{230}Th ^{231}Pa	Moore et al., 1981
57 DK 2	14°09'N,165°29.1W	3280	2.7 ± 0.4	^{230}Th ^{231}Pa	Halbach et al., 1983
111 DK 2	20°04'7N,170°38'5W	1240	1.2±0.06	"	"
76 DK	19°21'1N,170°59'W	1190	2.7±0.4	"	"
73 DK 7	09°24'1N,171°10'4W	2860	2.6±0.6	"	"
32 DK 2	9°08'5N,164°47'03W	1120	0.8±0.2	"	"
31 DK 5	9°08'9N,164°48'5W	2100	2.7±0.4	"	"

The South Pacific

1	2	3	4	5	6
V18-D 32	14°18'S,149°32'W	2000	3.1	^{230}Th	Ku and Broecker, 1969
V18-T119a,b	12°27'S,159°25'W	5000	3-6	^{230}Th,^{231}Pa	Ibid.
E 17-36	55°S,95°W	4700	3	^{230}Th,^{231}Pa	Ibid.
DWHD-47	41°51S,102°01'W	4240	0.5-6	K-Ar,^{230}Th	Barnes and Dymond, 1967

Table 128 (continued)

1	2	3	4	5	6
DW 72	21°31'S,85°14'W	~920	18	^{230}Th	Barnes and Dymond, 1967
2P-50	13°53'S,150°35'W	3695	1	^{230}Th	Krishnaswamy and Lal, 1972
TF-1.2	13°52'S,135°35'W	3623	1	^{230}Th	Bhat et al., 1973
E24-15	35°48'S,134°50'W	4696	6.4	^{230}Th	Kraemer and Shornick, 1974
DO 23 G	near Tuamotu Is.	1600	3.5	^{230}Th	Boulad et al., 1975
-	north of Tuamotu	4020	1-5	^{26}Al	Reyss and Yokoyama, 1976
-	same	4020	2	^{10}Be	Guichard et al., 1977
TECHNO 1	15°S,145°W	4020	2.8	^{10}Be	Guichard et al., 1978
Same	- " -	"	0.8-17	^{230}Th ^{231}Pa, ^{10}Be,^{26}Al, ^{234}U	Lalou et al., 1979b
Sonne	7°40S,91°59.5W	4240	162	^{230}Th	Reyss et al., 1982
TF - 5	13°53'S,135°35'W	3623	1.0-2.0	^{10}Be,^{10}Be ^{9}Be	Krishnaswamy et al., 1982
ANTP 58D	18°57'S,175°48E	3555	3.0-4.0	same	Sharma and Somayajulu, 1982
GEOSECS 1D	38°38'S,167°55'E	1418	1.0-1.7	^{10}Be,^{230}Th	-"-
-	South China Sea	4700	4.1	^{230}Th	Xia and Ming, 1984

The Atlantic

1	2	3	4	5	6
G 58-100	30°57'N,65°47'W	4800	4	^{230}Th ^{231}Pa	Ku and Broecker, 1969
A266-41	30°59'N,78°15'W	830	2	Same	- " -
G 74-2374	30°31'N,79°01'W	876	2	- " -	- " -
V 16-T3	10°04'S,24°41'W	4415	3	- " -	- " -
Lusiad AD4	6°03'N,32°22'W	1020	8-10	^{230}Th K-Ar	Barnes and Dymond, 1967
RC 15D5	48°28'S,55°14'W	2394	1.2-1.4	^{10}Be,^{230}Th ^{230}Th ^{232}Th	Sharma and Somayajulu, 1982
RC16D10	28°25'S,40°41'W	4388	1.3-3.5	Same and ^{10}Be,^{9}Be	- " -
ES-4	60°02', 67°15'	3475	4-9	^{230}Th	Kraemer and Shornik, 1974

The Indian Ocean

1	2	3	4	5	6
V16-T19	29°52'S,62°36'	4500	2.3-2.9 (3 samp.)	^{230}Th	Ku and Broecker, 1969
Dodo-66a	19°56'S,100°E	-	5.5	^{230}Th	Bhandari et al., 1971
M1	East of Madagascar	-	0	^{230}Th	Heye and Beiersdorf, 1973
RC 14D 4F	23°26'S,50°49'E	4052	1.3-1.9	^{10}Be,^{10}Be, ^{9}Be,^{230}Th, ^{230}Th,^{232}Th	Sharma and Somayajulu, 1982
RC 14D 4S	23°26'S,50°49'E	4052	1.3-3.3	same	- " -
	2°55'N,60°02'E	4000-	1.5-3.2	^{230}Th	Sharma and Somayajulu, 1983
		4500	(3 samp.)		

TABLE 129

Comparison of the dating results obtained by various methods:
manganese nodules and ore crusts

Sample	Method	Maximum age of the sample (yr)	Accretion rate of ore matter (mm/m yr)	References
1	2	3	4	5
nodule				
V21-D2	^{230}Th	-	4.0	Ku, 1977
35°54'N	^{231}Pa	-	4.3	Ku and Broecker, 1967
160°19'W	^{234}U/^{238}U	-	4.6	
5400 m	^{230}Th/^{232}Th	-	3.0-5.0	
Nodule DWHD 47	K-Ar(nucl)	-	0.1-1	Barnes and Dymond, 1967
41°51'S	2 samples			
102°01'W	^{230}Th	-	6	
4240 m				
Nodule Mn 139	^{10}Be	-	1.3	Ku et al., 1979
20°01'N	^{230}Th	-	4.1	
136°36'W	^{231}Pa	-	5.5	
3916 m				
Ore crust at basalt	^{230}Th	320 000	47.9	Burnett and Morgenstein,
Hess Deep	^{230}Th/^{232}Th	-	over 100	1976; Burnett and Piper,
2°16.8'N	Palagonitization			1977
101°1.5'W	of basalts	19 000	816	
3150 m	^{230}Th excessive	18 500	ca. 1000	
Ore crust	^{231}Pa	320 000	2.5	Lalou et al., 1979a
at basalt	^{230}Th	550 000	2.3	
13°09'S	^{10}Be	890 000	2.8	
148°62'W	^{26}Al	1.09×10^6	2.3	
4020 m	^{234}U	$\sim 1 \times 10^6$	2.5	
	Track		0.8-2.5	
Ore crust	Age of oceanic	5×10^6	8	Lalou et al., 1979b
at basalt	crust			
9°08.65'N	Age of sea	4×10^6	10	- " -
105°12.94'W	mountains			
3214-3265m	Age of	1.7×10^6	23	- " -
	nucleus (K-Ar)	$(330-570) \times 10^3$	67-114	- " -
	Track Palagonitization			
	of basalt	120 000	333	- " -
	by ionium			
	excess	5 000	8 000	- " -
Nodule	Racemization of	8.7×10^6	0.6-1.2	Bada, 1972
with bone	amino acids	-	1.2	
in nucleus				
19°30'N				
168°50'W				
Bones of				
whale in	Racemization			
nodule	of amino acids:			
nucleus	alloisoleuicine/	$(1.6-5.8) \times 10^6$	0.4-2.0	Kvenvolden and Blunt,
7-28°N	leuicine			1979
148°-163°W	D/L asparagine			
8 samples	acid	$(0.39-1.3) \times 10^6$	1.3-10.3	- " -

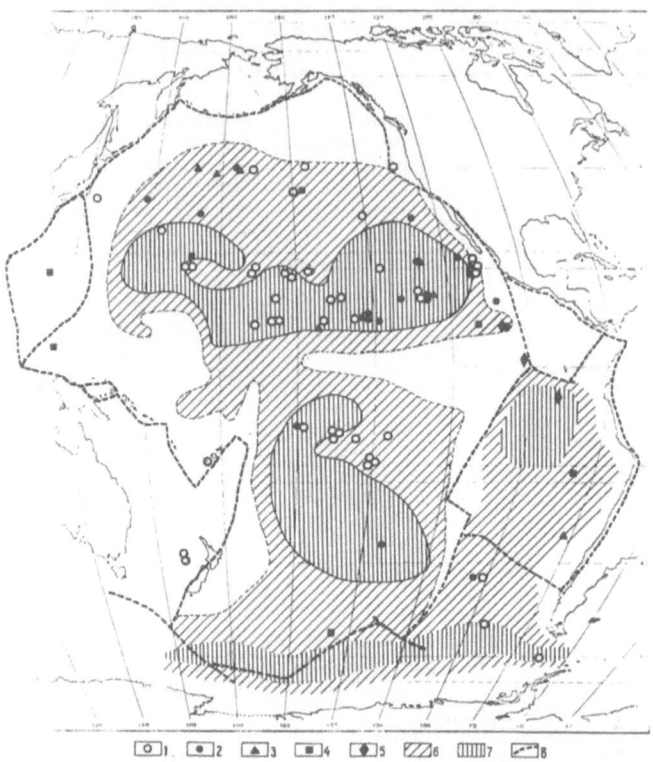

Fig. 59. Accretion rates of the Pacific nodules (data from Table 128), mm per million
years. (1) 0.8 to 4; (2) 4 to 8; (3) 8 to 12; (4) 12 to 40; (5) over 40;
(6) zones of occurrence of nodules; (7) zones of maximum distribution
of nodules; (8) rift zones.

particularly of the radiolarian zone. Similar rates were determined by radiometric
techniques for the nodules of the Atlantic and Indian oceans (Table 128).

According to the radiometric data, the accretion rates of the nodule tops and
bottoms are consistent though apparently different. In two nodules from the radiolarian zone, the rate at which their top parts grew is 1.3-1.7 times higher contrast
to the growth rate of their bottom parts. In the third sample analysed, the picture
is the opposite, the bottom part growing at least 3 times faster than the top one
(Table 130). The latter result was obtained by ionium and protactinium techniques
and was then verified by the beryllium technique in one nodule sample from the
southern Indian Ocean (23°26'S; 50°49'E): its top part grew at a rate of 1.4-1.9
mm/m yr, its bottom one grew at a rate of 4.0-5.2 mm/m yr (Sharma and Somayajulu, 1982).

The problem of the variations in nodule accretion rate has been discussed in
numerous works (Mero, 1965; Bezrukov, 1976; Heye, 1975; Cronan, 1980, etc.) on
the basis of structural-textural composition of ore matter and also from considerations of the radioisotope behaviour. However, layer-by-layer age determinations are
scarce. In one nodule sample from the Antarctic zone of the Pacific, the accretion
rates of individual layers were 4 to 19 mm/m yr by the ionium-thorium technique
(Kraemer and Shornick, 1974). In two other nodule samples from the North and
South Pacific, the accretion rates of individual layers vary from 0.78 to 7.31 mm/m
yr. The rates determined for 0.193-to-0.332 mm microlayer by the ionium-thorium
technique turned out to be 461 mm/m yr, i.e. 2 orders higher in contrast to the

TABLE 130
Comparative analysis of accretion rates in nodule tops and bottoms

Sample	Layer	Accretion rates (mm/10^6 yr)				References
		^{230}Th	^{230}Th/^{232}Th	^{231}Pa	^{10}Be	
A47-16-1	Top	3.9-4.0	4.3-4.6	3.7	-	Krishnaswamy and Cochran, 1978
9°2.3'N 151°11.4'W 5040 m	Bottom	-	2.5-3.3	-	-	
A47-16-2	Top	1.9-2.4	2.0-2.5	2.8-3.0	-	
same location	Bottom	1.5-2.5	-	-	-	
Mn7601-20-2	Top	4.7	5	5.4	-	Moore et al., 1981
11°07'N 140°05'W 4722 m	Bottom	12	16	13-29	-	
RC 14 D4 F	Top	-	-	-	1.4-1.9	Sharma and Somayajulu, 1982
23°26'S 50°49'E 4050 m	Bottom	-	-	-	4.0-5.2	

TABLE 131
Irregular accretion rates of various layers in nodules and ore crusts

Sample	Layer, mm	Accretion rate (mm/m yr)				
		^{230}Th	^{230}Th/^{232}Th	^{231}Pa	^{10}Be	^{10}Be/^{9}Be
Nodule* E 5-4 60°02'S 67°15'W	0-1.5	-	4	-	-	-
	1.5-5.5	-	19	-	-	-
	5.5-9	-	8	-	-	-
Ore crust** 13°09'S 148°62'W	0-1	2.3+0.8	-	-	-	-
	1-4	13+7	-	-	-	-
	0-0.5	-	-	2.5+0.9	-	-
	0.5-2.5	-	-	17+10	-	-
	0-20	-	-	-	2.8+0.6	-
Nodule*** A47-16-4 9°2.3'N 151°11.4'W	0.036-0.193	2.33	3.55		-	-
	0.193-0.332	7.31	461	-	-	-
	0.332-0.729	5.2	3.60	-	-	-
	0.036-0.729	5.12	5.06	-	-	-
	0.135-0.394	-	-	6.85	-	-
	0.4-6.6	-	-	-	2.0	2.4
Nodule*** FT 5 19°53'S 135°35'W 3623 m	0.0125-0.272	1.4	1.4	-	-	-
	0.273-0.494	0.78	0.81	-	-	-
	0.0125-0.494	1.1	1.1	-	-	-
	0.0125-0.210	-	-	2.1	-	-
	0.95-7.1	-	-	-	1.0	1.0
	0.95-13.0	-	-	-	2.0	1.3

From: *Kraemer and Shornick, 1974; **Lalou et al., 1979b; ***Krishnaswamy et al., 1982.

typical rate for nodules from the radiolarian zone (Krishnaswami et al., 1982). In
the ore crusts from the Tuamotu Ridge, accretion rates determined for their individual
layers are 2.3 to 17 mm/m yr (Lalou et al., 1979b; Table 131).

Radial cross-sections of alpha-track distribution, which are often of a step-
like geometry, also demonstrate the irregular character of nodule accretion. The age
intervals between these individual steps may correspond to 40 000 years and more
(Krishnaswami et al., 1979).

RELIABILITY OF NODULE DATING

As was briefly mentioned at the beginning of this chapter, the first results of
nodule dating by the radium method drew some criticism. The radium technique was
based on the supposition that radium is supplied to nodules directly from sea water
(Kurbatov, 1936). Later, other researchers pointed out that accretion rates obtained
by the radium technique could be considerably overestimated in contrast to real
rates since radium is the daughter product of ionium decay, and radium decays
50 times faster. Further researches showed that radium distribution in nodules is
really controlled by ionium distribution. Besides, radium is significantly mobile.
So, it does not seem to be worth consideration for the dating of nodules (Ku, 1977;
Krishnaswami and Cochran, 1978).

The technique for dating of nodules based on other isotopes (Table 128),
though widely applied, have also been criticised. The arguments arise from the
following considerations:

(1) Nodules may grow at a high rate and then, when they stop growing,
radioelements precipitating out of sea water are accumulated at their surface.
Diffusion of these elements to the inner parts of nodules results in the observed
pattern of their distribution (Arrhenius, 1967).

(2) Nodules may grow at a high rate due to supply of volcanic material and
they may accumulate volcanic radioisotopes in random proportions that does not in
principle allow their dating by radiometric techniques (Bonatti and Nayudu, 1965;
Cherdyntsev et al., 1971; Lalou and Brichet, 1976).

(3) The simple fact that nodules rest on a sediment surface is sufficient to
consider them as younger formations in contrast to the underlying substrate; any
radiometric data on the inverse age relations between them should be ignored (Lalou
and Brichet, 1972; Volkov, 1979).

Recent marine geology and geochemistry would provide representative data
enough to oppose these arguments:

(1) Should the behaviour of radioisotopes in nodules be controlled by the
diffusion mechanism, the radial cross-sections of their distribution in nodules would
follow the laws of diffusion. However, the radial cross-sections have an entirely
different geometry (Ku et al., 1979). The data on concentration and distribution of
radioisotopes in sea water show that a period of about tens and hundreds of
thousands of years is necessary to accumulate the observed amount in the surficial layer of nodules (Ku, 1977; Ku et al., 1979).

The period needed to achieve equilibrium in the system: adsorption-of-isotope-
diffusion-decay would additionally increase the time of nodule exposition at least by
10 times (Krishnaswamy et al., 1982). Thus, the diffusion hypothesis proposed to
"rejuvenate" nodules cannot cope with this problem and leads to the opposite result.

(2) Volcanic and hydrothermal processes, no doubt, affect oceanic sedimentation
and marine geochemistry of manganese. This effect can be estimated quantitatively
when analysing the distribution of sediments on the scale of the entire ocean or its
separate large regions (Skornyakova, 1964; Bostrom and Petterson, 1966; Lisitsin,
1979; 1983). However, considering the behaviour of manganese in a sea environment
this effect is of a local scale (Beiersdorf et al., 1982) which is demonstrated by the
map of nodule growth (Figure 59). So, the conclusion about volcanic origin of all
oceanic nodules contradicts the actual data.

The inference made by Cherdyntsev et al. (1971) about the irregular
distribution of radioisotopes in nodules does not agree with a great amount of the
data obtained which verify their regular distribution within nodules, and all methods
of absolute geochronology of nodules are based on these data. When analysing
nodules, the authors mentioned did not seem to regard either the specification of

the material considered, or the pattern of radioisotope distribution determined earlier. In contrast to other researchers, who analysed thin layers of nodules (several decimals of millimeter), they analysed layers of several centimeters thick, lump samples (crusts, nuclei) or nodules as a whole with inclusions of basalts, tuffs and phosphate rocks.

In numerous works by Lalu et al. (1972-1980), data are given on the excess of "young" isotopes in the inner nodule parts. This can be logically explained by the occurrence of cracks in nodules through which clayey material of the host sediments, rich in corresponding isotopes, is injected into the nodules (Heye, 1975; 1979; Ku and Knauss, 1979).

(3) The contradiction between the ancient age of nodules and the young age of the underlying sediments determined by the same radiometric techniques remained unresolved for a long time.

One of the opinions was that zones of nodule occurrence correspond to the zones of minimum, zero or even negative sedimentation (Bezrukov, 1976). Such zones do exist in the ocean, but many regions exist where nodules rest on the young (Holocene and Pleistocene) sediments which seems to determine the age of nodules. However, some additional data exist that testify against the young age of nodules. In contrast to sediments, nodules are rich in magnetic spherules of a cosmic nature and they accumulate at an extremely low rate (Finkelman, 1970). In contrast to sediments, organic matter in nodules is metamorphosed to a greater degree (Romankevich, 1977), which testifies to a longer period of exposition of nodules with respect to sediments.

In the nodule interior no deficit of ionium and protactinium, with respect to radioactive equilibrium, has been recognized. Such a deficit is an indispensable condition for the young age of nodules (Ku et al., 1979).

In the cracks that transect nodules, younger clayey material (as compared to that of the nodules themselves) was detected, but even this material is old enough, about 300 000 years (Ku and Knauss, 1979). However, sometimes it may result from the old age of the terrigenous portion of sediments. Finally, the coincidence in the dating of nodules by several independent techniques (Table 129) is sufficient evidence of their ancient age, despite the more rapid accumulation of the underlying sediments. We shall discuss this paradox in the next chapter.

The fact that nodules can grow under experimental conditions at the same rate as that in the pelagic environment of the ocean testifies in favour of the age-dating results obtained by radiometric techniques (Table 132).

Meanwhile, many contradictions in the age-dating results still remain unresolved, since all varieties of the radiometric techniques are based on four major assumptions: (1) the radiochemical composition of sea water is homogeneous and constant; (2) sea water is the source of radioisotopes in nodules; (3) the time of radioisotope accumulation in nodules is short as compared to the age of the latter; (4) nodules become an isolated system after their formation is completed.

The recent data demonstrate that assumptions (2) and (4) are only partly valid if at all. The bulk or the great part of nodule matter is supplied to them out of host sediments, not out of sea water. Of great doubt is the assumption that nodule interiors are isolated from the sea since nodules are characterized by great porosity, the share of natural pore water in them reaching 30 to 40%. Besides, diagenetic transformations are sure to occur in nodules, thus affecting their mineral, chemical and, consequently, their radiochemical composition (Burns and Burns, 1978a; Sterenberg et al., 1986).

It is obvious that further efforts are needed to improve the theoretical basis of radiometric methods applied to nodule dating, consistent with recent developments in the geochemistry of marine sedimentation and diagenesis.

TABLE 132
Time at which 1 mm of Mn hydrooxide film is formed:
with respect to concentration of dissolved Mn in water, nodule radius and
temperature of the medium (Krylov et al., 1981; Tikhomirov and Gromov, 1983)

Mn concentration in water, µg/l	Time of 1mm-layer formation (yr)			
	R = 0.25 cm	R = 0.5 cm	R = 1 cm	R = 5 cm
t = +2°C				
0.1	$(8.8 \pm 1.7) \times 10^5$	$(1.1 \pm 0.2) \times 10^6$	$(1.2 \pm 0.2) \times 10^6$	$(1.3 \pm 0.2) \times 10^6$
1	$(7.5 \pm 1.5) \times 10^3$	$(8.6 \pm 1.5) \times 10^3$	$(9.6 \pm 0.6) \times 10^3$	$(1.0 \pm 0.1) \times 10^4$
2	$(1.6 \pm 0.4) \times 10^3$	$(2.0 \pm 0.3) \times 10^3$	$(2.3 \pm 0.3) \times 10^3$	$(2.4 \pm 0.4) \times 10^3$
5	$(2.4 \pm 0.3) \times 10^2$	$(3.0 \pm 0.4) \times 10^2$	$(3.4 \pm 0.5) \times 10^2$	$(3.6 \pm 0.5) \times 10^2$
10	55 ± 7	70 ± 9	77 ± 9	88 ± 11
20	13 ± 1	16 ± 2	18 ± 2	19 ± 2
t = 19°C				
0.1	$(3.7 \pm 0.7) \times 10^5$	$(4.6 \pm 0.9) \times 10^5$	$(5.1 \pm 0.9) \times 10^5$	$(5.5 \pm 1.0) \times 10^5$
1	$(2.9 \pm 0.5) \times 10^3$	$(3.6 \pm 0.6) \times 10^3$	$(4.0 \pm 0.7) \times 10^3$	$(4.3 \pm 0.7) \times 10^3$
2	$(6.8 \pm 1.0) \times 10^2$	$(8.4 \pm 0.2) \times 10^2$	$(9.4 \pm 1.4) \times 10^2$	$(1.0 \pm 0.1) \times 10^3$
5	$(1.0 \pm 0.1) \times 10^2$	$(1.2 \pm 0.2) \times 10^2$	$(1.4 \pm 0.2) \times 10^2$	$(1.5 \pm 0.2) \times 10^2$
10	23 ± 3	29 ± 4	32 ± 4	35 ± 4
20	5.3 ± 6	6.6 ± 0.7	7.4 ± 0.8	8.0 ± 0.8
t = 30°C				
0.1	$(2.1 \pm 0.4) \times 10^5$	$(3.0 \pm 0.5) \times 10^5$	$(3.3 \pm 0.6) \times 10^5$	$(3.6 \pm 0.7) \times 10^5$
1	$(1.9 \pm 0.3) \times 10^3$	$(2.4 \pm 0.4) \times 10^3$	$(2.6 \pm 0.4) \times 10^3$	$(2.8 \pm 0.5) \times 10^3$
2	$(4.4 \pm 0.6) \times 10^2$	$(5.5 \pm 0.8) \times 10^2$	$(6.1 \pm 0.9) \times 10^2$	$(6.0 \pm 1.0) \times 10^2$
5	64 ± 9	80 ± 11	88 ± 12	96 ± 13
10	15 ± 2	19 ± 2	21 ± 3	23 ± 3
20	3.5 ± 0.4	4.2 ± 0.5	4.8 ± 0.5	5.2 ± 0.6

CHAPTER XII

PROBLEMS OF ORIGIN OF MANGANESE NODULES

The problem of the origin of nodules, as well as of any economic minerals, concerns the source of ore matter, the manner of its migration and supply, and the mechanism of concentration and favourable environmental factors.

To solve the problem of nodule origin one should utilize the sum of total knowledge on marine geology, geochemistry and, to a certain degree, on biogeochemistry. This chapter is devoted to some of these aspects.

SOURCE OF ORE MATTER AND MANGANESE BALANCE IN THE OCEAN

When analysing the origin of ore matter in nodules, one should consider two things, namely, direct sources of ore material for individual nodule formation and sources of ore material supply to the ocean as a whole. We shall start from the second aspect of the problem, which was already being debated in the last century.

First Gümlel (1878), and then Murray and Irvine (1894) supposed that the main ore material supply to the ocean comes from terrigenous and volcanic, and also hydrothermal material, though the role they play in different parts of the ocean may be divergent.

Later, Bonatti and Nayudu (1965) and Zelenov (1972) verified the decisive role of volcanism in the supply of ore matter to the ocean; at the same time, Strakhov (1960; 1976) expressed - with no less confidence - his belief in the decisive role of terrigenous material. However, at present, the role of hydrothermal supply of ore matter is obvious, though the share of its components is not unanimously agreed upon resulting in various balance estimates.

Goldschmidt (1954) was the first to open the problem of balance for discussion. He voiced the opinion that absolute masses of some metals in the ocean exceed their supply from continental rock weathering over geological time.

With a continuous inflow of new data on input and outflow quantities, the problem of the balance of elements in the ocean has become one of the permanent problems in geochemistry, the interest in which is sustained by the problem of manganese nodule origin and also by the problem of hydrothermal matter supply to the ocean. The number of publications devoted to various aspects of manganese balance in the ocean continues to grow (Horn and Adams, 1966; Vinogradov, 1967; Bostrom, 1967; Krishnaswamy and Lal, 1972; Lyle, 1976; Elderfield, 1976, 1977; Bender et al., 1977; Lisitsin, 1978, 1981; Edmond et al., 1979; Wedepohl, 1980; Volkov, 1981; Li, 1981, 1982; Whitfield and Turner, 1982; Baturin, 1983; Lisitsin et al., 1985).

Using the initial data published by the early 1960s, Horn and Adams (1966) computed the balance of some macro- and micro-elements in the ocean. These computations revealed an actual excess of manganese over its supply from continental sources. Such elements as chlorine, sulphur, bromine, iodine, molybdenum also turned out to be "excessive" (Table 133).

Later, manganese supply to the ocean from various sources was estimated by Elderfield (1976) who considered several different models. From these models, 2.2 to 4.0 million tonnes of dissolved manganese are supplied to the ocean and 0.7 - 7.4 million tonnes of manganese sink to the bottom annually (Table 134).

Lately, the supply of hydrothermal manganese to the ocean has become of

TABLE 133

Distribution of absolute masses of elements (in geograms, or 10^{20}g) in sediments,
water and magmatic rocks eroded during the geological time (Horn and Adams, 1966)

Element	Sediments		Oceanic		Relict water	Oceanic water	Eroded magmatic rocks	Difference $(1+2+3+4+5+6)-7$
	Continental	Shelf	Hemi-pelagic	Pelagic				
	1	2	3	4	5	6	7	
Cl	38.7	116′	74.5	45.6	27.8	267	9.79	559
S	3.17	8.96	9.20	5.52	1.32	12.6	10.6	30.2
B	0.130	0.429	0.346	0.206	0.00703	0.0673	0.245	0.940
Br	0.00571	0.0108	0.200	0.124	0.0953	0.912	0.639	1.28
I	0.00444	0.0149	0.000186	0.000144	73.3×10^{-6}	0.000701	0.0102	0.00102
Mo	0.00192	0.00594	0.0223	0.0137	14.7×10^{-6}	0.00014	0.0306	0.0134
Mn	0.573	1.44	27.2	16.1	22.93×10^{-6}	28.1×10^{-6}	22.4	22.9
Na	38.3	104	170	102	15.4	147	572	0
K	44.6	151	203	120	0.557	5.33	524	0
Ca	159	208	210	155	0.586	5.61	739	0
Fe	69.3	228	355	210	14.7×10^{-6}	0.00014	862	0
P	1.50	4.85	10.1	5.99	103×10^{-6}	0.000982	22.4	0
U	0.00774	0.0235	0.0153	0.00912	44×10^{-6}	0.000421	0.0561	0

TABLE 134
Fluxes of dissolved manganese in the ocean (m t/yr)
(Elderfield, 1976)

Supply		Deposition	
River discharge	0.2	(1) By comparison of sedimentation rates in the oceans	4.1
Leaching out of magmatic rocks	0.1-1.9	(2) By accumulation rate of hydrogenous Mn: a/ by partition geochemistry	1.2
Secondary alterations of basic rocks	1.1	b/ by correction to shales	7.4
		c/ by accretion rate of nodules	6.1
Halmyrolysis	0.8	(3) By Horn and Adams (1966)	0.7

special concern (see Chapter I). According to various authors, 0.5 to 10 million tonnes of hydrothermal manganese are supplied to the ocean yearly (Table 13).

Diagenetic manganese diffused into sea water out of sediments is involved in a closed cycle. However, it should be also regarded when considering the balance of dissolved manganese in the ocean: this causes additional difficulties.

Volkov (1981) has assessed the balance of manganese for three oceanic megazones - coastal, pelagic and ore-fields - using data on sedimentary material supply to the ocean from Lisitsin, 1974, 1978. For the "debit" part of the balance he proposed several ways to determine the areas and absolute mass of manganese; however, he assumed that the manganese content in pelagic sediments beyond ore fields was constant at 0.4%. The most reliable estimate, as Volkov believes, is given in Table 135. It shows that a given constant (manganese content in sediments beyond ore fields) causes the rise of supposed manganese concentration in the sediments that underlie manganese nodules to 6.7%. So, he concluded that it is precisely this excessive manganese which was used to form nodules, mainly during the Holocene.

Another method of estimating the balance using mainly the same initial data is given in Table 136. In the Table, absolute masses of manganese in nodules are determined by their growth rate, the latter being estimated by radiometric techniques (several millimeters per one million years).

In this case, one can see that "excessive" manganese dissipates in sediments which results from assuming the manganese average in pelagic sediments to be 1%. As such estimates are rather arbitrary, a special thorough investigation should be undertaken to detect the absolute rate of manganese accumulation in sediments. Such analyses were attempted earlier many times (Bender et al., 1970, 1971; Bostrom, 1973; Kraemer and Shornick, 1974; Lisitsin, 1978) and a map was compiled for absolute manganese mass in the ocean (Figure 60), though no differentiated analysis of these masses was made for individual oceanic regions. Nevertheless, the total absolute manganese mass was assessed as 8.35 m t/yr for the pelagic zone (beyond the 3000 m isobath) (Lisitsin, 1981). This estimate is very close to that reported by Elderfield (Table 134) - 7.4 m t/yr, and coincides with that in Table 136 - 8 m t/yr.

Our existing knowledge of manganese balance in the ocean is still insufficient - and that concerns both its "debit" and "credit" - in particular, through hydrothermal and diagenetic flows.

According to different estimates, the supply of labile and dissolved manganese to the ocean is 0.2-0.4 m t/yr through river discharge, 0.05-0.3 by atmospheric material, 1-10 by hydrothermal material, and 0.4-4 m t/yr by diagenetic flow. So, the amplitude of minimum-maximum summary estimates is 1.2-14.7 m t/yr. The accumulation rate of manganese should be studied in individual oceanic regions with due concern to their peculiarities, so that a true view of its actual amount can emerge.

Attempts were also made to estimate the absolute mass of ore matter accumulated in individual nodules.

TABLE 135

Manganese balance in the ocean (Volkov, 1981)

Supply				Deposition				
Source	Matter mass (b t/yr)	Mn content (%)	Mn mass (m t/yr)	Zone	Area (10^6/km^2)	Sedimentary mass (b t/yr)	Mn content (%)	Mn mass (m t/yr)
River suspension	18.5	0.11	20.35	Coastal	87.6	23.37	0.07	16.36
Glacial material	1.5	0.07	1.05	Pelagic	225.5	1.672	0.4	6.69
				Ore fields	47.1	0.058	6.7	3.85
Eolian material	1.6	0.07	1.12	Total				26.9
Coastal abrasion	0.5	0.07	0.35					
Volcanism	3.0	0.12	3.60					
River water	43×10^3/km^3	10/μg/l	0.43					
Total			26.9					

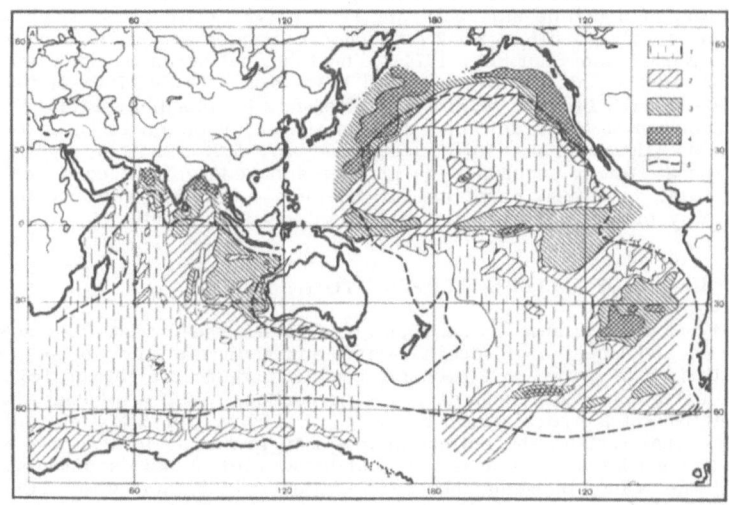

Fig. 60. Accumulation rate of Mn in the Pacific and Indian ocean bottom sediments, µg/cm²/1000 yr (Lisitsin, 1978). (1) less 0.5; (2) 0.5 to 2; (3) 2 to 5; (4) over 5; (5) boundary between oxidized and reduced sediments.

TABLE 136
Deposition of manganese in the ocean (Baturin, 1983)

Zone	Area (m km²)	Deposition to the bottom		
		abiogenic matter	Mn content (%) in sediments and in nodules	Mn*
Peripheral	87	$\dfrac{23 \times 10^9}{25}$	0.07	$\dfrac{16 \times 10^6}{0.0175}$
Hemipelagic and pelagic	220	$\dfrac{2 \times 10^9}{1}$	0.4	$\dfrac{8 \times 10^6}{0.0040}$
Ore fields	50	$\dfrac{50 \times 10^6}{0.1}$	1	$\dfrac{0.5 \times 10^6}{0.0010}$
Nodules	5 - 10	$\dfrac{(50-500) \times 10^3}{0.0005-0.005}$	20	$\dfrac{(10-100) \times 10^3}{0.0001-0.0010}$

* Over the line: t/yr;
  Under the line: mg/cm² yr.

From the radiometric techniques, manganese accumulation rates (with due concern of the total nodule surface) vary from 0.372 to 2.6 g/cm^2/m yr; iron accumulation rates are 0.516 to 2.572 g/cm^2/m yr (Bender et al., 1970; Somayajulu et al., 1971; Kraemer and Shornick, 1974). The following accumulation rates were obtained for other metals (in mg/cm^2/my): 0.15 to 9.4 Cu; 0.17 to 24 Ni; 0.019 to 37 Co; 0.004 to 21.2 Pb; 2.0 to 5.6 Zn (Kraemer and Shornick, 1974).

A comparison of these data and data on the accumulation rates of metals in host sediments (Heath et al., 1970; Bostrom et al., 1973; Lisitsin, 1978) shows that, in general, metals accumulate in nodules and sediments at consistent rates, except for iron, vanadium, chromium and titanium - they are supplied to sediments at considerably higher rates in contrast to nodules (Somayajulu et al., 1971).

DIRECT SOURCES OF MATERIAL SUPPLY TO NODULES

The majority of ore elements that are supplied to the ocean from various sources are dispersed and mixed up in oceanic water and sediments and do not preserve any genetic mark in them. Most probably oceanic nodules are formed from the ore material of polygenic origin, though some researchers would emphasise the predominant role of a hydrothermal sources (Chukhrov, 1978; Lisitsin, 1978).

Nodules that rest on the bottom surface may grow due to a supply of material both from above (out of oceanic water) and below (out of host sediments). In particular, the manner of manganese precipitation from oceanic water (Chapter IV) and manganese behaviour in sediments and pore waters (Chapters V, VI) demonstrate the influence of the two supply sources. On this basis, Skornyakova (1984, 1986) and other investigators tried to divide nodules into two major types, namely, hydrogenous and diagenetic. Bonatti et al. (1972) were the first to attempt a genetic classification of iron-manganese formations in the ocean on the basis of the major ore elements (manganese, iron and base metals) they contain. Three major types were distinguished, namely, hydrogenous, i.e., formed due to slow deposition of metals out of sea water and characterized by a high concentration of base metals and varying Mn/Fe ratios (from 0.5 to 5); hydrothermal, i.e., rich in iron and depleted of other metals; diagenetic, characterized by high Mn/Fe ratios and relatively low concentration of base metals.

The classification in question, and a three-component diagram of ore composition based on it, were modified many times by various researchers on the basis of more representative and extensive data. In particular, it appeared that hydrothermal formations have an extremely wide range of Mn/Fe ratios (Cronan, 1980).

Ore crusts on the rocks of sea mountains, characterized by minimum rate of accretion, are considered as a standard of hydrogenous manganese ore in the ocean. Manganese formations in the direct vicinity of hydrothermal sources are assumed to be a standard for hydrothermal ore. Nodules of hemipelagic regions resting on sediments enriched in organic matter are considered as a typical diagenetic formation.

So, the question may be raised: what is the relative role of hydrogenous ious and diagenetic processes in the formation of manganese nodules in oceanic pelagic regions in various sedimentary environments.

To answer this question Dymond (1981; Dymond et al., 1984) suggested a normative model (Figure 61) accounting for three accretion processes, namely, hydrogenous, suboxic diagenetic and oxic diagenetic. The latter was proposed earlier as an independent type of the diagenetic processes on the basis of a detailed geochemical and lithological study of pelagic oceanic sediments (Logvinenko, 1974; Volkov et al., 1976). Later, it appeared that a considerable supply of organic matter to the bottom of oceanic pelagic zones exists (Honjo, 1978, 1980, 1982; Suess, 1980; Reimers, Suess, 1983; Savenko, 1985), playing an important role in the latter process.

A special analysis, based on the model in question, was performed to find out the contribution of the three accretion processes to the supply of metals to the nodules in the eastern equatorial hemipelagic zone of the Pacific. It revealed that 57.8% of the bulk matter is supplied by oxic diagenesis, 40.7% by suboxic diagenesis and only 1% by hydrogenous accretion (Table 137).

It seems that the major part of the material of pelagic nodules, which was traditionally believed to be hydrogenous, is mobilized from sediments due to oxic diagenesis.

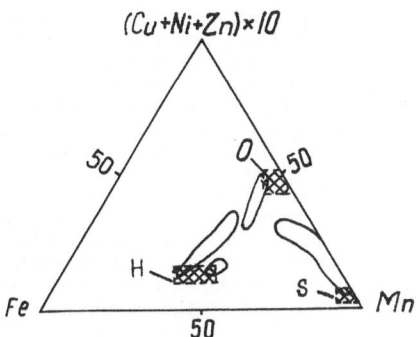

Fig. 61. A triplot of (Cu + Ni + Co) x 10 versus Mn versus Fe contents in nodules of Sites H, S and R (MANOP) (Dymond et al., 1984). Dashed boxes show fields of hydrogenous (H), oxic (O) and suboxic (S) accretion.

TABLE 137

Sources of nodule material supply in the eastern equatorial hemipelagic zone of the Pacific (Dymond et al., 1984)

Element	Sources		
	Hydrogenous	Oxic diagenetic	Suboxic diagenetic
Na	0.7	40.8	58.5
Mg	0.8	69.7	29.5
Al	0.9	82.9	16.2
Si	1.9	82.1	16.0
K	1.0	64.6	34.4
Ca	2.8	61.5	35.7
Mn	0.9	47.9	51.2
Fe	8.6	85.4	6.0
Co	10.1	82.7	7.2
Ti	6.7	81.0	12.3
Ni	1.2	74.0	24.8
Cu	0.2	88.2	11.6
Zn	0.5	61.4	38.1
Ba	0.7	75.2	24.1
Average	1.5	57.8	40.7

    Hydrogenous ore crusts are believed to be formed due to the accumulation of dissolved metals at electrochemically active surfaces of the solid phase, whereas formation of diagenetic nodules is associated with colloidal-chemical processes at the sea water/bottom interface (Goldberg and Arrhenius, 1958; Sevastyanov and Volkov, 1965, 1967; Burns and Burns, 1975).

WHY DO MANGANESE NODULES REST ON THE SEDIMENT SURFACE?

An obvious contradiction between the low growth rates of manganese nodules (millimeters per million years) and sedimentation rates that are three orders higher (millimeters per thousand years) remained unexplained for many years. It was used as the main argument against the ancient age of nodules. The very occurrence of dense nodules on the surface of semi-liquid sediments was a puzzle. However, special analyses of the physical properties of sediments, in particular their shear strength, revealed that sediments are dense enough to support nodule bodies at the bottom surface and to prevent their sinking (Piper and Fowler, 1980).

Similar results were obtained by Goriyainov and Goriyainova (1983) who estimated the forces pushing nodules out of sediments. When a nodule rests on the surface of a dense sediment, it experiences a uniform hydrostatic pressure (Figure 62a). As the nodule sinks into the mud, its bottom segment experiences greater compression in contrast to its top parts, which meet only hydrostatic pressure (Figure 62b). In this case, if an upward constituent of the pressure that affects the nodule exceeds its weight and adhesive strength, it will be squeezed upward by the sediments as their density becomes greater until the oppositely-directed forces reach their equilibrium; the nodule will come to the surface and its position will be higher in contrast to the initial moment (Figure 62c). In the general case, a nodule can be assumed to be an ellipsoid of revolution with normal compression f affecting its bottom part; it can be decomposed into horizontal (fr) and vertical (fb) constituents. The pushing force will be the resultant of the two vertical constituents (Figure 62d).

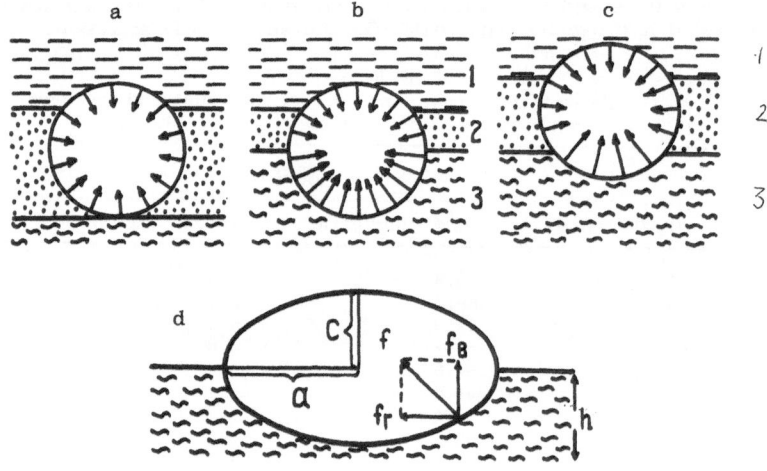

Fig. 62. Scheme illustrating the forces that push nodules out of sediments (Goriyainov and Goriyainova, 1983). (1) sea water; (2) bottom suspension; (3) sediments. Other notations are given in the text.

Another mechanism can be found to explain the occurrence of nodules at the bottom surface. It results from the activity of bottom fauna which was detected in pelagic regions of the ocean even at maximum depth (Figure 63). In a search for food, microorganisms inhabiting the bottom surface of nodules, for instance, benthic dwellers can push nodules out of the mud or even turn them upside down. Nodules would remain on the surface if they are pushed upward once every several thousand years (Von Stackelberg, 1985). Such episodes in a nodule's "biography" were detected by the radiometric age-dating of the bottom and top parts of nodules (Krishnaswamy and Cochran, 1978). (It is worth mentioning that Darwin wondered how earth worms can move rather large stones.)

Fig. 63. Gigantic enteropneust worm (~ 1 m long) and its spiral tracks in surficial
sediments of the South Pacific (29°40S; 176°43W; depth 4735 m) (Bourne
and Heezen, 1965).

According to another hypothesis, benthic fauna stimulate the injection of sedimentary matter from the surface to the passages they make, the nodules remaining on the surface (Piper and Fowler, 1980). Traces of an active bioturbation revealed in the upper 10 to 20 cm of pelagic sediments and verified by radiometric data testify to this hypothesis (Berger and Killingley, 1982; Sanderson, 1985).

BIOGENIC ASPECTS IN MANGANESE NODULE GENESIS

Though manganese does not belong to the major biogenic elements, it is in charge of a number of important beiogenic functions (Chapter III). The more important thing, however, is that the biogenic factor determines to a considerable extent the manganese destiny on its way to the ocean. The entire fine terrigenous material and the greater part of hydrothermal material that enter into the ocean pass repeatedly through the biogenic filter before they reach the bottom (Lisitsin, 1983). Manganese species in water and suspension are associated with the occurrence of organic matter and activity of sea bacteria (Chapters II, IV).

At the oceanic bottom, manganese and other metals are also affected by the biogenic factor, beginning with the downward flux of organic matter and terminating with benthic activity (from microorganisms to macrofauna). Bacterial activity is of especial concern as it may affect the growth, structure and composition of nodules.

Specific sea bacteria that accumulate iron and manganese were discovered in water, suspension, sediments and manganese nodules of the ocean. In the waters of the northeastern Pacific, they comprise to 5% of the total microbial inhabitants of suspension (Cowen, 1982). In bottom sediments, both metal-oxidizing and reducing bacteria were detected (Ehrlich, 1963-1980; Krumbein, 1971; Schuett and Ottow, 1977).

Butkevich (1928) was the first one to report on the specific iron bacteria in manganese nodules recovered from the Arctic basin bottom. He discovered Gallionella in the jars containing sea water and nodules. Later, in nodules from the Kara Sea, other species of microorganisms were found which were capable of accumulating iron and manganese (Kalinenko, 1946; Kalinenko et al., 1962). The microbiological analysis of manganese nodules from the northeastern and southern Pacific revealed to 80 mln./cm³ of bacteria in them, though no specific iron bacteria were detected (Sorokin, 1971, 1972). Other similar analyses had more success - they revealed abundant iron bacteria. In the nodules from the Central Pacific, a specific bacterial microflora appeared to be abundant in contrast to underlying sediments (Ehrlich, 1972; Ehrlich et al., 1974). Similar results were obtained in the microbiological analysis of nodules from the southeastern Pacific, the density of microbe inhabitants in nodules being 30 mln/g; in particular, 10 thousands of specific manganese oxidizing bacteria per gram. The number and composition of microflora in bottom, equatorial and top parts of nodules appeared to be similar (Schuett and Ottow, 1977). Some data on microbe inhabitants of the Pacific nodules are given in Table 138.

A number of other papers reported on the occurrence of new species of bacteria, bacterial spores and bacterial ferments that inhabit sea water and are capable of oxidizing and depositing manganese and iron (Kepkay and Nielson, 1982; Rosson and Nielson, 1982; Schuett, 1982; Cowen and Bruland, 1985).

The function of microorganisms in manganese and iron oxidation and the mechanism of oxidation are still under debate.

Vinogradsky (1952) believed that bacteria oxidizing iron are inorgoxidative since they use the energy of oxidation for their growth and practically do not need any organic matter as they consume carbon dioxide.

Kalinenko (1946) supposed that iron and manganese oxidation by bacteria is their accessory function, having no connection with their life cycle and being caused by pH increase. It was also supposed that an iron coating over bacteria may protect them from Protozoans grazing or that metal oxidation is somehow connected with heterotrophic organisms (Zavarzin, 1972).

Various aspects of oxidation-reduction functions of marine bacteria detected in manganese nodules were considered in many works by Ehrlich and other researchers (Ehrlich, 1963, 1964, 1966, 1968, 1971, 1972, 1974, 1976, 1980; De Castro and Ehrlich, 1970; Ehrlich et al., 1972, 1973; Giorze, 1977; Arcuri and Ehrlich, 1979, 1980; Marshall, 1979, etc.). The data considered confirm that occurrence of various

TABLE 138

Density, composition and activity of bacterial inhabitants of manganese nodules, host sediments and bottom water: The Pacific

Group of bacteria	Sediments	Nodules			Sea water	References
		Top	Bottom	Nucleus		
Total number of bacteria, per cm^3	$(14-37)$x10^6	$(31-37)$x10^6	$(15-21)$x10^6	-	-	Sorokin, 1971
in particular: heterotrophic	10-220	123-3600	0-52	-	-	
Ratio: $\dfrac{\text{productivity x 10}^3}{\text{biomass}}$	5-18	50-280	2-120	-	-	
Total number of aerobic, per cm^3	$(1.6-2.9)$x10^8	$(0.5-3.1)$x10^6	$(0.9-5.4)$x10^6	$(0.1-5.4)$x10^6	$(4-9)$x10^6	Schutt and Ottow, 1977
in particular, denitrificators:	0	$(0.1-5.4)$x10^3	$(0.2-3)$x10^3	$(0.1-5.4)$x10^3	0-7, 6x10^5	
iron (III) depositing	92-7,6x10^5	0-120	0-31	0-120	$(0.9-40)$x10^3	
manganese (II) depositing	10^3-10^7	10^4-10^5	10^4	10^3-10^6	10^5-10^6	
Total number of anaerobics	-	$(1.2-54)$x10^3	$(0.1-12)$x10^3	$(0.5-31)$x10^3	-	

bacteria, in particular, specific iron bacteria in deep-sea manganese nodules is a fact. Now, the question under debate is how great and independent their role is in nodule formation.

Electron microscopy of manganese nodules from numerous regions of the World Ocean revealed various biomorphic ultramicrostructures in them that are sometimes predominant and seem to be bacteriogenic.

In the Pacific nodules, dendritic filaments, tubes and rods of vernadite and other manganese minerals were recognized, that might appear as the result of the activity of microorganisms similar to Metallogenium or Kuznetsovia specimens (Andruschenko and Skornyakova, 1967; Andruschenko, 1976; Chukrov, 1978; Chukrov et al., 1978).

In the Indian ocean nodules, cocci-like and filamentous formations of iron and manganese hydrooxides and also spheric bodies with thorns and hyphae were identified as bacterial spores (Kalinenko et al., 1962; Lazurenko and Khoruzhyi, 1980; Lazurenko, 1982).

In the Atlantic nodules of the Caribbean basin, biomorphic ultramicrostructures were also detected of rod-rounded and star-like shapes (Chugunnyi and Demenko, 1973). Xavier (1976) analysed a representative collection of the North Pacific nodules using the electron microscope and inferred that all ultramicrostructures of manganese and iron hydrooxides of spheric, oval and rod-like shape are of bacterial nature in the bulk on the nodules analysed.

Such a conclusion might seem extremely explicit, but recent data testify to the occurrence of bacteriogenic structures in nodules (Burnett and Nielson, 1981, 1983; Baturin and Dubinchuk, 1983, 1986).

In nodules of the Pacific and Indian oceans which we analysed, the most typical biomorphic ultramicrostructures are cocci, filaments, tubes and chains.

Cocci-like structures are formed of rounded or slightly oval and lens-like bodies of 1.0 to 1.5 μm in diameter. They occur as individual formations or as dense aggregates. Sometimes they are coated by iron and manganese hydrooxides which hide the details of their original microstructure.

Tubular structures are straight or slightly curved tubes 1 to 10 μm long and 0.1 to 0.3 μm in outer diameter, thickness of its walls being about 0.01 μm. The diameter of individual tubes is more or less constant: however, alteration of slight bulging and narrowing may be observed along their axes. The structures in question are entirely, only rarely partly, filled with iron and manganese hydrooxides, as was discoverd by the microdiffraction technique.

Filament structures are formed by dense aggregates of parallel fibres 2 to 5 μm long and to 0.2- μm thick. Individual fibres are no more than 0.001 μm thick. Aggregates are usually placed chaotically, rarely parallel to each other. Filament structures are associated with cocci and tubular ones. From the microdiffraction data they are mainly composed of iron and manganese hydrooxides.

Chain-like aggregates are formed of bodies joined into linear or bifurcating chains. Figure 64 shows the most typical of them.

The ultramicrostructures described are spread irregularly within various regions and within separate sites; however, they may predominate in one sample or another.

According to their morphology, they can be correlated with the known water (in particular, sea water) forms of the bacterial flora and microflora described in many works (Kholodnyi, 1952; Kalinenko, 1962; Zavarzin, 1972; Balashova, 1974; Stanier et al., 1976; Artemchuk, 1981, etc.).

Coccoid formations resemble in shape and size various species of spores, cocci, marine yeasts and fungi, some of which can be also compared with small coccolithophorids that are rock-forming components in oceanic deposits. They are also similar to microorganisms depositing iron and manganese (Siderocapsa, Caulococcus, Ochrobium, Siderococcus).

Tubular structures are also extremely widespread in the microbial world. In particular, they are typical of Sphaerotilus and Leptothrix species that accumulate (or deposit) iron and manganese. Besides, they resemble stems of iron bacteria Gallionella impregated by iron.

Filament forms may be compared with actinomycetic hyphae, marine fungal mycelium, and also filaments of iron bacteria Gallionella and Toxothrix.

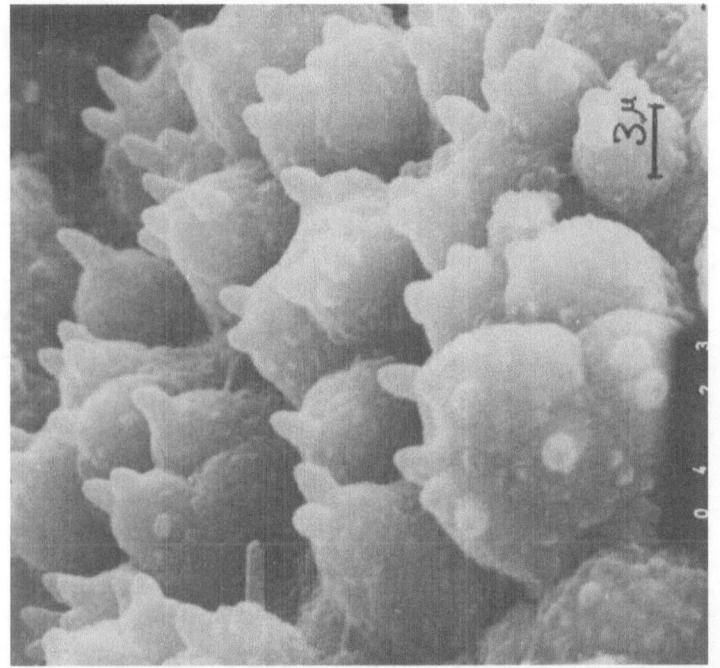

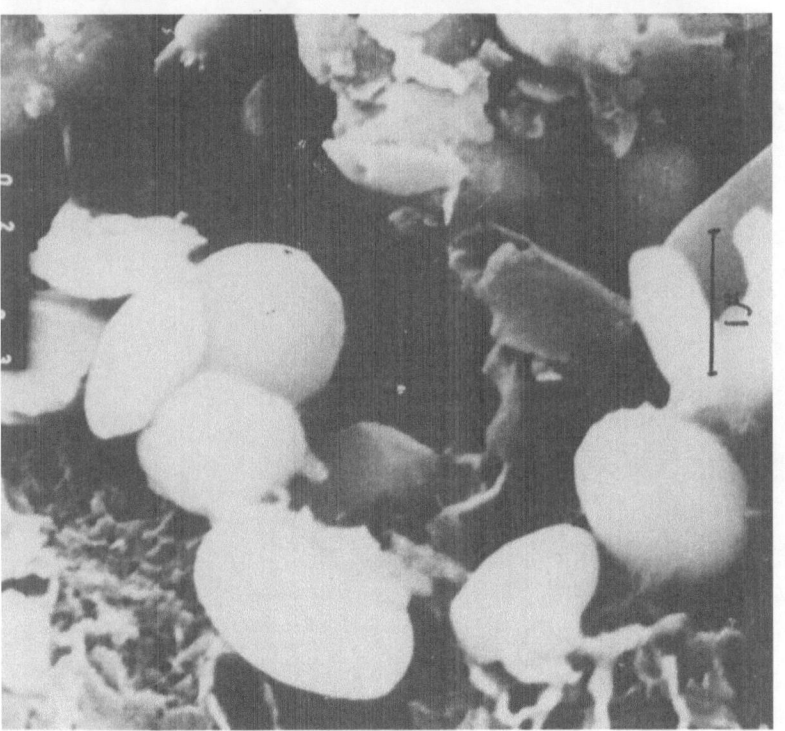

Fig. 64. Bacterial remains in manganese nodules: (a) Cocci-like formations: the Pacific radiolarian zone. (b) Rounded mammiliferous bodies: the South Pacific. (c) Rounded knobbed bodies: the south-eastern Indian ocean. (d) Chain aggregates: the south-eastern Indian ocean.

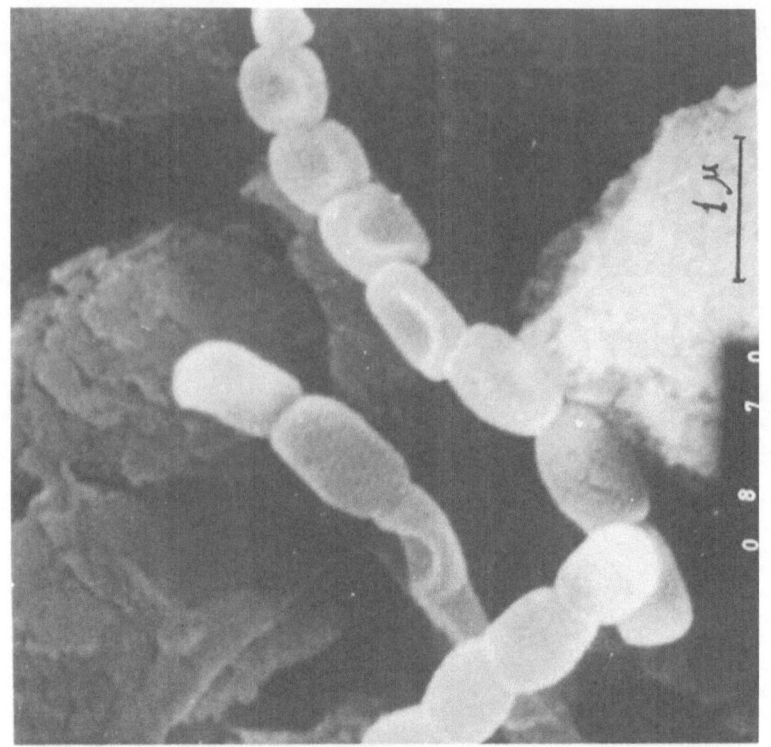

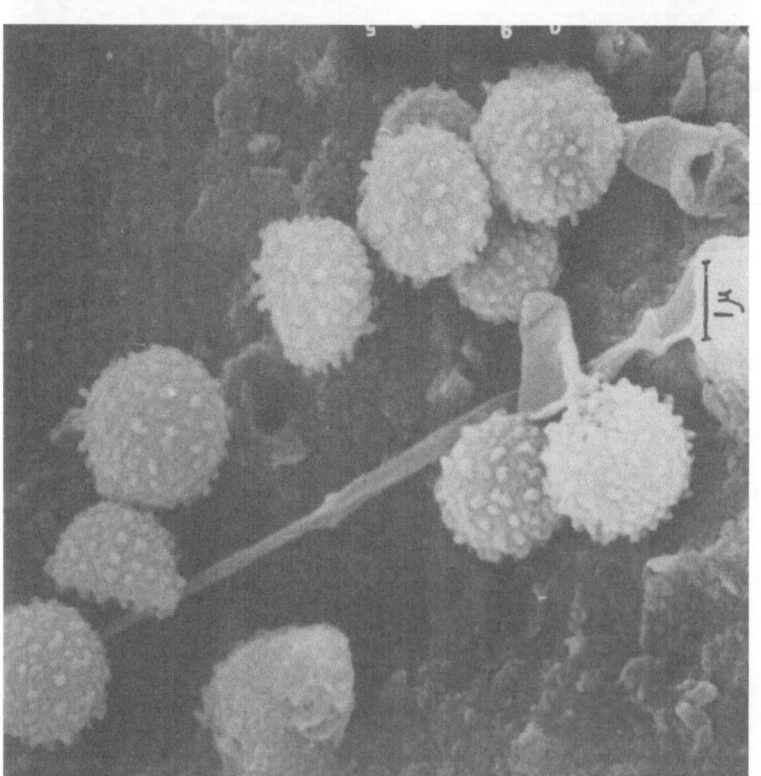

Fig. 64. Bacterial remains in manganese nodules: (a) Cocci-like formations: the Pacific radiolarian zone.
(b) Rounded mammiliferous bodies: the South Pacific.
(c) Rounded knobbed bodies: the south-eastern Indian ocean.
(d) Chain aggregates: the south-eastern Indian ocean.

GENESIS OF SULPHIDE MINERALS AND NATIVE METALS IN MANGANESE NODULES

The occurrence of sulphide minerals, native metals, in particular aluminium and iron and intermetallic compounds in pelagic nodules under extremely oxic conditions is a paradox. Nevertheless, these findings are not isolated (Müller, 1979; Os.wald, 1983; Baturin and Dubinchuk, 1983a, 1984b, c; Baturin et al., 1984; Yushko-Zakharova et al., 1984) and are mainly limited by the availability of highly sensitive instruments. Manganese nodules are only one of the new natural formations that turned out to contain native metals under perverse conditions from the classical standpoint of geochemistry. Earlier, native metals were recoverd in the lunar samples, various magmatic and hydrothermal rocks, oceanic sediments, under conditions far from suitable for their formation (Oleinikov et al., 1978; Siesser, 1978; Ashikhmina et al., 1979, 1981; Novgorodova, 1979, 1983; Shterenberg et al., 1980, 1981; Maury et al., 1980; Lazur et al., 1984).

The origin of the minerals considered in nodules might be cosmogenic, magmatogenic, volcanic-hydrothermal, terrigenous, diagenetic and biogenic.

The cosmogenic material comes to the Earth's surface in various forms, particularly as magnetic spherules, discovered in pelagic sediments and manganese nodules as early as in 1884 by Murray and Renard and then by many others (Finkelman, 1970; Jebwab, 1975; Müller, 1979). Wüstite and taenite were also detected in magnetic spherules (Finkelman, 1970) which seems to testify to their cosmic nature. However, in manganese nodules, wüstite and taenite were detected not in magnetic spherules themselves but were dispersed in hydrooxide matter, the former as a finely-dispersed inclusions, the latter as microgranules of indefinit habit.

In oceanic basaltoids, dispersed sulphide mineralization was recorded as the finest inclusions, rarely as veinlets, of chalcopyrite, bornite and pyrite (Vakhrushev and Prokoptsev, 1969; Frolova et al., 1979). However, manganese nodule fields are not confined to basaltoid outcrops.

The supposition on volcanogenic-hydrothermal nature of the minerals considered does not seem to be true either, as nodule fields are usually at a great distance from active volcanic sources and active tectonic rift zones. Therefore, a finely-dispersed volcanic and hydrothermal material should have resisted a long exposure to an oxidizing sea environment before it comes to the nodule field. Hydrothermal sulphides cannot pass this test. In the areas of recent hydrothermal activity, sulphide minerals are preserved at the bottom and in near-bottom suspension exclusively in the direct vicinity of hydrothermal springs and then they change to hydrooxides as they move off only at several meters or tens of meters (Hekinian et al., 1980, 1983).

The idea of the terrigenous nature of these minerals is also problematic as they occur neither in river nor in eolian suspensions. Ostwald (1983), who studied manganese nodules of the South Pacific, supposed that copper and nickel sulphides are formed in nodules due to "local enrichment by these elements above that which could be accomodated in the todorokite structure. The excess metals were trapped, however, by the fortuitous presence of sulphur species, resulting in the development of fine inclusions of sulphides within manganese oxide matrix". However, such mechanism seems dubious since Chukrov et al. (1983) detected a high isomorphic capacity of manganese minerals with respect to nickel. So, the occurrence of these metals in nodules seems to be better explained by the biochemical mechanism. This supposition is testified to by the data on active participation of organic components in oceanic geochemistry of manganese, copper, nickel and other metals.

Detailed study of the suspended matter from the World Ocean pelagic zones revealed authigenic ultramicroparticles of native sulphur, pyrite, covellite, and sphalerite; their genesis can be explained only by biochemical activity of organic matter and low oxidation rate of these formations which are associated with organic aggregates (Jedwab, 1979, 1980). It is not clear, yet, whether these particles can reach the bottom and be preserved in sediments, as their oxidation must be already commencing in the water column. However, the flux of organic matter to the bottom surface is intensive enough to create at least a local and temporary reducing microenvironment (Romankevich, 1977; Bishop et al., 1977a, b, 1980; Suess, 1980; Reimers and Suess, 1982; Baturin, 1983, 1984; Lisitsin, 1983; Savenko, 1984).

This supposition is supported by denitrification in oxidized oceanic sediments (Wilson, 1978) and formation of authigenic native sulphur and gypsum in nodules (Kirchner et al., 1978; Xavier and Klemm, 1979).

This very flux seems to support the existence of benthic microfauna on manganese nodules (Greenslate, 1974; Wendt, 1974; Dugolinsky, 1977; Thiel, 1978; Von Stackelberg, 1985) and a considerable population of viable heterotrophic microorganisms, among which specific iron bacteria, and even anaerobic organisms were detected, the latter being able to provide for sulphate reduction under corresponding conditions (Ehrlich, 1972, 1974; Ehrlich et al., 1972; Schuett and Ottow, 1977; Marshall, 1979; Schuett, 1982).

The appearance of sulphide mineral phases in nodules possibly results from the life activity processes of these assemblages.

To verify this concept, complex biochemical analyses of both nodules and sulphide minerals proper are necessary, including the determination of isotopic composition of sulphur that preserves the biological marker in itself.

It is not clear whether the occurrence of native metals such as aluminium and iron in nodules may also be explained by biogeochemical processes. In any case, it is only a part of the general problem of the genesis of these minerals in various sedimentary formations, both in the oceans and on land. To solve it, new theoretical and experimental approaches should be conceived.

CONCLUSIONS

The present work was based partly on published and partly on original data. Its aim was to review the recent progress on the knowledge of nodule formation in the ocean. It is our hope that we have coped with the problem - the last most extensive review of this kind was published in 1977 by Glasby.

The monograph is a collection of the data available on the marine geochemistry of dispersed manganese and on the geochemistry of manganese nodules; in particular the behaviour of some rare elements in them. The latter only became possible thanks to the recent progress in marine geology and geochemistry: for the last ten years the collection and processing of information has entered a quantitatively and qualitatively new stage.

Like many such comprehensive reviews, the present work dwells upon a wide range of problems; however, not all of them are considered equally. Some problems are traditional in the general topic of nodules, some are considered for the first time and deserve further analysis.

We see the general pattern of manganese nodule formation in the ocean as follows.

Manganese and other metals are supplied to the ocean from various sources and in various forms: mainly solid species (terrigenous and volcanic material) and as mainly dissolved species (hydrothermal material). The former are deposited in the vicinity of their source, the latter may be transported to a greater distance which is limited by the high reactivity of manganese. The third, though internal, source of ore matter supply in the ocean is a diagenetic flux from hemipelagic sediments.

Despite its short residence time in oceanic water (several hundred years), manganese from these source is well mixed by currents, advection and biological activity, and is deprived of any markers to indicate its initial origin.

Manganese and associated metals are precipitated out of sea water as the result of oxidation, coagulation, sorption and as a component of biogenic detritus. One of the possible additional modes of manganese extraction out of sea water is its sorption by the surface of bottom sediments.

In the surficial sediments and in the bottom nepheloidal layer, the initial stage of diagenetic transformation of sedimentary matter occurs which leads to a partial dissolution of biogenic components, and enrichment of residual fractions by base and other metals. At this stage, numerous ultramicroscopic accretions of manganese oxides are formed as a result of colloidal chemical processes. Some of these accretions may serve as nuclei of micronodules along with other sedimentary materials. A diffusive flux of metals from the underlying sediments may supply additional reserves of metals for such accumulation to the extent depending on the given facial environment.

In general, the accretion rates of nodules are 2-3 orders lower than sedimentation rates. Nodules rest on the bottom surface either due to the plastic properties of sediments, or due to bioturbation producing a "sieve" effect: surficial sediments and also micronodules sieve through the holes made by bottom fauna, whereas large nodules remain on the surface. Lithification of underlying sediments (for instance, during long intervals of non-deposition) results in the subsequent burial of nodules under new portions of sedimentary material.

To specify and clarify this picture further investigations are required.

Among the most pressing problems of geochemistry of manganese and manganese nodule-ore formation in the ocean are the following:

The share of various sources in the total ore matter supply to the pelagic zone of the ocean. In particular, it is important to determine more precisely the possible share of hydrothermal manganese in pelagic nodules. Therefore it is necessary to improve the models of identification of the sources of ore matter in the ocean (Bonatti, 1972; Dymond, 1981, etc.), to define to what degree they are informative, and the range of their application.

The role played by the diffusion of metals from hemipelagic sediments in their total balance in pelagic sedimentation and ore formation. Judging by its scale, this process does not seem to fall behind the terrigenous and hydrothermal supply of metals.

The global balance of metals in the ocean and the relative share of ore formation in it. The available maximum estimates of terrigenous, hydrothermal and diffusive supply of manganese each exceed the corresponding absolute amount of manganese supplied to sediments, whereas the share of the manganese expended on nodule formation varies from several hundredths of a percent (Baturin, 1983) to a few percent (Volkov, 1981).

Now, oxic oceanic diagenesis is believed to play an important role in nodule formation; however no plausible interpretation of its mechanism exists and the problem requires further lithological, geochemical and thermodynamic basement. Thus, the question of the proportions of hydrogenous and diagenetic supply of metals to those nodules, which earlier were considered exclusively "sedimentary" ones, is still open.

The high relative concentration of mobile forms of hydrolizate metals (aluminium titanium, zirconium) in manganese nodules causes one to raise the question on the manner of their transformation in the process of nodule formation, the sources and forms of their supply and on their possible application as geochemical indices of nodule genesis.

The behaviour of such elements as selenium, tellurium, bismuth, tantalum, platinum and metals of the platinum group in nodules and in the marine environment as a whole is poorly investigated. Of particular concern is the high platinum to palladium ratio in nodules.

The origin of sulphides and native metals in nodules goes beyond the theme of manganese nodules and is one of the general problems of sedimentary geochemistry. Any success achieved in deciphering their occurrence in nodules will contribute to geochemistry as a whole.

The biogenic aspect of nodule formation has acquired a special meaning since new inferences were made on the dynamics of organic matter supply to the bottom, due to the sedimentary trap technique. Suspensions rich in organic matter reach the bottom surface, the nodule surface included. Utilization of this organic matter is at a maximum in pelagic sediments which contain a minimum of organic matter. Microorganisms participating in the process affect it geochemically, including mobilization and demobilization of manganese, iron and other metals. Despite a great number of works devoted to the problem in question, many of its aspects are not clear and demand further detailed field, experimental and theoretical researches.

The progress in determining the age and accretion rates of nodules is impressive.

However, those results are overshadowed by the evident violation of some basic prerequisites of radiometric dating methods such as the participation of diagenetic flux of metals in the formation of nodules and their postdepositional alterations implying further surch for new independent methods of dating and interpreting of results.

This list of acute problems for the geochemistry of nodules in the ocean is not complete; however, it illustrates that a wide sphere of activity is open for modern and future researches.

REFERENCES

Adams, F., Von Craen, M., Van Espen, P., and Andreuzzi, D., 1980, 'The elemental composition of atmospheric aerosol particles at Chacoltaya, Bolivia', Atm. Envir., 14, N 8, 879-893.

Addy, S.K., 1979, 'Rare earth elements patterns in manganese nodules and micronodules from northwest Atlantic', Geochim, et Cosmochim. Acta, 43, N 7. 1105-1115.

Addy, S.K., Presley, B.J. and Ewing, M., 1976, 'Distribution of manganese, iron and other trace metals in a core from northeast Atlantic', J. Sed. Petrol., 46, N 4, 813-818.

Afanasyev, Yu. A., Eremin, V.P., Osadchuk, T.M., and Ryabinin, A.I., 1979, 'Analysis of pressure effect upon solubility of iron (III) and manganese (IV) hydrooxides', Zhurnal fiz. khimii, 53, N 8, 1960-1962 (in Russian).

Afanasyev, Yu.A., Eremin, V.P. and Ryabinin, A.I., 1982, 'On the hydrostatic pressure effect upon solubility of major nodule components', Okeanologia, 22, N 3, 420-422 (in Russian).

Agadi, V.V., Bhosle, N.B. and Untawabe, A.G., 1978, 'Metal concentration in some seaweeds of Goa (India)', Botan. Marina, 21, 247-250.

Agiorgitis, G. and Gundlach, H., 1978, 'Platin-Gehalte in Tiefsee-Manganknollen', Naturwiss., B.65, N 10, 534.

Ahrens, L.H., Willis, J.P. and Oosthuizen, C.O., 1967, 'Further observations on the composition of manganese nodules, with particular reference to some of the rare elements', Geochim. et cosmochim. acta, 31, N 11, 2169-2180.

Alekin, O.A. and Brazhnikova, L.V., 1964, 'Discharge of dissolved matter from the USSR territory', Moscow, Nauka, 144 pp. (in Russian).

Alekseev, V.V. and Lisitsina, K.N., 1974, 'Discharge of suspended load', in The world water balance and water resources of the Earth, Leningrad, Gidrometeoizdat, pp. 510-517 (in Russian).

Allen, J.A., 1960, 'Manganese deposition on the shells of living mollusks', Nature, 185, N 4709, 336-337.

Amari, S. and Ozima, M., 1985, 'Search for the origin of exotic helium in deep-sea sediments', Nature, 317, N 6037, 520-522.

Amdurer, M., Adler, D. and Santschi. P.H., 1983, 'Studies of the chemical forms of trace elements in sea water using radiotracers', in C.S. Wong et al. (eds.) Trace metals in sea water, N.Y. Plenum, pp. 537-562.

Amiruddin, A. and Ehmann, W.D., 1962, 'Tungsten abundances in meteoritic and terrestrial materials', Geochim. et Cosmochim. acta, 26, 1011-1022.

Anders, E. and Ebihara, M., 1982, 'Solar-system abundances of the elements', Geochim. et Cosmochim. Acta, 46, N 11, 2363-2380.

Andersen, M.E. and Macdougall, J.D., 1977, 'Accumulation rates of manganese nodules and sediments: an alpha track method', Geophys. Res. Lett., 4, 351-353.

Andreae, M.O., 1982, 'Marine aerosol chemistry at Cape Grim, Tasmania, and Townsville, Queensland', J.Geophys. Res., 87, N c11, 8875-8885.

Andreev, S.I., Kazmin, Yu.B., Egiazarov, B.Kh., Korsakov, O.D., Lygina, T.I. and Mirchink, I.M., 1984, 'Distribution of manganese nodules' in Manganese nodules of the World Ocean, Moscow, Nedra, pp. 18-61 (in Russian).

Andrews, J.E., 1972, 'Distribution of manganese nodules in the Hawaiian archipelago. Manganese nodule deposits', in Pacific Symposium/Workshop Proc.. Honolulu. Hawaii, pp. 61-65.

Andruschenko, P.F., 1976, 'Mineral composition and texture of manganese nodules ,
    in Manganese nodules of the Pacific, Moscow, Nauka, pp. 123-167 (in Russian).
Andruschenko, P.F., Gradusov, B.P., Eroschev-Shak, V.A., Yashina, R.S. and
    Borisovskiy, S.E., 1975, 'Composition and structure of metamorphozed
    manganese nodules, of newly formed vein manganese hydrooxides and of
    pelagic host sediments in the South Pacific', Izv. AN SSSR, ser. geol., N 1,
    91-111 (in Russian).
Andruschenko, P.F. and Skornyakova, N.S., 1967, 'The Pacific manganese nodules:
    composition, structure and formation', in Manganese ore fields in the USSR,
    Moscow, Nauka, pp. 94-116 (in Russian).
Andruschenko, P.F. and Skornyakova, N.S., 1969, 'Texture and mineral composition
    of manganese nodules in the South Pacific', Okeanologia, 9, N 2, 282-294 (in
    Russian).
Anikeeva, L.I., Andreev, S.I., Kazmin, Yu.B., Korsakov, O.D., Egiazarov, B.Kh.,
    Lygina, T.I., Mirchink, I.M., 1984, 'Morphology of manganese nodules', in
    Manganese nodules of the World Ocean, Moscow, Nedra, pp. 62-104 (in
    Russian).
Aplin, A.C., 1985, 'Rare earth element geochemistry of Central Pacific
    encrustations', Earth and Planet. Sci. Lett., 71, N 1, 13-22.
Aplin, A.C. and Cronan, D.S., 1985a, 'Ferromanganese oxide deposits from the
    Central Pacific Ocean. 1. Encrustations from the Line Island Archipelago',
    Geochim. et Cosmochim. Acta, 49, N 2, 427-436.
Aplin, A.C. and Cronan, D.S., 1985b, 'Ferromanganese oxide deposits from the
    Central Pacific Ocean. 2. Nodules and associated sediments', Geochim. et
    Cosmochim. Acta, 49, N 2, 437-451.
Arcuri, E.J. and Ehrlich, H.L., 1979, 'Cytochrome involvment in Mn (II) oxidation
    by two marine bacteria', Appl. Microbiol., 37, N 5, 916-923.
Arcuri, E.J. and Ehrlich, H.L., 1980, 'Electron transfer coupled to Mn(II) oxidation
    in two deep-sea Pacific Ocean isolates', in Biogeochem. Ancient and Modern
    Envir. Berlin, Springer, pp. 339-344.
Arrhenius, G., 1963, 'Pelagic sediments' in The sea, 3, N.Y., Interscience, 655-
    727.
Arrhenius, G., 1967, 'Deep-sea sedimentation: a critical revue of U.S. work',
    Trans. Am. Geophys. Un., 48, 604-631.
Artemchuk, N.Y., 1981, Microflora of the seas of the USSR, Moscow, Nauka,
    192 pp. (in Russian).
Ashikhmina, N.A., Bogatikov, O.A., Gorshkov, A.I. et al., 1979, 'First findings
    of metallic aluminium particles in the lunar ground', Dokl. AN SSSR, 246, N 4,
    958-961 (in Russian).
Ashikhmina, N.A., Bogatikov, O.A., Stepanchikov, V.A., Frikh-Khar, D.I., 1981,
    'On the findings of zincous copper in the lunar ground', Dokl. AN SSSR, 256,
    N 5, 1212-1214 (in Russian).
Aumento, F., 1969, 'The Mid-Atlantic Ridge near 45°N; V: Fission track and
    ferromanganese chronology', Can. J. Earth Sci., 6, 1431-1440.
Bacon, M.P. and Anderson, R.F., 1982, 'Distribution of thorium isotopes between
    dissolved and particulate forms in the deep sea', J. Geophys. Res., 87,
    2045-2056.
Bacon, M.P., Brewer, P.G., Spencer, D.W., Murray, J.W., Goddard, J., 1980,
    'Lead-210, polonium-210, manganese and iron in the Cariaco trench',
    Deep-Sea Res., 27, N 2A, 119-135.
Bada, J.L., 1972, 'The dating of fossil bones using the racemization of isoleucine',
    Earth and Planet. Sci. Lett., 15, 223-231.
Bada, J.L., Kvenvolden, K.A. and Peterson, E., 1973, 'Racemization of animo
    acids in bones', Nature, 245, 308-310.
Baker, E.T., Feely, R.A. and Takahashi, K., 1979, 'Chemical composition, size
    distribution and particle morphology of suspended particulate matter at DOMES
    sites A, B and C: relationships with local sediment composition', in
    J.L. Bischoff and D.Z. Piper (eds.), Marine Geology and Oceanography of the
    Pacific Manganese Nodule Province, New York, Plenum, pp. 163-201.

Balashov, Yu.A., 1976, 'Geochemistry of rare-earth elements', Moscow, Nauka, 267 pp. (in Russian).

Balashova, V.V., 1974, **Micoplasms and iron-bacteria**. Moscow, Nauka, 65 pp. (in Russian).

Balistrieri, L.S., Brewer, P.G. and Murray, J.W., 1981, 'Scavenging residence times of trace metals and surface chemistry of sinking particles in the deep ocean', Deep Ses Res., 28, 101-121.

Balistrieri, L.S. and Murray, J.W., 1983, 'Apparent equilibrium constants for metal-solid interactions in the marine environment', Geochim. et Cosmochim. Acta, 47, 1091-1098.

Balistrieri, L.S. and Murray, J.W., 1984, 'Marine scavenging: trace metal adsorption by interfacial sediment from MANOP Site H', Geochim. et Cosmochim. Acta, 48, 921-929.

Baragar, W.R.A., Plant, A.G., Pringle, G.J. and Schou, M., 1977, 'The petrology of alteration in three discrete flow units of sites 332 and 335', in Init. Rep. DSDP, 37, Wash. D.C., pp. 811-819.

Barker, J.L. and Anders, E., 1968, 'Accretion rate of cosmic matter from iridium and osmium contents of deep-sea sediments', Geochim. et Cosmochim. Acta, 32, N 6, 627-645.

Barnes, S.S., 1967, 'Minor elements composition of ferromanganese nodules', Science, 157, N 3784, 63-65.

Barnes, S.S., 'Dymond J.R. 1967, Rates of accumulation of ferromanganese nodules', Nature, 213, N 5082, 1218-1219.

Baseline studies of pollutants in the marine environment, 1972, E. Goldberg (ed.), Background papers for a workshop at Brook-Haven Nat. Labor., 799 pp.

Baturin, G.N., 1975, 'Uranium in recent marine sedimentation', Moscow, Atomizdat, 152 pp. (in Russian).

Baturin, G.N., 1978, 'Phosphorites on the sea floor', Moscow, Nauka, 230 pp. (in Russian).

Baturin, G.N., 1983, 'Manganese balance and ore formation in the ocean', Dokl. AN SSSR, 268, N 1, 214-217 (in Russian).

Baturin, G.N., 1984, 'Manganese nodules in the ocean and biota', Priroda, N 9, 19-23 (in Russian).

Baturin, G.N., 1986a, 'Rare ore minerals', in Manganese nodules of the Central Pacific, Moscow, Nauka, pp. 192-200 (in Russian).

Baturin, G.N., 1986b, 'Rare metals in manganese nodules', ibid., pp. 201-210 (in Russian).

Baturin, G.N., 1986c, 'Age and accretion rates of manganese nodules', ibid., pp. 284-296 (in Russian).

Baturin, G.N., 1986d, 'Microelements the Pacific manganese nodules', Geokhimia, N 4, 489-501 (in Russian).

Baturin, G.N., Bekov, G.I., Egorov, A.S., Kurskii, A.N., Puchkova, T.V. and Radaev, V.N., 1984a, 'Tracing of platinum metals and gold in manganese nodules by the methods of atomic absorption spectroscopy and laser photoionization in Geology of oceans and seas, Abstracts of the reports to the 6th All-Union seminar on marine geology, Moscow, Nauka, pp. 13-15 (in Russian).

Baturin, G.N., Bekov, G.I., Egorov, A.S., Zubov, I.V., Kurskii, A.N., Pakhomov, D.Yu. and Radaev, V.N., 1986a, 'Determination of rhodium in marine samples by the method of laser photoionization analytical spectroscopy', in Geology of oceans and seas, Abstracts of the reports to the 7th All-Union seminar on marine geology, Moscow, Nauka (in Russian).

Baturin, G.N., Boiko, T.F., Miller, A.D., 1985a, 'Rhenium in manganese nodules and oceanic metalliferous sediments', Dokl. AN SSSR, 280, N 1, 211-215 (in Russian).

Baturin, G.N. and Dubinchuk, V.T., 1979, 'Microstructures of oceanic phosphorites', Moscow, Nauka, 200 pp. (in Russian).

Baturin, G.N. and Dubinchuk, V.T., 1983a, 'Sulphide minerals in the Pacific manganese nodules', Dokl. AN SSSR, 272, N 4, 950-953 (in Russian).

Baturin, G.N. and Dubinchuk, V.T., 1983b, 'Biomorphic ultramicroscopic structures in pelagic manganese nodules', Okeanologia, 23, N 6, 997-1000 (in Russian).

Baturin, G.N. and Dubinchuk, V.T., 1984a, 'Manganosite and wüstite in the
    Pacific manganese nodules', Okeanologia, 24, N 2, 311-315 (in Russian).
Baturin, G.N. and Dubinchuk, V.T., 1984b, 'Sulphide mineralization in pelagic
    manganese nodules', Lithol. i Polezn. iskop., N 2, 53-61 (in Russian).
Baturin, G.N. and Dubinchuk, V.T., 1984c, 'Nickel minerals in the Pacific
    manganese nodules', Dokl. AN SSSR, 278, N 4, 958-961 (in Russian).
Baturin, G.N. and Dubinchuk, V.T., 1986, 'On the biogenic influence upon
    formation of microstructures of deep sea manganese nodules', Geol. zhurnal,
    46, N 1, 109-116 (in Russian).
Baturin, G.N., Dubinchuk, V.T. and Shevchenko, A.Yu., 1984b, 'On native metals
    in oceanic manganese nodules', Okeanologia, N 5, 797-781 (in Russian).
Baturin, G.N., Fisher, E.I., Fisher, V.L.; 1984c, 'On gold concentration in oceanic
    manganese nodules', Dokl. AN SSSR, 275, N 2, 421-424 (in Russian).
Baturin, G.N., Fisher, E.I., Kurskii, A.N. and Puchkova, T.V.; 1985b, 'Metals of
    platinum group in deep-sea manganese nodules , Dokl. AN SSSR, 285, N 4,
    992-996 (in Russian).
Baturin, G.N., Fisher, E.I., Kurskii, A.N., Puchkova, T.V., Serebryanyi, B.L.
    and Shwartzman, S.I., 1986b 'Gold in deep-sea manganese nodules', Geokhimia,
    N6, 751-759 (in Russian).
Baturin, G.N., Gordeev, V.V. and Kosov, A.E.; 1986c, 'Metals in interstitial
    water', in Manganese nodules of the Central Pacific, Moscow, Nauka,
    pp. 251-269 (in Russian).
Baturin, G.N. and Isaeva, A.B., 1986, 'Tungsten in oceanic manganese nodules',
    Lithol. i Polezn. iskop., N 2, 17-24 (in Russian).
Baturin, G.N., Kochenov, A.V. and Dubinchuk, V.T., 1981, 'Uranium in
    phosphorites of the Pacific sea mountains', Lithol. i Polezn. iskop., N 1, 3-10
    (in Russian).
Baturin, G.N., Kochenov, A.V. and Dubinchuk, V.T., 1986d, 'Uranium and torium
    in oceanic manganese nodules', Lithol. i Polezn. iskop., N 6, 19-27 (in Russian).
Baturin, G.N., Kochenov, A.V. and Trimonis, E.S., 1969, 'On composition and
    origin of iron-ore sediments and hot brine in the Red Sea', Okeanologia, 9,
    N 3, 442-451 (in Russian).
Baturin, G.N., Lukashin, V.N. and Shevchenko, A.Ya., 1986e, 'On manganese
    nodules in the Antarctic Pacific', Tikhookeanskaya geologia, N 5, 23-28 (in Russian).
Baturin, G.N. and Oreshkin, V.N., 1984, 'Cadmium in manganese nodules', in
    Geology of oceans and seas, Abstr. of reports to the 6th. All-Union seminar
    on marine geology, Moscow, Nauka, 3, 12-13 (in Russian).
Baturin, G.N. and Oreshkin, V.N., 1986, 'Silver in the Pacific manganese nodules',
    Dokl. AN SSSR, 289, N1, 189-192 (in Russian).
Baturin, G.N., Oreshkin, V.N., Ashchyan, T.O. and Vnukovskaya, G.L., 1986f,
    'Mercury in oceanic ores', in Geology of oceans and seas, Abstr. of reports
    to the 7th. All-Union seminar on marine geology, Moscow, Nauka, 3,
    168-169 (in Russian).
Baturin, G.N., Oreshkin, V.N. and Skornyakova, N.S.; 1986g, 'Cadmium in the
    Pacific manganese nodules', Geokhimia, N 7, 1052-1055 (in Russian).
Baturin, G.N. and Savenko, V.S., 1984, 'Manganese in pore water of marine and
    oceanic sediments', Litol. i Polezn. iskop., N 1, 89-104 (in Russian).
Bazilevskaya, E.S., 1976, 'Chemical-mineralogical analysis of manganese ore',
    Trudy Geol. inst. AN SSSR, Moscow, is. 287, pp. 71-86 (in Russian).
Bazilevskaya, E.S., 1981, 'Ferromanganese ore manifestation in oceanic bottom
    sediments (data of deep-sea drilling)', Litol. i Polezn. iskop., N 5, 38-50 (in
    Russian).
Bazilevskaya, E.S., 1985, 'Role of diagenesis in formation of manganese nodules in
    the Clarion-Clipperton ore provinces', Tikhookeanskaya geologia, N 6, 60-70
    (in Russian).
Bazilevskaya, E.S., Iljicheva, L.V. and Stepanetz, M.I., 1979, 'On the formation
    mode of oceanic manganese nodules', Litol. i Polezn. iskop., N 4, 85-93 (in
    (Russian).
Beauford, W.J., Barber, J. and Barringer, A.R., 1975, 'Heavy metals release from
    plants into the atmosphere', Nature, 256, N 5512, 35-37.

Beauford, W., Barber, J. and Barringer, A.R., 1977, 'Release of particles containing metals from vegetation into the atmosphere', Science, 195, N 4278, 571-573.

Beiersdorf, H., Gundlach, H., Heye, D., Marchig, V., Meyer, H. and Schnier, C., 1982, 'Heated bottom water and associated Mn-Fe crusts from the Clarion fracture zone southeast of Hawaii', in K.A. Fanning and F.T. Manheim (eds.), The dynamic environment of the ocean floor. Toronto, Lexington Books, pp. 359-368.

Bekov, G.I., Letokhov, V.S., Radaev, V.N., Baturin, G.N., Yegorov, A.S., Kursky, A.N. and Narseyev, V.A., 1984, 'Ruthenium in the ocean'. Nature, 312, N 5996, 748-750.

Belov, N.A., Kulikov, N.N., Lapina, N.I. and Semenov, Yu.P., 1968, 'Distribution of iron, manganese and carbonates in deposits of the Arctic seas', Trudy Ark. i Antark. nauch.-issled.inst., N 285, 67-73 (in Russian).

Belov, N.A. and Lapina, N.I., 1961, 'Bottom sediments of the Arctic Basin', Leningrad, Morskoi transport, 152 pp. (in Russian).

Belyaev, G.M. and Glikman, L.S., 1965, 'Findings of abundance of sharks' teeth at the bottom of the Pacific and Indian oceans'. In Problems of Cenozoic stratigraphy, International geological congress, session 22, Dokl. sov. geologov, Moscow, Nedra, pp. 74-79 (in Russian).

Belyaev, G.M. and Glikman, L.S., 1970a, 'Sharks' teeth at the Pacific bottom', Trudy inst. okeanologii AN SSSR, 88, 252-276 (in Russian).

Belyaev, G.M. and Glikman, L.S., 1970b, 'On geological age-dating of the teeth of Megaselachus megalodon shark', Trudy inst. okeanologii AN SSSR, 88, 277-280 (in Russian).

Belyaev, L.I. and Ovsyanyi, E.I., 1969, 'Analysis of microelements in atmospheric precipitations in the off-shore region in respect to some problems of chemical oceanography', Gidrokhim. materialy, 51, 3-12 (in Russian).

Belyaeva, N.V., 1973, 'Peculiarities of chemical composition of plankton foraminifera tests', Okeanologia, 13, N 2, 303-306 (in Russian).

Bemmelen, R.V. von, 1949, 'The geology of Indonesia'. The Hague, Gov. Print.Off., 1-2, 732 pp.

Bender, M.L., 1971, 'Does upward diffusion supply the excess manganese in pelagic sediments?' J. Geophys. Res., 76, N 18, 4212-4215.

Bender, M.L., 1972, 'Mechanisms of trace metal removal from the oceans', in D.R. Horn (ed.), Ferromanganese deposits on the ocean floor, Wash. D.C., Nat. Sci. Found., pp. 73-80.

Bender, M.L., 1983, 'Pore water chemistry of the mounds hydrothermal field, Galapagos spreading center: results from Glomar Challenger piston coring', J. Geophys. Res., B 88, N 2, 1049-1056.

Bender, M.L., Broecker, W., Gornitz, V., Middel, U., Kay, R., Sun, S.S. and Biscaye, P., 1971, 'Geochemistry of three cores from the East Pacific Rise', Earth and Planet. Sci. Lett., 12, 425-433.

Bender, M.L., Klinkhammer, G.P. and Spencer, D.W., 1977, 'Manganese in seawater and the marine manganese balance', Deep-Sea Res., 24, N 9, 799-812.

Bender, M.L., Ku, T.L. and Broecker, W.S., 1966, 'Manganese nodules: their evolution', Science, 151, N 3708, 325-328.

Bender, M.L., Ku, T.L. and Broecker, W.S., 1970, 'Accumulation rates of manganese in pelagic sediments and nodules', Eart and Planet. Sci. Lett., 8, N 2, 143-148.

Bender, M. and Schulz, C., 1969, 'Distribution of trace metals in cores from a traverse across the Indian ocean', Geochim. et cosmochim. Acta, 33, N 2, 292-297.

Berg, W.W. and Winchester, J.W., 1978, 'Aerosol chemistry of the marine atmosphere', in J.P. Riley and R. Chester(eds.), Chemical Oceanography, 2 ed., 7, New York, Academic, pp. 173-231.

Berger, W.H. and Killingley, J.S., 1982, 'Box cores from the equatorial Pacific: [14]C sedimentation rates and benthic mixing', Mar. Geol., 45, N 1, 99-125.

Bertine, K.K. and Goldberg, E.D., 1971, 'Fossil fuel combustion and the major sedimentary cycle', Science, 173, N 3993, 233-235.

Bewers, J.M., Sundby, B. and Yeats, P.A., 1976, 'The distribution of trace metals in the western North Atlantic off Nova Scotia', Geochim. et Cosmochim. Acta, 40, N 6, 687-696.

Bewers, J.M. and Windom, H.L., 1983, 'Intercomparison of seawater sampling devices for trace metals', in C. Wong et al. (eds.), Trace metals in sea water. N.Y., Plenum, pp. 143-154.

Bezrukov, P.L., 1960, 'Bottom sediments of the Sea of Okhotsk', Trudy Inst. Okeanologii AN SSSR, 32, 15-95 (in Russian).

Bezrukov, P.L., 1970, 'General pattern of sedimentation in the Pacific', in Sedimentation in the Pacific. The Pacific, 6, Moscow, Nauka, pp. 301-321 (in Russian).

Bezrukov, P.L., 1976, 'Manganese nodules: stratigraphic position and age', in Manganese nodules of the Pacific, Moscow, Nauka, pp. 82-90 (in Russian).

Bhandari, N., Arnold, J.R. and Parkin, D.W., 1968, 'Cosmic dust in the stratosphere', J. Geophys. Res., 73, N 5, 1837-1845.

Bhandari, N., Bhat, S.G., Krishnaswami, S. and Lal, D., 1971, 'A rapid gamma-beta coincidence technique for determination of natural radionuclides in marine deposits', Earth and Planet. Sci. Lett., 11, 121-126.

Bhat, S.G., Krishnaswami, S., Lal, D., Rama and Somayajulu, B.L.K., 1970, 'Radiometric and trace elemental studies of ferromanganese nodules', in Proc. Symp. Hydrogeochem. and Biochem., Tokyo, Sept. 1, 1. Wash., The Clark Co., 1973, pp. 443-462.

Bischoff, J.L., 1969, 'Red Sea geothermal brine deposits: their mineralogy, chemistry and genesis', in E.T. Degens and D.A. Ross (eds.), Hot brines and recent heavy metal deposits in the Red Sea, N.Y., Springer, pp. 368-401.

Bischoff, J.L. and Dickson, F.W., 1975, 'Seawater-basalt interaction at 200 °C and 500 bars: implications for origin of sea-floor heavy metal deposits and regulation of seawater chemistry', Earth and Planet. Sci. Lett., 25, N 3, 385-397.

Bischoff, J.L., Heath, G.R., and Leinen, M., 1979, 'Geochemistry of deep-sea sediments from the Pacific manganese nodule province: DOMES site A, B and C', in J.L. Bischoff and D.Z. Piper (eds.), Marine Geology and Oceanography of the Pacific manganese nodule province, N.Y.-L., Plenum, pp. 397-436.

Bischoff, J.L. and Ku, T.L., 1971, 'Pore fluids of recent marine sediments. II. Anoxing sediments of 35° to 45° N, Gibraltar to Mid Atlantic Ridge', J. Sediment. Petrol., 41, N 4, 1008-1017.

Bischoff, J.L., Piper, D.Z. and Leong, K., 1981, 'The aluminosilicate fraction of North Pacific manganese nodules', Geochim. et Cosmochim. Acta, 45, 2047-2063.

Bischoff, J.L. and Seyfried, W.E., 1977, 'Seawater as a geothermal fluid: Chemical behavior from 25° to 350° C', Proc. 2nd. Int. Symp. Water-Rock Interact, Strasbourg, Sec. 4, pp. 165-172.

Bischoff, J.L. and Seyfried, W.E., 1978, 'Hydrothermal chemistry of seawater from 25° to 350° C', Amer. J. Sci., 278, N 6, 838-860.

Bischoff, J.L. and Sayles, F.L., 1972, 'Pore fluid and mineralogical studies of recent marine sediments: Bauer Depression region of East Pacific Rise', J. Sediment. Petrol., 42, 711-724.

Bishop, J.K., Collier, R.W., Ketten, D.R. and Edmond, J.M., 1980, 'The chemistry, biology and vertical flux of particulate matter from the upper 1500 m of the Panama Basin', Deep-Sea Res., 27 A, 615-640.

Bishop, J.K.B., Edmond, J.M., Ketten, D.R., Bacon, M.P. and Silker, W.B, 1977a, 'The chemistry, biology and vertical flux of particulate matter from the upper 400 m of the equatorial Atlantic ocean', Deep-Sea Res., 24, 511-548.

Bishop, J.K., Ketten, D.R., and Edmond, J.M., 1977b, 'The chemistry, biology and vertical flux of particulate matter from the upper 400 m of Cape Basin in the southeast Atlantic Ocean', Deep-Sea Res., 25, N 12, 1121-1161.

Black, W.A.P. and Mitchell, R.L., 1952, 'Trace elements in the common brown algae and sea water', J. Mar. Biol. Assn. U.K., 30, N 3, 575-584.

Blashchishin, A.I., Mitropolskiy, A.Yu. and Shtraus, A.D., 1982, 'Microelements in Recent sediments of the Baltic Sea', Kiev, Inst. geol. nauk. AN UkSSR, preprint 82-4, 66 pp. (in Russian).

Blazhchishin, A.I., 1976, 'Major chemical components in bottom sediments', in Geology of the Baltic Sea, Vilnius, Mokslas, pp. 255-287 (in Russian).

Blazhchishin, A.I., Gudelis, V.K. and Emelyanov, E.M. et al., 1976, 'Geology of the Baltic Sea', Vilnius, Mokslas, 383 pp. (in Russian).

Blimblcombe, P. and Hunter, K.A., 1977, 'Rock volatility and aerosol composition', Nature, 265, N 5996, 761-762.

Blokh, A.M. and Kochenov, A.V., 1964, 'Microelements in bone phosphate of fossil fishes. Geology of ore deposits of rare elements', Moscow, Nedra, 108 pp. (in Russian).

Bobrov, V.A., Bolotov, V.V., Gilbert, E.N., Parkhomenko, V.S. and Shipitsin, Yu.G., 1980, 'Analysis of the element composition of lunar regolith by instrumental neutron-activation technique', in Lunar soils in the Sea of Crises, Moscow, Nauka, pp. 345-351 (in Russian).

Boiko, T.F., Baturin, G.N. and Miller, A.D., 1985, 'On rhenium content in recent sediments of the ocean; Geokhimiya, N 11, 1662-1671 (in Russian).

Bolger, G.W., Betzer, P.R. and Gordeev, V.V., 1978, 'Hydrothermally-derived manganese suspended over the Galapagos Spreading Center', Deep-Sea Res., 25, 721-733.

Bonatti, E., 1981, 'Metal deposits in the ocean lithosphere', in C. Emiliani (ed.), The Sea, The Oceanic Lithosphere, 7, N.Y., Wiley, pp. 639-686.

Bonatti, E., Kraemer, T. and Rydell, H., 1972, 'Classification and genesis of submarine iron-manganese deposits', in D.R. Horn (ed.), Ferromanganese deposits on the ocean floor, Wash., Nat. Sci. Found., pp. 149-166.

Bonatti, E. and Nayudu, Y.R., 1965, 'The origin of manganese nodules on the ocean floor', Am. J. Sci., 263, N 1, 17-39.

Bonatti, E., Fisher, D.E., Joensuu, O. and Rydell, H.S., 1971, 'Postdepositional mobility of some transition elements, phosphorus, uranium and thorium in deep sea sediments', Geochim. et Cosmochim. Acta, 35, N 2, 189-201.

Borole, D.V., Krishnaswami, S. and Somayajulu, B.L.K., 1977, 'Investigations on dissolved uranium, silicon and on particulate trace metals in estuaries', Estuarine Coastal Mar. Sci., 5, N 6, 743-754.

Bostrom, K., 1967, 'The problem of excess manganese in pelagic sediments.' In P.H. Abelson (ed.), Researches in Geochemistry, 2, N.Y., Wiley, pp. 421-452.

Bostrom, K., 1972/1973, 'The origin and fate of ferromanganese active ridge sediments', Stockholm Contrib. Geology. Acta Universitatis Stockholmiensis, Stockholm, pp. 149-242.

Bostrom, K., 1980, 'The origin of ferromanganoan active ridge sediments', in P.A. Rona and R.P. Lowell (eds.), Seafloor spreading centers: Hydrothermal systems, Stroudsburg; Dowden, Hutchinson a.Ross, pp. 288-332.

Bostrom, K., Burman, J.O., Porter, C. and Ingri, J., 1981, 'Selective removal of trace elements from the Baltic by suspended matter', Mar. Chem., 10, N 4, 335-354.

Bostrom, K., Joensuu, O., Valdes, S. and Riera, M., 1972, 'Geochemical history of South Atlantic Ocean sediments since Late Cretaceous', Mar. Geol., 12, N 2, 85-121.

Bostrom, K., Kraemer, T. and Gartner, S., 1974, 'Provenance and accumulation rates of opaline silica, Al, Ti, Fe, Mn, Cu, Ni and Co in Pacific pelagic sediments. Chemical Geology', 11, N 2, 123-148.

Bostrom, K. and Peterson, M.N.A., 1966, 'Precipitates from hydrothermal exhalations on the East Pacific Rise', Econ. Geol., 61, N 7, 1258-1265.

Bostrom, K. and Peterson, M.N.A., 1969, 'The origin of aluminium-poore ferro-manganoan sediments in areas of high heat flow on the East Pacific Rise', Mar. Geol., 7, N 5, 427-448.

Bostrom, K., Wiborg, L. and Ingri, J., 1982, 'Geochemistry and origin of ferro-manganese concretions in the Gulf of Bothnia', Mar. Geol., 50, 1-24.

Boothe, P.N. and Knauer, G.A., 1972, 'The possible importance of fecal material in the biological amplification of trace and heavy metals', Limnol. Oceanogr., 17, N 2, 262-274.

Boudreau, B.P. and Scott, M.R., 1978, 'A model for the diffusion-controlled growth of deep-sea manganese nodules', Amer. J. Sci., 278, 903-929.

Boulad, A.P., Condomines, M., Bernat, M., Michard, G. and Allegre, C.J., 1975, 'Vitesse d'accrétion des nodules de manganèse des fonds océaniques, C. R. Acad. Sci. Paris, 280, Ser. D., N 21, 2425-2428.

Boulègue, J., Renard, D., Michard, G., Boulad, A.P. and Chantret, F., 1978, 'Manganese dioxide concretions on granite outcrops in an intertidal area (Cove of Belmont, France). II. Rapid manganese dioxide formation during the mixing of freshwater and seawater', Chem. Geol., 23, N 1, 41-63.

Bourne, D.W. and Heezen, B.C., 1965, 'A wandering Enteropneust from the abyssal Pacific, and the distribution of "spiral" tracks on the sea floor', Science, N 3692, 150, 60-63.

Boutron, C., 1979, 'Trace element content of Greenland snows along an eastwest transect', Geochim. et Cosmochim. Acta, 43, N 8, 1253-1258.

Boutron, C. and Lorius, C., 1979, 'Trace metals in Antarctic snow since 1914', Nature, 277, N 5697, 551-554.

Bowen, H.J.M., 1966, Trace elements in biochemistry', London, Academic Press, 367 pp.

Bowen, H., 1979, 'Environmental chemistry of the elements', London e.a., Academic Press, 333 pp.

Boyle, E.A., 1983, 'Manganese carbonate overgrowths on foraminifera tests', Geochim. et Cosmochim. Acta, 47, N 10, 1815-1819.

Brewer, P.G., Densmore, C.D., Munns, R. and Stanley, R.J., 1969, 'Hydrography of the Red Sea brines', in E.T. Degens and D.A. Ross (eds.), Hot brines and recent heavy metal deposits in the Red Sea, N.Y., Springer, pp. 138-147.

Brewer, P.G., Nozaki, Y., Spencer, D.W. and Fleer, A.P., 1980, 'Sediment trap experiments in the deep North Atlantic: Isotopic and elemental fluxes', J. Mar. Res., 38, N 4, 703-728.

Brewer, P.G. and Spencer, D.W., 1969, 'A note on the chemical composition of the Red Sea brines', in E.T. Degens and D.A. Ross (eds.), Hot brines and recent heavy metal deposits in the Red Sea, N.Y., Springer, pp. 174-179.

Brewer, P.G. and Spencer, D.W., 1974, 'Distribution of some trace elements in Black Sea and their flux between dissolved and particulate phases', in The Black Sea - geology, chemistry and biology, Tulsa, Oklahoma, Amer. Ass. Petrol. Geol., pp. 137-143.

Brewer, P.G., Wilson, R.S., Murray, J.W., Munns, R.G. and Densmore, C.D., 1971, 'Hydrographic observations on the Red Sea brines indicate a marked increase in temperature', Nature, 231, N 5297, 32-38.

Bricker, O., 1965, 'Some stability relations in the system $Mn-O_2-H_2O$ at 25° and one atmosphere total pressure', Amer. Mineralogist, 50, N 9, 1296-1354.

Broecker, W., 1974, 'The great manganese nodule mystery', in Chemical Oceanography, N.Y., Harcourt Brace Jovanovich, pp. 89-113.

Bronshten, V.A., 1975, 'Nature and origin of meteorite bodies', in Problems of origin of solar system bodies, Moscow-Leningrad, Izd. AN SSSR, pp. 265-301 (in Russian).

Brooks, P.P., Kaplan, I.R. and Peterson, M.N.A., 1969, 'Trace element composition of Red Sea geothermal brine and interstitial water', in E. Degens and D. Ross (eds.), Hot brines and Recent heavy metal deposits in the Red Sea, N.Y., Springer, pp. 180-203.

Brooks, R.R. and Rumsby, M.G., 1965, 'The biochemistry of trace element uptake by some New Zealand bivalves', Limnol. Oceanogr., 10, 521-527.

Brooks, R.R. and Rumsey, D., 1974, 'Heavy metals in some New Zealand commercial seafishes', N.Z.J. Mar. Freshwat. Res., 8, 155-166.

Brownlow, A.E., Hunter, W. and Parkin, D.W., 1966, 'Cosmic spherules in a Pacific core', Geophys. J. Roy. Astr. Soc., 12, 1-12.

Bruevich, S.V. and Vinogradova, E.G., 1949, 'Sedimentation in the Caspian Sea', Tr. Inst. Okeanologii AN SSSR, 3, 119-156 (in Russian).

Bruland, K.W., 1983, 'Trace elements in sea-water', in Chemical Oceanography, Vol. 8. London, Academic Press, pp. 157-220.

Bruland, K.W. and Franks, R.P., 1983, 'Mn, Ni, Cu, Zn and Cd in the western North Atlantic', in C.S. Wong et al. (eds.), Trace metals in sea water, N.Y., Plenum, pp. 395-414.

Brumsack, H.J. and Gieskes, J.M., 1983, 'Interstitial water trace-metal chemistry of laminated sediments from the Gulf of California, Mexico', Mar. Chem., 14, N 1, 89-106.

Bryan, G.W., 1973, 'The occurrence and seasonal variation of trace metals in the scallops Pecten maximums (L.) and Chlamys orepcularis (L.)', J. Mar. Biol. Assn. U.K., 53, 145-166.

Bryan, G.W., 1976, 'Heavy metal contamination in the sea', in R.J. Johnston (ed.), Marine pollution, L.: Acad. Press.

Bryan, G.W. and Hummerstone, L.G.. 1973, 'Brown seaweed as an indicator of heavy metals in estuaries in south west England', J. Mar. Biol. Assn. U.K., 53, 705-720.

Bryan, G.W. and Hummerstone, L.G., 1977, 'Indicators of heavy metal contamination in the Loo estuary (Cornwall) with particulate regard to silver and lead', J. Mar. Biol. Assn. U.K., 57, 75-92.

Bryan, G.W. and Ward, E., 1965, 'The absorption and loss of radioactive and non-radioactive manganese by the lobster, Homarus vulgaris', J. Mar. Biol. Assn. U.K., 45, 65-95.

Buat-Menard, P. and Arnold, M., 1978, 'The heavy metal chemistry of atmospheric particulate matter emitted by Mount Etna volcano', Geophys. Res. Lett., 5, N 4, 245-248.

Buat-Menard, P. and Chesselet, R., 1979, 'Variable influence of the atmospheric flux on trace metal chemistry of suspended matter', Earth and Planet. Sci. Lett., 42, N 3, 399-411.

Buddhue, J.D., 1950, 'Meteoritic dust', Univ. New Mexico. Publ. Meteoritics, N 2, 102 pp.

Bumbu, Y.V., 1976, Microelements in phytoplancton life, Kishinev, Shteenza, 115 pp. (in Russian).

Burdige, D.J. and Gieskes, J.M., 1983, 'A pore water/ solid phase diagenetic model for manganese in marine sediments', Am. J. Sci., 283, 29-47.

Burnett, B.R. and Nielson, K.H., 1981, 'Organic films and microorganisms associated with manganese nodules', Deep-Sea Res., 28, N 6A, 637-645.

Burnett, B.R. and Nielson, K.H., 1983, 'Energy dispersive X-ray analysis of the surface of a deep-sea ferromanganese nodule', Mar. Geol., 53, N 4, 313-329.

Burnett, W.C. and Morgenstein, M., 1976, 'Growth rates of Pacific manganese nodules as deduced by uranium-series and hydration-rind dating techniques', Earth and Planet. Sci. Lett., 33, N 2, 208-218.

Burnett, W.C. and Piper, D.Z., 1977, 'Rapidly-formed ferromanganese deposit from the Hess Deep, eastern Pacific', Nature, 265, N 5595, 596-600.

Burns, R.G. and Burns, V.M., 1977, 'Mineralogy of manganese nodules', in G.P. Glasby (ed.), Marine manganese deposits, N.Y., Elsevier, pp. 185-248.

Burns, V.M. and Burns, R.G., 1978a, 'Post-depositional metal-enrichment processes inside manganese nodules from the north equatorial Pacific', Earth and Planet. Sci. Lett., 39, 341-348.

Burns, V.M. and Burns, R.G., 1978b, 'Diagenetic features observed inside deep-sea manganese nodules from the north equatorial Pacific', Scanning Electron Microscopy, pp. 245-252.

Burns, V.M. and Burns, R.G.. 1978c, 'Authigenic todorokite and phillipsite inside deep-sea manganese nodules', Am. Mineral., 63, 827-831.

Burns, R.G. and Burns, V.M., 1979a, 'Manganese oxides', in R.G. Burns (ed.), Marine minerals, Wash., Mineral. Soc. Amer., pp. 1-46.

Burns, R.G. and Burns, V.M., 1979b, 'Authigenic oxides', in C. Emiliani (ed.), The Sea, N.Y., John Wiley, Vol. 7.

Burns, R., Burns, V. and Stockman, H., 1983, 'A review of the todorokite-buserite problem: implication to the mineralogy of marine manganese nodules', Am. Mineral., 6, 972-980.

Burns, R.G. and Fuerstenau, D.W., 1966, 'Electron probe determination of inter-element relationships in manganese nodules', Am. Mineral., 51, N 56, 895-902.

Burton, J.D., 1975, 'Radioactive nuclides in the marine environment', in Chemical Oceanography. London, Academic Press, Vol. 4, pp. 203-285.

Burton, J.D., Maher, W.A. and Statham, P.J., 1983, 'Some recent measurements of
    trace metals in Atlantic ocean waters', in C.S. Wong et al. (eds.), Trace
    metals in sea water, N.Y., Plenum, pp. 415-426.
Buser, W. and Grütter, A., 1956, 'Über die Natur der Manganknollen', Schweiz. Min.
    Petr. Mitt., B. 36, S. 49-62.
Butkevich, V.S., 1928, 'Formation of marine iron-manganese deposits and
    participating microorganisms', Tr. Mor. nauch. Inst., 3, N 3, 1-81 (in Russian)
Butuzova, G.Yu., 1969, 'Recent volcanic-sedimentary iron-ore process in the
    Santorini caldera (the Aegean Sea) and its effect on sediment geochemistry',
    Moscow, Nauka, 109 pp. (in Russian).
Butuzova, G.Yu., 1984, 'Mineralogy and some aspects af genesis of metalliferous
    sediments in the Red Sea', Litologia i polezn. iskop., N 4, pp. 11-32 (in
    Russian).
Butuzova, G.Yu., 1985, 'Peculiarity of hydrothermal-sedimentary ore genesis in the
    Red Sea rift zone', Litologia i polezn. iskop., N 5, pp. 39-55 (in Russian).
Butuzova, G.Yu., Dvoretskaya, O.A. and Stepanetz, M.I., 1967, 'Experience of
    application of chloridized alcohol for extraction of free iron oxides and hydro-
    oxides out of Recent sediments from the Santorini volcano zone', Litologia i
    polezn. iskop., N 4, 130-136 (in Russian).
Butuzova, G.Yu. and Lisitsina, N.A., 1983, 'Metalliferous sediments of the Red Sea
    deep basins', Litologia i polezn. iskop., N 3, pp. 16-32 (in Russian).
Cadle, R., 1975 'Volcanic emissions of halides and sulfur compounds to the troposphere
    and stratosphere', J. Geophys. Res., 80, N 12, 1650-1652.
Callender, E. and Bowser, C.J., 1980, 'Manganese and copper geochemistry of
    interstitial fluids from manganese nodule-rich pelagic sediments of the north-
    eastern equatorial Pacific Ocean', Am. J. Sci., N 10, 1063-1096.
Calvert, S.E., 1978, 'Geochemistry of oceanic ferromanganese deposits', Phil. Trans.
    R. Soc. London, 290, 43-73.
Calvert, S.E. and Piper, D.Z., 1984, 'Geochemistry of ferromanganese nodules from
    DOMES Site A, Northern Equatorial Pacific: Multiple diagenetic metal sources
    in deep sea', Geochimica et cosmochim. Acta, 48, 1913-1928.
Calvert, S.E. and Price, N.P. 1970, 'Composition of manganese modules and manganese
    carbonates from Loch Fyne, Scotland', Contrib. Mineral. Petrol., 29, 215-233.
Calvert, S.E. and Price, N.B., 1972, 'Diffusion and reaction profiles of dissolved
    manganese in pore waters of marine sediments', Earth and Planet. Sci. Lett.,
    16, 245-249.
Calvert, S.E. and Price, N.B., 1977, 'Geochemical variation in ferromanganese
    nodules and associated sediment from the Pacific ocean', Marine Chemistry,
    5, N 1, 43-74.
Calvert, S.E. and Price, N.B., 1977, 'Shallow-water continental margin and
    lacustrine nodules: distribution and geochemistry', in G.P. Glasby (ed.),
    Marine manganese deposits, Amsterdam, Elsevier, pp. 45-86.
Calvert, S.E., Price, N.B., Heath, G.R. and Moore, T.C., 1978, 'Relationship
    between ferromanganese nodule composition and sedimentation in a small survey
    area of the equatorial Pacific', J. Mar. Res., 36, N 1, 161-183.
Cambray, R.S., Jeffris, D.F. and Topping, G., 1975, 'An estimate of the input of
    atmospheric trace elements into the North Sea and the Clyde Sea', -U.K. Atom.
    Energy Auth. Harwell Rep., N 7733, 30 pp.
Cameron, A.G.W., 1982, 'Elementary and nuclidic abundances in the solar system',
    in C.A. Barnes, D.N. Schramm and D.D. Clayton (eds.), Essays in Nuclear
    Astrophysics, Cambridge Univ. Press, pp. 23-43.
Capelli, R., Contardi, V., Cosma, B., Minganti, V. and Zanicchi, G., 1983, 'A
    four-year study on the distribution of some heavy metals in five marine
    organisms of the Ligurian Sea', Mar. Chem., 12, N 4, 281-293.
Carbonnel, J.P. and Meybeck, M., 1975, 'Quality variation of the Mekong River at
    Pnom Penh', J. Hydrol., 27, 249-265.
Carmichael, N.G., Squibb, K.S. and Fowler, B.A., 1979, 'Metals in the molluscan
    kidney: A comparison of two closely related species of bivalve species
    (Argopecten) using X-ray microanalysis and atomic absorption spectroscopy',
    J. Fish. Res. Bd. Canada, 36, 1149-1155.

Carpenter, R., 1960, 'Factors controlling the marine geochemistry of fluorine', Geochim. et cosmochim. acta, Vol. 33, N 10, pp. 1153-1168.

Carpenter, R., 1983, 'Quantitative electron spin resonance (ESR) determinations of forms and total amounts of Mn in aqueous environmental samples', Geochim. et Cosmochim. Acta, 47, N 5, 875-885.

Carpenter, R., Johnson, H.P. and Twiss, E.S., 1972, 'Thermomagnetic behaviour of manganese nodules', J. Geophys. Res., 77, 7163-7174.

Carroll, D., 1958, 'Role of clay minerals in the transportation of iron', Geochim. et Cosmochim. Acta, 14, N 1, 1-27.

Charlot, G. and Bezier, D., 1957, Quantitative inorganic analysis, New York, Wiley, 535 pp.

Chelishev, N.F., 1985, 'The role of ion exchange in deep oceanic mineral formation', Geokhimiya, N 4, pp. 540-547 (in Russian).

Chelishev, N.F. and Gribanova, N.K., 1985, 'Exchange reactions and forms of metal occurrence in manganese nodules', Mineral. zhurnal, 7, N 4, 3-10 (in Russian).

Cherdyntsev, V.V., Kadyrov, N.V. and Novichkova, N.V., 1971, 'Origin of the Pacific manganese nodules from radioisotropic data', Geokhimiya, N 3, pp. 339-354 (in Russian).

Chernoff, B. and Dooley, J.K., 1978, 'Heavy metals in relation to the biology of the mummichog, Fundulus heteroclitus', J. Fish Biol., 14, 309-328.

Chester, R., 1972, 'Geological, geochemical and environmental implications of the marine dust veil', in D. Dyrssen and D. Jagner (eds.), The changing Chemistry of the Oceans, Nobel Symp. 20, Stockholm, pp. 299-307.

Chester, R., Aston, S.R., Stoner, I.H. and Bruty, D., 1974, 'Trace metals in soilsized particles from the lower troposphere over the world ocean', J. de Recherches atmospheriques, 8, N 3-4, 777-789.

Chester, R. and Hughes, M., 1967, 'A chemical technique for separation of ferro-manganese minerals, carbonate minerals and absorbed trace elements from pelagic sediments', Chem. Geol., 2, N 3, 249-262.

Chester, R. and Johnson, L.R., 1971, 'Trace element geochemistry of North Atlantic aeolian dust', Nature, 231, N 5299, 176-178.

Chester, R. and Messiha-Hanna, R.C., 1970, 'Trace elements partition patterns in North Atlantic deep-sea sediments', Geochim. et Cosmochim. Acta, 34, N 10, 1121-1128.

Chester, R. and Stoner, J.H., 1974, 'The distribution of zinc, nickel, manganese, cadmium, copper and iron in some surface waters from the World Ocean', Marine Chem., 2, N 1, 17-32.

Chester, R. and Stoner, J., 1974, 'The distribution of Mn, Fe, Cu, Ni, Co, Ga, Cr, V, Ba, Sr, Sn, Zn and Pb in some soilsized particulates from the lower troposphere over the world ocean', Marine Chem., 2, N 3, 157-188.

Chipman, W. and Thommeret, J., 1970, 'Manganese content and the occurence of fallout ^{54}Mn in some marine benthos of the Mediterranean', Bull. Inst. Ocean., 69, 1-15.

Chow, T.J. and Patterson, C.C., 1959, 'Lead isotopes in manganese nodules', Geochim. et Cosmochim. Acta, 17, N 1-2, 21-31.

Chow, T.J. and Patterson, C.C., 1962, 'The occurence and significance of lead isotopes in pelagic sediments', Geochim. et Cosmochim. Acta, 26, N 2, 263-308.

Chowdhury A.N. and Pal, J.C., 1983, 'Determination of platinum in USGS manganese nodule reference samples by a fire assay-spectrographic method', Geostand. Newslett., 7, N 2, 279-280.

Chudaev, O.V., 1986, 'Manganese minerals', in Ferromanganese nodules of the Central part of the Pacific Ocean, Moscow, Nauka, pp. 186-192 (in Russian).

Chudaev, O.V., Skornyakova, N.S., Pushcharovsky, D.Yu., Kholodkevich, I.V. and Khudoloshkin, V.M., 1983, 'Mineral composition of manganese nodules in the Central Pacific', Dokl. AN SSSR, 269, N 6, 1444-1448 (in Russian).

Chudaeva, V.A., 1978, 'Trace elements in river discharge in the western Sea of Japan', in Migration processes of ore matter in coastal area, Vladivostok, pp. 51-55 (in Russian).

Chudaeva, V.A., Gordeev, V.V. and Fomina, L.S., 1982, 'Speciation of elements in suspension in some rivers in the Sea of Japan region', Geokhimiya, N 4, pp. 585-596 (in Russian).

Chugunnyi, Yu.G., 1982, 'Manganese nodules of the Caribbean Sea and some peculiarities in their formation', Geol. zhurnal, N 82, pp. 78-84 (in Russian).

Chugunnyi, Yu.G. and Demenko, D.P., 1973, 'On the role of bacterial activity in formation of manganese nodules in recent sea basins', in Nodules and nodule analysis, Kharkov, pp. 155-156 (in Russian).

Chugunnyi, Yu.G. and Demenko, D.P., 1977, 'On the role of biogenetic activity in formation of manganese nodules in recent sea basins', in Nodules and nodule analysis, Moscow, Nauka, pp. 176-180 (in Russian).

Chukrov, F.V., 1978, 'On common source of metals in deep oceanic ores and sediments', Geologiya rudnykh mestorozhdenii, N 3, pp. 3-15 (in Russian).

Chukrov, F.V., Gorshkov, A.I., Dritz, V.A., Shterenberg, L.E., Sivtsov, A.V. and Sakharov, B.A., 1983a, 'Mixed-layer minerals asbolan-buserite and asbolans in oceanic iron-manganese nodules', Izv. AN SSSR, ser. geol., N 5, pp. 91-99 (in Russian).

Chukrov, F.V., Gorshkov, A.I., Ermilova, L.P., Berezovskaya, V.V. and Sivtsov, A.V., 1981a, 'Mineral forms of Mn and Fe occurrence in oceanic sediments', Izv. AN SSSR, ser. geol., N 4, pp. 5-21 (in Russian).

Chukrov, F.V., Gorshkov, A.I., Rudnitskaya, E.S., Beresovskaya, V.V. and Sivtsov, A.V., 1978a, 'On vernadite', Izv. AN SSSR, ser. geol., N 6, pp. 5-19 (in Russian).

Chukrov, F.V., Gorshkov, A.I., Rudnitskaya, E.S. and Sivtsov, A.V., 1978b, 'On birnessite characteristics', Izv. AN SSSR, ser. geol., N 9, pp. 67-75, (in Russian).

Chukrov, F.V., Gorshkov, A.I., Rudnitskaya, E.S. and Berezovskaya, V.V., 1979a, 'New data on natural todorokite', Nature, 278, 631-632.

Chukrov, F.V., Gorshkov, A.I., Rudnitskaya, E.S. and Berezovskaya, V.V., 1979b, 'A new 14 $\overset{\circ}{A}$ mineral of the birnessite group in deep-sea micronodules', Nature, 280, N 5718, 136-137.

Chukrov, F.V., Gorshkov, A.I., Sivtsov, A.V. and Berezovskaya, V.V., 1979c, 'On new mineral phases of oceanic manganese micronodules', Izv. AN SSSR, ser. geol., N 1, pp. 83-90 (in Russian).

Chukrov, F.V., Gorshkov, A.I., Sivtsov, A.V., Berezovskaya, V.V. and Rudnitskaya, E.S., 1979d, 'On ranceite nature', Izv. AN SSSR, ser. geol., N 11, pp. 71-81 (in Russian).

Chukrov, F.V., Gorshkov, A.I. and Sivtsov, A.V., 1981b, 'New structural modification of todorokite', Izv. AN SSSR, ser. geol., N 5, pp. 88-91 (in Russian).

Chukrov, F.V., Gorshkov, A.I., Vitovskaya, I.V., et al., 1980a, 'Crystallochemical nature of Co-Ni asbolans', Izv. AN SSSR, ser. geol., N 6, pp. 73-81 (in Russian).

Chukrov, F.V., Gorshkov, A.I., Zvyagin, B.B. and Ermilova, L.P., 1980b, 'Iron oxides as minerals of sedimentary units and chemogenic eluvium', in I.M. Varentsov and G.Y. Grasselly (eds.), Geology and geochemistry of manganese. Budapest, Akad. Kiado, pp. 231-257.

Chukrov, F.V., Shterenberg, L.E., Gorshkov, A.I., Dritz, V.A., et al., 1983b, 'On 10-$\overset{\circ}{A}$ manganese mineral nature in iron-manganese nodules', Litol. i polezn. iskop., pp. 33-41 (in Russian).

Chukrov, F.V., Zvyagin, B.B., Ermilova, L.P. and Gorshkov, A.I., 1976a, 'Feroxyhyte, new modification of FeO (OH)', Izv. AN SSSR, ser. geol., N 5, pp. 5-24 (in Russian).

Chukrov, F.V., Zvyagin, B.B. and Yermilova, L.P., 1976b, 'Mineralogical criteria in the origin of marine iron-manganese nodules', Min. Deposita, 11, 24-32.

Clarke, F.W., 1924, 'The data of geochemistry', 5th. ed. Bull. U.S. Geol. Surv., N 770, 841 pp.

Coastal upwelling, its sediment record (Ed. J. Thiede, E. Suess). N.Y. and London, Plenum Press, 1983, Pt. A, 580 pp., Pt. B, 610 pp.

Cochran, J.K. and Krishnaswamy S., 1980, 'Radium, thorium, uranium and ^{210}Pb in deep-sea sediments and sediment pore waters from the north equatorial Pacific', Amer. J. Sci., 280, 849-889.

Collier, R.W. and Emond, J.M., 1983, 'Plancton composition and trace element fluxes from the surface ocean', in C.S. Wong et al. (eds.), Trace metals in sea water, N.Y., Plenum, pp. 789-809.

Collier, R. and Edmond, J., 1984, 'The trace element geochemistry of marine biogenic particulate matter', Prog. Oceanogr., 13, N 2, 113-199.

Colly, S., Thompson, J., Wilson, T.R.S. and Higgs, N.C., 1984, 'Post-depositional migration of elements during diagenesis in brown clay and turbidite sequences in the North East Atlantic', Geochim. et Cosmochim. Acta, 48, N 6, 1223-1235.

Corliss, J., 1971, 'The origin of metal-bearing submarine hydrothermal solutions', J. Geophys. Res., 76, N 33, 8128-8138.

Corliss, J.B., Dymond, J., Gordon, L.I., Edmond, J.M., Von Herzen, R.P., Ballard, R.D., Green, K., Williams, D., Bainbridge, A., Crane, K. and Von Andel, T.H., 1979, 'Submarine thermal springs on the Galapagos Rift', Science, 203, N 4385, 1073-1083.

Corliss, J.B., Lyle, M., Dymond, Y. and Crane, K., 1978, 'The chemistry of hydrothermal mounds near the Galapagos Rift', Earth and Planet. Sci. Lett., 40, 12-24.

Correns, C.W., 1937, 'Die Sedimente des equatorialen Atlantischen Ozeans', in Dtsch. Atlant. Exped. "Meteor" 1925-1927, Wiss. Erg., 1937, Bd. 3, Tl. 3, S. 135-298.

Courtois, C. and Clauer, N., 1980, 'Rare earth elements and strontium isotopes of polymetallic nodules from the southeast Pacific Ocean', Sedimentology, 1980, 27, N 6, 687-695.

Cowen, J.P., 1982, 'Iron and manganese depositing bacteria in suspended particles', Eos, 63, N 45, 960.

Cowen, J.P. and Bruland K.W., 1985, 'Metal deposits associated with bacteria: implications for Fe and Mn marine biogeochemistry', Deep-sea Res., 32, N 3, 253-272.

Cowen, J.P. and Silver, M.W., 1984, 'The association of iron and manganese with bacteria on marine microparticulate material', Science, 224, 1340-1342.

Cox, M.E. and McMurtry, G.M., 1981, 'Vertical distribution of mercury in sediments from the EPR', Nature, 289, N 5 800, 789-792.

Crecelius, E.A., Carpenter, R. and Merrill, R.T., 1973, 'Magnetism and magnetic reversals in ferromanganese nodules', Earth and Planet. Sci. Lett., 17, 391-396.

Crerar, D.A. and Barnes, H.Z., 1974, 'Deposition of deep-sea manganese nodules', Geochim. et Cosmochim. Acta, 38, N 2, 279-300.

Crocket, J.H. and Kuo, H.Y., 1979, 'Sources for gold, palladium and iridium in deep-sea sediments', Geochim. et Cosmochim. Acta, 43, N 6, 831-842.

Cronan, D.S., 1969, 'Average abundances of Mn, Fe, Ni, Co, Cu, Pb, Mo, V, Cr, Ti and P in Pacific pelagic clays', Geochim. et Cosmochim. Acta, 33, N 12, 1562-1565.

Cronan, D.S., 1972, 'Regional geochemistry of ferromanganese nodules in the world ocean', in D.R. Horn (ed.), Ferromanganese deposits on the ocean floor, N.Y.: Columbia Univ. Palisades, pp. 19-30.

Cronan, D.S., 1975, 'Manganese nodules and other ferromanganese oxide deposits from the Atlantic ocean', J. Geophys. Res., 80, N 27, 3831-3837.

Cronan, D.S., 1980, 'Underwater Minerals', London, Academic, 362 pp.

Cronan, D.S. and Moorby, S.A., 1981, 'Manganese nodules and other ferromanganese oxide deposits from the Indian Ocean', J. Geol. Soc. London, 38, 527-539.

Cronan, D.S. and Tooms, J.S., 1967a, 'Sub-surface concentrations of manganese nodules in Pacific sediments', Deep-Sea Res., 14, N 1, 117-119.

Cronan, D.S. and Tooms, J.S., 1967b, 'Geochemistry of manganese nodules from the N.W. Indian Ocean', Deep-Sea Res., 14, N 2, 239-249.

Cronan, D.S. and Tooms, J.S., 1968, 'A microscopic and electron probe investigation of manganese nodules from the Northwest Indian Ocean', Deep-Sea Res., 15, N 2, 215-223.

Cronan, D.S. and Tooms, J.S., 1969, 'The geochemistry of manganese nodules and associated pelagic deposits from the Pacific and Indian oceans', Deep-Sea Res., 16, N 4, 335-359.

Davidson, C.J., Chu, L., Grimm, T.C., Nasta, M.A. and Quamoos, M.P., 1981, 'Wet and dry deposition of trace elements onto the Greenland ice sheet', Atm. Envir., 15, N 8, 1429-1437.

de Baar, H.J.W., Bacon, M.P., Brewer, P.G. and Bruland, K.W., 1985, 'Rare earth elements in the Pacific and Atlantic oceans', Geochim. et Cosmochim. Acta, 49, N 9, 1943-1959.

De Castro, A.F. and Ehrlich, H.L., 1970, 'Reduction of iron oxide minerals by a marine Bacillus', Antonie van Leeuwenhoek, 36, 317-327.

Defant, A., 1961, Physical oceanography, N.Y., Pergamon Press, v. 1, 729 pp., v. 2, 598 pp.

Demina, L.L., 1982, 'Forms of migration of heavy metals in the ocean', Moscow, Nauka, 120 pp (in Russian).

Demina, L.L., Gordeev, V.V. and Fomina, L.S., 1978, 'Forms of Fe, Mn, Zn and Cu in river water and suspension and their transformation in the river and sea water mixing area: rivers of the Black, Azov, and Caspian Sea basins', Geokhimiya, N 8, pp. 1211-1229 (in Russian).

Demina, L.L., Shumilin, E.N. and Tambiev, S.B., 1984, 'Forms of metals in suspension of the Indian ocean surficial waters', Geokhimiya, N 4, pp. 565-576 (in Russian).

Depetris, P.J. and Griffin, J.J., 1968, 'Suspended load in the Rio de la Plata drainage basin', Sedimentology, 11, 53-60.

Desai, M.V.M. and Ganguly, A.K., 1980, 'Organo-metallic interactions of manganese and other heavy metals in the marine environment', in Geology and Geochemistry of manganese, vol. 1, Budapest, Kiado, pp. 399-410.

Description of manganese nodule processing systems for environmental studies. Progr. Rept., vol. 1, Rockville, U.S. Dept. of Commerce, Office of Marine Minerals, 1977.

Description of manganese nodule processing activities for environmental impact studies, v. 3. Processing systems technical analyses. US Dept. Comm., NOAA, Off. Mar. Minerals, 1979, pp. 14-60.

Dickson, F.W., 1977, 'The role of rhyolite-seawater reaction in the genesis of Kuroko ore deposits', Proc. 2 nd. Int. Symp. Water-Rock Interact, Strasbourg, Sec. 4, pp. 181-190.

Diem, D. and Stumm, W., 1984, 'Is dissolved Mn^{2+} being oxidized by O_2 in absence of Mn-bacteria or surface catalysts?', Geochim. et Cosmochim. Acta, 48, N 7, 1571-1573.

Discharge of selected rivers of the world. Vol. III. Mean monthly and extreme discharges (1972-1975). Paris, UNESCO, 1979, 104 pp.

Disnar, J.R., 1981, 'Etude expérimentale de la fixation de métaux par un matérial sédimentaire actuel de l'origine algaire. II. Fixation "in vitro" de UO_2^{2+}, Cu^{2+}, Ni^{2+}, Zn^{2+}, Pb^{2+}, Co^{2+}, Mn^{2+}, ainsi que de VO_3^-, MoO_4^{2-} et $GeO_2^{2'}$, Geochim. et Cosmochim. Acta, 45, N 3, 363-379.

Dritz, V.A., Petrova, V.V., Gorshkov, A.I., Svalnov, V.N., Sokolova, A.L., Sivtsov, A.V. and Karpova, G.V., 1985, 'Manganese minerals of Mn-Fe micro-nodules of the Central Pacific sediments and their post-depositional transformations', Lithol. i polezn. iskop., N 3, pp. 17-39 (in Russian).

Drozdova, V.M. and Makhon'ko, E.P., 1970, 'Content of trace elements in precipitation', J. Geophys. Res., 75, N 18, 3610-3612.

Duce, R.A., Arimoto, R., Ray, B.J., Unni, C.K. and Harder, P.J., 1983, 'Atmospheric trace elements at Enivetak atoll: concentrations, sources and temporal variability', J. Geophys. Res., 88, N C9, 5321-5342.

Duce, R.A., Hoffman, G.L., Ray, B.J., Fletcher, J.S., Piotrowicz, S.R., Hoffman, E.J., Miller, J.M. and Heffter, J.L., 1976, 'Trace metals in the marine atmosphere: sources and fluxes', in H. Windom and R.A. Duce (eds.), Marine Pollutant transfer, D.C. Heath, Lexinton, Mass., pp. 77-119.

Duce, R.A., Hoffman, G.L. and Zoller, W.H., 1975, 'Atmospheric trace metals at remote northern and southern hemisphere sites: pollution or natural?', Science, 187, N 4171, 59-61.

Duce, R.A., Unni, C.R., Ray, B.J., Prospero, J.M. and Merrill, J.T., 1980, 'Long range atmospheric transport of soil dust from Asia to the tropical North Pacific: temporal variability', Science, 209, N 4464, 1522-1524.

Dudley, W.C., 1976, 'Cementation and iron concentration in foraminifera on
    manganese nodules', J. Foraminifer. Res., 6, 202-207.
Dudley, W.C., 1979, 'Biogenic influence on the composition and structure of marine
    manganese nodules', Proc. Colloq. Internatl. C.N.R.S., Paris, pp. 227-232.
Dugolinsky, B.K., 1976a, 'Chemistry and morphology of deep-sea manganese
    nodules and the significance of associated encrusting protozoans on nodule
    growth'. Ph. Diss., Univ. of Hawaii, Honolulu, 228 pp.
Dugolinsky, B.K., 1976b, 'Theory of formation: life forms on manganese nodules',
    Ocean Industry, 11, 88-90.
Dugolinsky, B.K., Margolis, S.V. and Dudley, W.C., 1977, 'Beogenic influence on
    growth of manganese nodules', J. Sed. Petrol., 47, N 1, 428-445.
Duinker, J.C., Hillebrant, M.T.J. and Nolting, R.F., 1979 'Organochlorines and metals
    in harbour seals (Dutch Wadden Sea)', Mar. Poll. Bull., 10, 360-364.
Duinker, J.C. and Nolting R.F., 1976, 'Distribution model for particulate trace
    metals in the Rhine estuary, southern bight and Dutch Wadden Sea', Neth. J.
    Sea. Res., 10, N 1, 71-102.
Dunham, A.C. and Glasby, G.P., 1974, 'Petrographic and electron microprobe
    investigation of some deep-sea and shallow-water manganese nodules', N.Z.J.
    Geol. Geophys., 17, 929-935.
Dvoretskaya, O.A., 1971, 'Sediments of the Indian ocean south of Ceylon', Litol. i
    Polezn. iskop., N 4, 16-33 (in Russian).
Dvoretskaya, O.A. and Boiko, T.F., 1979, 'Distribution of rare alkaline elements
    and thallium in surficial sediments along the profile across the Pacific', Litol. i
    polezn. iskop., N 4, pp. 95-104 (in Russian).
Dymond, J., 1981, 'Geochemistry of Nazca plate surface sediments: An evaluation
    of hydrothermal, biogenic, detrital, and hydrogenous sources', in L.D. Kulm
    et al. (eds.), Nazca Plate: Crustal formation and Andean convergence,
    Geol. Soc. Amer. Memoir, 154, 133-173.
Dymond, J. and Eklund, W., 1978, 'A microprobe study of metalliferous sediment
    components', Earth and Planet. Sci. Lett., 40, N 2, 243-251.
Dymond, J., Fischer, K., Clanson, M., Cobler, R., Gardner, W., Richardson,
    M.J., Berger, W., Soutar, A. and Dunbar, R., 1981, 'A sediment trap inter-
    comparison study in the Santa Barbara Basin', Earth and Planet. Sci. Lett.,
    53, N 3, 403-418.
Dymond, J., Lyle, M., Finny, B., Piper, D.Z., Murphy, K., Conard, R. and
    Pisias, N., 1984, 'Ferromanganese nodules from MANOP Sites H, S and R -
    control of mineralogical and chemical composition by multiple accretionary
    processes', Geochim. et Cosmochim. Acta, 48, N 5, 931-949.
Edmond, J.M., Measures, C., McDuff, R.E., Chan, L.H., Collier, R.and Grant, B,
    1979, 'Ridge crest hydrothermal activity and the balances of the major and
    minor elements in the Ocean: The Galapagos data', Earth and Planet. Sci. Lett.,
    46, N 1, 1-18.
Edmond, J.M., Von Damm, K.L., McDuff, R.E. and Measures, C.J., 1982,
    'Chemistry of hot springs on the East Pacific Rise and their effluent dispersal',
    Nature, 297, N 5863, 187-191.
Edzwald, J.K., Upchurch, J.B. and O'Melia, C.R., 1974, 'Coagulation in estuaries',
    Environ. Sci. Technol., 8, 58-63.
Efimova, E.I. and Nikolaev, D.S., 1965, 'Radiochemical composition of manganese
    nodules and manganese ores', Radiokhimiya, N 5, pp. 603-610 (in Russian).
Egorov, V.V., Zhigalovskaya, T.N. and Malakhov, S.C., 1970, 'Microelement content
    of surface air above the continent and the ocean', J. Geophys. Res., 57, N 18,
    3650-3656.
Ehrlich, A.M., 1968, 'Rare earth abundances in manganese nodules', Ph. D. Thesis,
    Massachusetts Institute of Technology, Cambridge, Mass., 225 pp.
Ehrlich, H.L., 1963, 'Bacteriology of manganese nodules', J. Appl. Microbiol., 11,
    15-19.
Ehrlich, H.L., 1964, 'Bacterial release of manganese from Mn nodules', Bacteriol.
    Proc. Abstr., G 156, 42-43.
Ehrlich, H.L., 1966, 'Reactions with manganese by bacteria from marine ferro-
    manganese nodules', Dev. Ind. Microbiol., 7, 279-286.

Ehrlich, H.L., 1968, 'Bacteriology of manganese nodules. II. Manganese oxidation by cell-free extract from a manganese nodule bacterium', Appl. Microbiol., 16, 197-202.

Ehrlich, H.L., 1971, 'Bacteriology of manganese nodules. V. Effect of hydrostatic pressure on bacterial oxidation of Mn (II) and reduction of MnO_2', Appl. Microbiol., 4, N 2, 306-310.

Ehrlich, H.L., 1972, 'The role of microbes in manganese nodules genesis and degradation', in Ferromanganese deposits on the ocean floor. Wash. D.C., Nat. Sci. Found., pp. 63-70.

Ehrlich, H.L., 1974, 'The formation of ores in the sedimentary environment of the deep sea with microbial participation: the case for ferromanganese concretions', Soil. Sci., 119, N 1, 36-41.

Ehrlich, H.L., 1976, 'Manganese as an energy source for bacteria', in J.O. Nriagu (ed.), Environ. Biogeochem., v. 2. Metals transfer and ecological mass balances. Ann Arbor Sci. Publ., Ann Arbor, Mich., pp. 633-644.

Ehrlich, H.L., 1980, 'Different forms of microbial manganese oxidation and reduction and their environmental significance', in Biogeochem. Ancient and mordern Environment. Berlin, Springer, pp. 327-332.

Ehrlich, H.L., Giorse, W.S. and Johnson, G.L., 1972, 'Distribution of microbes in manganese nodules from the Atlantic and Pacific oceans', Dev. Ind. Microbiol., 13, 57-65.

Ehrlich, H.L., Yang, S.H. and Mainwaring, J.D., 1973, 'Bacteriology of manganese nodules. VI. Fate of copper, nickel, cobalt and iron during bacterial and chemical reduction of the manganese (IV)', Z. Allg. Mikrobiologie, 13, N 1, 39-48.

Eisler, R., 1977, 'Toxicity evaluation of a complex metal mixture to the soft shell clam , Mya arenaria'. Mar. Biol., 43, 265-276.

Eisler, R., 1981, Trace metal concentrations in marine organisms. N.Y. e.a., Pergamon Press, 687 pp.

Eisler, R., Barry, M.M., Lapan, R.L., Telek, G., Davey, E.W. and Soper, A.E., 1978, 'Metal survey of the marine clam Pitar morrhuana collected near a Rhode Island (USA) electroplating plant', Mar. Biol., 45, 311-317.

Eittreim, S.L., 1970, 'Suspended particulate matter in the deep waters of the northwest Atlantic Ocean', Ph. D. Thesis, Lamont-Doherty Geol. Observ., Columbia Univ.

Elderfield, H., 1976, 'Manganese fluxes to the ocean', Mar. Chem., 4, N 2, 103-132.

Elderfield, H., 1981, 'Metal-organic associations in interstitial waters of Narrangansett Bay sediments', Amer. J. Sci., 281, N 9, 1184-1196.

Elderfield, H. and Greaves, M.J., 1981, 'Negative cerium anomaly in the rare elements patterns of oceanic ferromanganese nodules', Earth and Planet. Sci. Lett., 55, N 1, 163-170.

Elderfield, H. and Greaves, M.J., 1982, 'The rare elements in seawater', Nature, 296, N 5854, 214-219.

Elderfield, H., Gunnlaugsson, E., Wakefield, S.J. and Williams, P.T., 1977, 'The geochemistry of basalt-sea-water interactions: evidence from Deception Island, Antarctica, and Reykyanes Island', Miner. Mag., 41, 217-226.

Elderfield, H., Hawkesworth, C.J., Greaves, M.J. and Calvert, S.E., 1981a, 'Rare earth geochemistry of oceanic ferromanganese nodules and associated sediments' Geochim. et Cosmochim. Acta, 45, N 4, 513-528.

Elderfield, H., Hawkesworth, C.J., Greaves, M.J. and Calvert, S.E., 1981b, 'Rare earth element zonation in Pacific ferromanganese nodules', Geochim. et Cosmochim. Acta, 45, N 7, 1231-1234.

El Ghobary, H., 1982, 'Fe, Mn, Cu and Zn in interstitial water of near-shore sediments', Mar. Geol., 47, N 1-2, M11-M20.

Ellis, A.J., 1968, 'Natural hydrothermal systems and the experimental hot water-rock interaction: reactions with NaCl solutions and trace metal extraction', Geochim. et Cosmochim. Acta, 32, N 12, 1356-1363.

El Wakeel, S.K. and Riley, J.P., 1961, 'Chemical and mineralogical studies of deep-sea sediments', Geochim. et Cosmochim. Acta, 25, 110-146.

Emelyanov, E.M., 1975, 'Fe, Mn and Ti in the Atlantic sediments', Litol. i polezn. iskop., N 3, pp. 3-19 (in Russian).

Emelyanov, E.M., 1977 'Formation of chemical composition of the Atlantic suspension (from the data on Fe, Al, Ti, Mn)', Geokhimiya, N 4, pp. 565-577 (in Russian)

Emelyanov, E.M., 1981, 'Alumosilicate carbonate-manganese lithologic-geochemical region of the Gotland and Landsort depressions', in Sedimentation in the Baltic Sea, Moscow, Nauka, pp. 136-180 (in Russian).

Emelyanov, E.M., 1982, 'Sedimentation in the Atlantic Basin', Moscow, Nauka, 190 pp. (in Russian).

Emelyanov, E.M., Baturin, G.N., Vlasenko, N.B. and Orlova, S.A., 1973, 'Iron, manganese and copper in pore waters of the Atlantic bottom sediments', in Study of the World Ocean rift zones, Moscow, Nauka, 3, pp. 187-198 (in Russian).

Emelyanov, E.M., Lisitsin, A.P., Shimkus, K.M. et al., 1978, 'Geochemistry of Late Cenozoic sediments of the Black Sea, Leg 42 B.', in Ross, D.A., Neprochnov, Yu.P. et al., Initial Rep. of the DSDP, 42, pt. 2, Wash. D.C., US Gov. Print. Off., p. 543-606.

Emelyanov, E.M., Mitropolskiy, A.Yu., Shimkus, K.M. and Mussa, A.A., 1979, 'Geochemistry of the Mediterranean Sea', Kiev, Naukova Dumka, 132 pp. (in Russian).

Emelyanov, E.M., Pilipchuk, M.F., Volostnykh, B.V. et al., 1982, 'Iron and manganese forms in sediments along the geochemical profile across the Baltic Sea', Baltica, Vilnius, Mokslas, N 7, pp. 110-120 (in Russian).

Emelyanov, E.M. and Pustelnikov, O.S., 1976, 'Suspended matter, its composition and sediment material balance in the Baltic sea water', in Gudelis, V., Emelyanov, E.M., (eds.), Geology of the Baltic Sea, Vilnius, Mokslas pp. 159-186 (in Russian).

Emelyanov, E.M. and Stryuk, V.L., 1981, 'Lithology, geochemistry and stratigraphy of the Baltic Sea sediments', in Sedimentation in the Baltic Sea, Moscow, Nauka pp. 79-106 (in Russian).

Emelyanov, E.M., Trimonis, E.S. and Shimkus, K.M., 1976, 'Quantitative distribution and absolute masses of Fe, Al, Ti and Mn in the Black Sea suspension', Geokhimiya, N 9, 1375-1390 (in Russian).

Emelyanov, E.M., Trimonis, E.S. and Shimkus, K.M., 1980, 'Geochemical analysis of Upper Cenozoic deposits', in Geological history of the Black Sea from the deep sea drilling data, Moscow, Nauka, pp. 87-100 (in Russian).

Emerson, S., Kalhorn, S., Jacobs, L., Tebo, B.M., Nealson, K.H. and Rosson, R.A., 1982, 'Environmental oxidation rate of manganese (II): Bacterial catalysis', Geochim. et Cosmochim. Acta, 46, N 6, 1073-1080.

Emiliani, C., 1955, 'Mineralogical and chemical composition of the tests of certain pelagic foraminifera', Micropaleontology, 1, N 4, 377-380.

Eremenko, V.Ya., 1964, 'Forms of heavy metals in natural water', Gidrokhim. Materialy, 36, 61-73 (in Russian).

Eremenko, V.Ya., 1966, 'Technique for analysing heavy metal species in natural waters', Gidrokhim. materialy, 41, 153-158 (in Russian).

Eustace, I.J., 1974, 'Zinc, cadmium, copper and manganese in species of finfish and shellfish caught in the Derwent Estuary, Tasmania', Austral. J. Mar. Freshwat. Res., 25, 209-220.

Evans, D.W., Cutshall, N.H., Cross, F.A. and Wolf, D.A., 1977, 'Manganese cycling in the Newport River estuary, North Carolina', Estuarine Coastal Mar. Sci., 5, N 1, 71-80.

Exon, N.F., 1983, 'Manganese nodule deposits in the central Pacific ocean and their variation with latitude', Mar. Mining, 4, N 1, 79-107.

Fairhall, L.T., 1969, 'Industrial toxicology', N.Y., Hafner Publ. Co, 376 pp.

Ferguson, J. and Lambert, I.B., 1972, 'Volcanic exhalations and metal enrichments at Matupi Harbor, New Britain, T.P.N.G.', Econ. Geol., 67, N 1, 25-37.

Fesenkov, V.G., 1965, 'Achievements in meteoritics', Meteoritika, 26, 3-16 (in Russian).

Fewkes, R.H., 1973, 'External and internal features of marine manganese nodules as seen with the SEM and their implications in nodule origin', in M. Morgenstein (ed.), The origin and distribution of manganese nodules and prospects for exploration, Hawaii Inst. Geophys., pp. 21-29.

Fife, W.S. and Lonsdale, P., 1981, 'Ocean floor hydrothermal activity', in C.E. Emiliani (ed.), The Sea, v.7. The Oceanic Lithosphere, N.Y., Wiley, pp. 589-638.

Finkelman, R.B., 1970, 'Magnetic particles extracted from manganese nodules: suggested origin from stony and iron meteorites', Science, 167, N 3920, 982-984.

Finney, B., Heath, G.R. and Lyle, M., 1984, 'Growth rates of manganese-rich nodules at MANOP Site H (Eastern North Pacific)', Geochim. et Cosmochim. Acta, 48, N 5, 911-919.

Fisher, E.I. and Fisher, V.L., 1983, 'On platinum concentration in oceanic manganese nodules', Izv. AN Latv. SSR, N 5, pp. 515-522 (in Russian).

Fisher, E.I. and Fisher, V.L., 1984, 'The role of humic acids in gold sorption by marine sediments', Litol. i polezn. iskop., N 5, pp. 77-82 (in Russian).

Flanagan, F.J. and Gottfried, D., 1980, 'USGS rock standards, III: manganese-nodule reference samples USGS-Nod-A-1 and USGS-Nod-P-1', U.S. Geol. Surv. Prof. Pap., N 1155, 39 pp.

Fleet, A.J., 1984, 'Aqueous and sedimentary geochemistry of the rare earths', in P. Henderson (ed.), Rare earth element geochemistry, Amsterdam, Elsevier, pp. 343-373.

Fleischmann, W. and von Heimendal, M., 'Electron microscopic investigation on the Pacific manganese nodules', Min. Deposita, 12, N 2, 155-162.

Fomina, L.S., 1966, 'Accumulation and redistribution of rare-earth elements in formation process of oceanic manganese nodules', Dokl. AN SSSR, 170, N 5, 1181-1184 (in Russian).

Förstner, U. and Stoffers, P., 1981, 'Chemical fractionation of transition elements in Pacific pelagic sediments', Geochim. et Cosmochim. Acta, 45, N 7, 1141-1146.

Fortescue, J., 1980, Environmental geochemistry, N.Y., Springer, 347 pp.

Foster, A.R., 1970, 'Marine manganese nodules: nature and origin of internal features', M.S. Thesis, Wash. State Univ., 131 pp.

Foster, P., 1976, 'Concentrations and concentration factors of heavy metals in brown algae', Environ. Poll., 10, 45-53.

Fowler, S.W., 1977, 'Trace elements in zooplancton particulate products', Nature, 269, N 5629, 51-53.

Fowler, S.W. and Oregioni, B., 1976, 'Trace metals in mussels from the North West Mediterranean', Mar. Poll. Bull., 7, 26-29.

Frakes, L.A., 1982, 'Metal chemistry of manganese nodules from the Cape Leeuwin field, southeast Indian Ocean', Mar. Geol., 47, N 1-2, M1-M10.

Frakes, L.A., Exon, N.F. and Granath, J.W., 1977, 'Preliminary studies of the Cape Leeuwin manganese nodules deposit off Western Australia', BMR Jour. Austral. Geol. Geophys., 2, N 1, 66-69.

Frakes, L.A., Exon, N.F. and Granath, J.W., 1977, 'Chemistry of manganese nodules from the Cape Leeuwin field off Western Australia', BMR Jour. Austral. Geol. Geophys., 2, 232-233.

Franclin, M.L. and Morse, J.W., 1983, 'The interaction of manganese (II) with the surface of calcite in dilute solution and seawater', Mar. Chem., 12, N 4, 241-254.

Frank, D. et al., 'Ferromanganese deposits of the Hawaiian archipelago', Hawaii Instit. Geophys., N 14, 1-69.

Frazer, J.Z. and Fisk, M.B., 1977, 'Occurrence and abundance of ferromanganese nodules in the northeast equatorial Pacific', Map, Bureau of Mines, N 77-16.

Frazer, J.Z. and Wilson, L.L., 1980, 'Manganese nodule resources in the Indian Ocean', Marine Mining, 2, N 3, 257-292.

Frazier, J.M., 1975, 'The dynamics of metals in the American oyster Crassostrea virginica. I. Seasonal effects', Chesapeake Sci., 16, 162-171.

Friedrich, G. and Plüger, W., 1974, 'Die Verteilung von Mangan, Eisen, Kobalt, Nikel, Kupfer und Zink in Manganknollen verschiedener Felder',

Meerestechnik, B.5, N 6, S. 203-206.

Friedrich, G., Rosner, B. and Demirsoy, S., 1969, 'Erzmikroskopische und mikro-analytische Untersuchungen an Manganerzkonkretionen aus dem Pazifischen Ozean', Miner. Dep., 4, 298-307.

Friedrich, G. and Schmitz-Wiechowsky, A., 1980, 'Mineralogy and chemistry of a ferromanganese crust from a deep-sea hill, central Pacific, Valdivia cruise VA 13-2', Mar. Geol., 37, N 1-2, 71-90.

Frolova, T.I., Zharikova, E.N., Zolotarev, B.P., Kashintsev, G.L., Pechersky, D.M. and Starostin, V.I., 1979, 'Magmatic rocks of the south-eastern Pacific and their secondary transformation with regard to the origin of metalliferous sediments', in Metalliferous sediments of the south-eastern Pacific, Moscow, Nauka, pp. 48-71 (in Russian).

Frondel, C., Marvin, U.B. and Ito, Y., 1960, 'New data on birnessite and hollandite', Am. Mineral., 45, 871-875.

Fukai, R., Oregioni, B. and Vas, D., 1978, 'Interlaboratory comparability of measurements of trace elements in marine organisms: results of intercalibration exercise on oyster homogenate', Ocean. Acta, 1, 391-396.

Furbusch, W.J. and Schraeder, E.L., 1977, 'Secondary biogenic textures in some iron-manganese nodules from the Blake Plateau, Atlantic Ocean', Mar. Geol., 25, N 4, 343-354.

Galloway, J.N., Thornton, J.D., Norton, S.A., Volchok, H.L. and McLean, R.A.N. 1982, 'Trace metals in atmospheric deposition: a review and assessment', Atmos. Envir., 16, N 7, 1677-1700.

Galtsoff, P.S., 1942, 'Accumulation of manganese and the sexual cycle in Ostrea virginica', Physiol. Zool., 15, 210-215.

Galtsoff, P.S., 1953, 'Accumulation of manganese, iron, copper and zinc in the body of American oyster, Crassostrea (Ostrea) verginica', Anat. Res., 117, 601-602.

Garanzha, A.P. and Konovalov, G.S., 1977, 'Technique of extraction of colloidal mobile form of microelements', in II All-Union meeting on techniques of natural and waste water analysis, Moscow, Inst. of Geochemistry (in Russian).

Gebbing, J., 1909, 'Chemische Untersuchungen von Meeresboden, Meereswasser und Luftproben der Deutschen Südpolar Expedition 1901-1903', Berlin, S. 87.

Geology and Geochemistry of Manganese, 1980, Varentsov, I.M. and Grasselly, G.Y. (ed.), Budapest, Akad. Kiado, Vol. 1-3.

George, S.G., 1980, 'Correlation of metal accumulation in mussels with the mechanisms of uptake, metabolism and detoxification: a review', Thalassia Jugosl., 16, N 2-4, 347-365.

George, S.G., Pirie, J.S. and Coombs, T.L., 1980, 'Isolation and elemental analysis of metal-rich granules from the kidney of the scallop, Pecten maximus (L.)', J. Exp. Mar. Biol. Ecol., 42, 143-156.

Georgescu, I.I. and Lupan, S., 1971, 'Contributions to the study of the ferro-manganese concretions of the Black Sea', Rev. Roum. Geol. Geophys. Geogr. Sci., Geol., 15, 157-163.

Gershanovich, D.E., 1862, 'New data on Recent deposits of the Bering Sea deposits' Tr. Vses. Inst. mor. ryb. khoz-va i okeanografii, 46, 128-164 (in (Russian).

Ghiorse, W.C., 1977, 'Enzymatic manganese reduction by manganese nodule bacteria (abstr.)', Third Intl. Symp. Environ. Biogeochem. Univ. Oldenburg, Environ. Lab. Occ. Publ. 1, p. 50.

Ghiretti, F., Salvato, B., Carlucci, S. and Pieri, R., 1972, 'Manganese in Pinna nobilis', Experimentia, 28, 232-233.

Gibbs, R.J., 1967, 'The geochemistry of the Amazon river system. Part I. The factors that control the salinity, composition and concentration of the suspended matter', Bull. Geol. Soc. Amer., 78, 1203-1232.

Gibbs, R.J., 1973, 'Mechanisms of trace metal transport in rivers', Science, 180, N 4081, 70-73.

Gibbs, R.J., 1977, 'Transport phases of transition metals in Amazon and Yukon rivers', Bull. Geol. Soc. Amer., 88, N 6, 829-843.

Gieskes, J. and Reese, H., 1980, 'Interstitial water studies, Legs 51-53', Init. Rep.
Deep-Sea Drill. Project, Wash. D.C., US Gov. Print. Off., Legs 51-53, Pt. 2,
pp. 747-751.

Giovanoli, G., 1980, 'On natural and synthetic manganese nodules', in
I.M. Varentsov and G.Y. Grasselly (eds.), Geology and geochemistry of
manganese, Budapest, Akad. Kiado, pp. 159-202.

Giovanoli, R., Stahli, E. and Feitkecht, W., 1970, 'Über Oxidhydroxide des
vierwertigen Mangans mit Schichtengitter. 1. Natrium-mangan (II, III)
manganat (IV)', Helv. Chim. Acta, 53, 209-220.

Girin, Yu.P., Gordeev, V.V., Gurvich, E.G., Lukashin, V.N., Migdisov, A.A.
and Barskaya, N.V., 1979, 'Chemical composition of manganese nodules from
metalliferous and common pelagic sediments of the southeastern Pacific',
Geokhimia, N 1, 90-107 (in Russian).

Glagoleva, M.A., 1959, 'Forms of element migration in river water', in To cognition
of sediment diagenesis, Moscow, Nauka, pp. 5-28 (in Russian).

Glagoleva, M.A., 1961, 'On geochemistry of the Black Sea sediments', in Recent
sediments of seas and oceans, Moscow, Izd. AN SSSR, pp. 448-476 (in Russian)

Glagoleva, M.A., 1972, 'Regularities in variations of chemical composition of nodules
from the northwestern Pacific', Litologia i polezn. iskop., N 4, 40-49 (in
Russian).

Glagoleva, M.A., 1979, 'Elements of iron-manganese group', in Lithology and
geochemistry of the Pacific sediments (profile across the ocean), Moscow,
Nauka, 143-202 (in Russian).

Glagoleva, M.A., Volkov, I.I., Sokolov, V.S. and Yagodinskaya, T.A., 1975,
'Chemical elements in the Pacific sediments along the profile from the Hawaiian
is. to Mexico coast', Litologia i polezn. iskop., N 5, 16-28 (in Russian).

Glagoleva, M.A. and Turovskii, D.S., 1975, 'To cognition of geochemistry of
microelements in the Mid-Caspian basin sediments', Litologia i polezn. iskop.,
N 1, 20-29 (in Russian).

Glasby, G.P., 1972a, 'The mineralogy of manganese nodules from a range of marine
environments', Marine Geology, 13, N 1, 57-72.

Glasby, G.P., 1972b, 'The nature of the iron oxide phase in marine manganese
nodules', N.Z.J. Sci., 15, 232-239.

Glasby, G.P., 1973, 'Mechanisms of enrichment of the rare elements in marine
manganese nodules', Mar. Chem., 1, N 2, 105-125.

Glasby, G.P., 1976, 'Manganese nodules in the South Pacific: A review', N.Z. J.
Geol. Geophys., 19, 707-736.

Glasby, G.P., 1977, 'Why manganese nodules remain at the sediment-water
interface', New Zealand Journal of Science, 20, N 2, 187-190.

Glasby, G.P., 1978, 'Deep-sea manganese nodules in the stratigraphic record:
evidence from DSDP cores', Mar. Geol., 28, N 1, 51-64.

Glasby, G.P., 1981, 'Manganese nodules studies in the Southwest Pacific, 1975-1980,
a review', South Pacific Mar. Geol. Notes, 2, N 3, 37-46.

Glasby, G.P., Bäcker, H. and Meylan, M.A., 1975, 'Metal contents of manganese
nodules from the Southwestern Pacific basin', Erzmetall, B. 28, H. 7/8,
S. 340-342.

Glasby, G.P., Friedrich, G., Thijssen, T., Plüger, W.L., Kunzendorf, H.,
Ghosh, A.K. and Roonwal, G.S., 1982, 'Distribution, morphology and
geochemistry of manganese nodules from the Valdivia 13/2 Area, equatorial
north Pacific', Pacific Sci., 36, N 2, 241-264.

Glasby, G.P., Keays, R.R. and Rankin, P.C., 1978, 'The distribution of rare
earth, precious metals and other trace elements in Recent and fossil deep-sea
manganese nodules', Geochem. J., 12, N 4, 229-243.

Glasby, G.P. and Lawrence, P., 1974, 'Manganese deposits in the South Pacific
ocean. Maps for Ni, Cu, Co and Mn'. N.Z. Oceanogr. Inst.

Glasby, G.P., Meylan, M.A., Margolis, S.V. and Bäcker, H., 1980, 'Manganese
deposits of the Southwestern Pacific basin', in Geology and geochemistry of
manganese, 3, Budapest, Acad. Kiado, pp. 137-183.

Glasby, G.P., Stoffers, P., Sioulas, A., Thijssen, T. and Friedrich, G., 1982,
'Manganese nodule formation in the Pacific ocean: a general theory', Geo.-
Marine Lett., 2, 47-53.

Glasby, G.P. and Thijssen, T., 1982, 'The nature and composition of the acid-insoluble residue and hydrolisate fraction of manganese nodules from selected areas in the equatorial and S.W. Pacific', Tschermarks Mineral. Mitt., 30, 205-225.

Glover, E.D., 1977, 'Characterization of marine birnessite', Am. Mineral., 62, N 3-4, 278-285.

Glover, J.W., 1979, 'Concentrations of arsenic, selenium and ten heavy metals in school shark, Galeorhinus australis (Macleay), and gummi shark, Mustelus antarcticus Gunther from southeast Australian waters', Austral. J. Mar. Freshwat. Res., 30, 505-510.

Goldberg, E.D., 1954, 'Marine geochemistry, I. Chemical scavengers of the sea', J. Geol., 62, 249-265.

Goldberg, E.D. and Picciotto, E., 1955, 'Thorium determinations in manganese nodules', Science, 121, N 3147, 613-614.

Goldberg, E., 1957, 'Biogeochemistry of trace metals', in J. Hedgpeth (ed.), Treatise on marine ecology and paleoecology, Geol. Soc. Amer. Mem. 67, 345-357.

Goldberg, E.D., 1961a, 'Chemistry in the oceans', in M. Sears (ed.), Oceanography. Am. Assoc. Adv. Sci. Publ. 67, pp. 583-597.

Goldberg, E.D., 1961b, 'Chemical and mineralogical aspects of deep-sea sediments', in Phys. Chem. Earth, 4, pp. 281-302.

Goldberg, E.D., 1963, 'Mineralogy and chemistry of marine sedimentation', in F.P. Shepard (ed.), Submarine geology. Harper and Row, N.Y., 2 ed., pp. 436-446.

Goldberg, E.D., 1965, 'Minor elements in sea water', in J.P. Riley and G. Skirrow (eds.), Chemical Oceanography, London, Academic Press, 1965, 1, pp. 163-196.

Goldberg, E.D., 1971, 'Atmosphere dust, the sedimentary cycle, and man.', Geophysics, 1, 117-132.

Goldberg, E.D., 1976, 'Rock volatility and aerosol composition', Nature, 260, N 5547, 128-129.

Goldberg, E.D. and Arrhenius, G., 1958, 'Chemistry of Pacific pelagic sediments', Geochim. et Cosmochim. Acta, 13, N 2/3, 153-212.

Goldberg, E.D., Bowen, V.T., Farrington, J.W., Harvey, G., Martin, J.H., Parker, P.L., Risenbrough, R.W., Robertson, W., Schneider, E. and Gamble, E., 1978, 'The mussel watch', Envir. Conserv., 5, 101-125.

Goldberg, E.D., Koide, M., Schmitt, R.A. and Smith, R.H., 1963, 'Rare earth distributions in the marine environments', J. Geophys. Res., 68, N 14, 4209-4217.

Goldberg, E.D., Somayajulu, B.L., Galloway, J., Kaplan, J.R. and Faure, G., 1969, 'Differences between barites of marine and continental origins', Geochim. et Cosmochim. Acta, 33, N 2, 287-289.

Goldschmidt, V.M., 1937, 'The principles of distribution of chemical elements in minerals and rocks', J. Chem. Soc., 655-672.

Goldschmidt, V.M., 1958, Geochemistry. Oxford, Clarendon Press, 730 pp.

Goldschmidt, V.M. and Peters, C., 1931a, 'Zur Geochemie des Galliums', Nachr. Ges. Wiss. Göttingen, Math. phys., B. III, N IV, 177.

Goldschmidt, V.M. and Peters, C., 1931b, 'Zur Geochemie des Scandiums', Nachr. Ges. Wiss. Göttingen, Math. Phys., B. III, N IV, 286.

Goldschmidt, V.M. and Peters, C., 1932a, 'Zur Geochemie des Bors', Nachr. Ges. Wiss. Göttingen, Math. Phys., B. III, N IV, 404.

Goldschmidt, V.M. and Peters, C., 1932b, 'Zur Geochemie der Edelmetalle', Nachr. Ges. Wiss. Göttingen, Math. Phys. Kl., B. III, N IV, pp. 377.

Goldstein, S.L. and O'Nions, R.K., 1981, 'Nd and Sr isotopic relationships in pelagic clays and ferromanganese deposits', Nature, 292, N 5821, 324-327.

Golenetskiy, S.P., Malakhov, S.G. and Stepanok, V.V., 1981, 'On nature of global atmospheric aerosols', Astronom. vestnik, 15, N 4, 226-233 (in Russian).

Golenetskiy, S.P., Stepanok, V.V., Kolesnikov, E.M. and Murashov, D.A., 1977, 'On chemical composition and nature of Tungus cosmic body', Astronom. vestnik 11, N 3, 126-136 (in Russian).

Golenetskiy, S.P., Stepanok, V.V., Malakhov, S.G., Zhigalovskaya, T.N. and
    Mazin, L.A., 1982, 'On comet nature of global atmospheric aerosols', in
    Cosmic matter on the Earth, Kiev, Naukova Dumka, pp. 133-141 (in Russian).
Goncharov, G.N., Kalyamin, A.V. and Lurye, B.G., 1973, 'Iron-manganese
    concretions from the Pacific Ocean studied by a nuclear -resonance method',
    Dokl. Akad. Nauk SSSR, 212, N 3, 720-723 (in Russian).
Goodell, H.G., Meylan, M.A. and Grant, B., 1971, 'Ferromanganese deposits of the
    South Pacific Ocean, Drake passage and Scotia sea', Antarct. Res., 15,
    27-92.
Gordeev, V.V., 1981, 'New estimates of surficial discharge of dissolved and
    suspended matter to the ocean', Dokl. AN SSSR, 262, N 5, 1227-1230 (in
    Russian).
Gordeev, V.V., 1983, 'River discharge into the ocean and its geochemical parameters',
    Moscow, Nauka, 160 pp. (in Russian).
Gordeev, V.V., 1984, 'Geochemical parameters of river discharge to the ocean',
    Litol. i polezn. iskop., N 5, 29-50 (in Russian).
Gordeev, V.V., 1986, 'Forms of elements in manganese nodules and host sediments',
    in Manganese nodules of the Central Pacific, Moscow, Nauka, pp. 211-238 (in
    Russian).
Gordeev, V.V. and Demina, L.L., 1979, 'Direct observations of hydrotherms at the
    Pacific bottom (Galapagos active zone, Hess depression)', Geokhimia, N 6,
    902-917 (in Russian).
Gordeev, V.V. and Khandros, G.S., 1976, 'Concentration of micro-elements in the
    Pacific water and suspension', Okeanologicheskie issledovaniya, N 29,
    83-105 (in Russian).
Gordeev, V.V. and Lisitsin, A.P., 1978, 'Average chemical composition of the World
    river system suspensions and supply of river sedimentary material to the
    ocean', Dokl. AN SSSR, N 1, 225-228 (in Russian).
Gordeev, V.V. and Lisitsin, A.P., 1979, 'Microelements', in Oceanology. Chemistry
    of the ocean, V.I. Chemistry of oceanic waters, Moscow, Nauka, pp. 338-375
    (in Russian).
Gordeev, V.V., Miklishanskiy, A.Z. and Tambiev, S.B., 1984, 'Geochemistry of
    suspension and water in the Riga Bay', in Geological history and geochemistry
    of the Baltic Sea, Moscow, Nauka, pp. 18-32 (in Russian).
Gordeev, V.V., Mitropolskiy, A.Yu. and Turkina, O.V., 1983, 'Metal forms in the
    Ganges-Brahmaputra suspension', Geokhimiya, N 10, 1461-1467 (in Russian).
Gordon, C.M., Carr, R.A. and Larson, R.E., 1970, 'The influence of environ-
    mental factors on the sodium and manganese content on barnacle shells',
    Limnol. Oceanogr., 15, 461-466.
Gorshkova, T.I., 1931, 'Chemical and mineralogical analyses of sediments from the
    Barentz and Bering Seas', Tr. Gos. okeanogr. Inst., N 2-3, 83-127 (in
    Russian).
Gorshkova, T.I., 1957, 'The Kara Sea sediments', Tr. Vses. gidrobiol. obsch, N 8,
    68-99 (in Russian).
Gorshkova, T.I., 1966, 'Manganese in bottom deposits of the northern USSR seas
    and its biological value', Tr. Vses. n.-issl. Inst. ryb. khozyaistva i
    okeanografii, 60, 89-102 (in Russian).
Gorshkova, T.I., 1967, 'Manganese in bottom deposits of the northern seas', in
    Manganese deposits in the USSR, Moscow, Nauka, pp. 117-135 (in Russian).
Gorshkova, T.I., 1970, 'Pore solutions of the Baltic Sea and Riga Bay', in
    Chemical resources of seas and oceans, Moscow, Nauka, pp. 67-78 (in Russian).
Goryainov, I.N. and Goryainova, G.I., 1983, 'On the buoyancy of manganese
    nodules', Dokl. AN SSSR, 272, N 2, 432-437 (in Russian).
Graham, D.L., 1972, 'Trace metal levels in intertidal molluscs of California', Veliger,
    14, N 4, 365-372.
Graham, W., Bender, M.L. and Klinkhammer, G., 1976, 'Manganese in Narrangasett
    Bay', Limnol. Oceanogr., 21, N 5, 665-673.
Graham, J.W. and Cooper, S.C., 1959, 'Biological origin of manganese-rich deposits
    of the sea floor', Nature, 183, N 4667, 1050-1051.
Graham, W.F. and Duce, R.A., 1979, 'Atmospheric pathways of the phosphorus
    cycle', Geochim. et Cosmochim. Acta, 43, N 8, 1195-1208.

Gramm-Osipov, L.M. and Kiseleva, O.A., 1970, 'Distribution of Cu, Zn, Pb, Mo and W in the Pacific sediments', in Geology, geochemistry and metallogeny of the northwestern Pacific belt, Vladivostok, pp. 216-218 (in Russian).

Gramm-Osipov, L.M., Repechka, M.A., Volkova, T.I., Pliss, S.G. and Chernykh, V.N., 1973, 'On geochemistry of the Sea of Japan sediments', in Geology of the Sea of Japan, Vladivostok, 91-114 (in Russian).

Gramm-Osipov, L.M. and Shulga, Yu.M., 1980, 'Elements of manganese balance in oceanic water', Geokhimiya, N 8, 1222-1228 (in Russian).

Grant, J.B., 1967, 'A comparison of the chemistry and mineralogy with the distribution and physical aspects of marine manganese concretions of the southern oceans'. Contr. Sediment. Res. Lab. Fla. State Univ., 19, 99 pp.

Graybeal, A.L. and Heath, G.R., 1984, 'Remobilization of transition metals in surficial pelagic sediments from the eastern Pacific', Geochim. et Cosmochim. Acta, 48, 965-975.

Green, D.H. and Ringwood, A.E., 1973, 'Significance of a primitive lunar basaltic composition present in Apollo 15 soils and breccias', Earth and Planet. Sci. Lett., 19, N 1, 1-8.

Greenslate, J.L., 1974, 'Microorganisms participate in the construction of manganese nodules', Nature, 249, N 5453, 181-183.

Greenslate, J.L., 1978, 'Marine manganese concretion growth rates: non-radiometric considerations', Geophys. Res. Lett., 5, N 4, 237-239.

Greig, R.A., Adams, A. and Wenzloff, D.R., 1977a, 'Trace metal content of plancton and zooplancton collected from the New York Bight and Long Island Sound'. Bull. Environ. Contamin. Toxicol., 18, 3-8.

Greig, B.H. and Wenzloff, D.R., 1977b, 'Trace metals in finfish from the New York Bight and Long Island Sound', Mar. Poll. Bull., 8, 198-200.

Greig, R.A., Wenzloff, D.R., Adams, A., Nelson, B. and Shelpuc, C., 1977c, 'Trace metals in organisms from ocean disposal sites of the middle eastern United States', Arch. Environ. Contamin. Toxicol., 395-409.

Grill, E.V., 1978, 'The effect of sediment-water exchange on manganese deposition and nodule growth in Jervis Inlet, British Columbia', Geochim. et Cosmochim. Acta, 42, N 5, 485-494.

Grill, E.V., 1982, 'Kinetic and thermodynamic factors controlling manganese concentrations in oceanic waters', Geochim. et Cosmochim. Acta, 46, N 12, 2435-2446.

Grill, E.V., Murray, J.W. and MacDonald, R.D., 1968, 'Todorokite in manganese nodules from a British Columbia fjord', Nature, 219, 358-359.

Gromov, V.V., 1974, 'Analysis of microelements' behaviour in the ocean by desorption technique', Radiokhimiya, 18, N 6, 757-760 (in Russian).

Gromov, V.V., 1975, 'Sorption of elements of iron group by the Pacific bottom sediments', Okeanologiya, 15, N 1, 59-65 (in Russian).

Gromov, V.V. and Spitsin, V.I., 1975, 'Artificial radionuclides in marine environment', Moscow, Atomizdat, 224 pp. (in Russian).

Gromov, V.V. and Surikov, V.V., 1976, 'Analysis of the forms of elements of iron group in sea water by ultrafiltration technique', Okeanologicheskie issledovaniya 29, 106-110 (in Russian).

Grönvold, F., Haraldsen, H. and Kjekshus, A., 'On the sulfides, selenides and tellurides of platinum', Acta chem. Scand., 14, N 9, 1879-1893.

Grütter, A. and Buser, W., 1957, 'Untersuchungen an Mangansedimenten', Chimia, 11, 132-133.

Gryzhankova, L.N., Laktionova, V.N., Boichenko, E.A. and Karyakin, A.V., 1973a, 'Distribution of polyvalent metals in various algae types', Okeanologiya, 13, is. 4, 611-614 (in Russian).

Gryzhankova, L.N., Saenko, G.N., Karyakin, A.V. and Laktionova, N.V., 1973b, 'Concentrations of some metals in the Sea of Japan algae', Okeanologiya, 13, N 2, 259-267 (in Russian).

Gryzhankova, L.N., Udelnova, T.M. and Boichenko, E.A., 1975, 'Iron, manganese and titanium content in marine plants', Izv. AN SSSR, ser. biol., N 1, pp. 76-81 (in Russian).

Guichard, F., Reyss, J.L. and Yokohama, Y., 1978, 'Growth rate of manganese nodule measured with ^{10}Be and ^{26}Al', Nature, 272, N 5649, 155-156.

Guichard, F., Reyss, J.L., Yokohama, Y. and Lalou, C., 1977, 'Etude radiochimique d'une croute de manganese par ^{10}Be et ^{26}Al: les nodules de manganèse croissant - its vite ou lantement', 5e Réun. annu. Sci. terre, Rennes, p. 260.

Gümbel, W., 1878, 'Über die im Stillen Ozean auf dem Meeresgrunde vorkommenden Manganknollen', Sitzungsber. Bayer. Akad. Wiss. Math., Phys. K., B. 8, S. 189-209.

Gundlach, H., Jung, W. and Schweisfurth, R., 'Manganese-oxidizing microorganisms and their importance for the genesis of manganese ore deposits (abstr.) Abstr.', 25th Intl. Geol. Congr., 3, p. 773.

Gunglach, H., Marchig, V. and Schnier, C., 1977, 'Zur Geochemie von Manganknollen aus dem Zentralpazific und ihrer Sedimentunterlage. 2. Porewasser und Meerwasser', Geol. Jb., D 23, S. 67-90.

Gurvich, E.G., Lisitsin, A.P. and Kurinov, A.D., 1980a, 'Hafnium'. In Geochemistry of hydrolizating elements, Moscow, Nauka, pp. 181-200 (in Russian).

Gurvich, E.G., Lukashin, V.N., Lisitsin, A.P. and Kurinov, A.D., 1980b, 'Rare-earth elements and yttrium', in Geochemistry of hydrolizating elements, Moscow, Nauka, pp. 71-116 (in Russian).

Gurvich, E.G. and Schurygina, E.V., 1982, 'Selective leaching of oxide minerals of tetravalent manganese from oceanic bottom sediments and manganese nodules'. In Geology of oceans and seas. Abstracts of the reports to the 5th All-Union seminar on marine geology, Moscow, Nauka, 1, p. 130 (in Russian).

Guseva, K.A., 1937, 'Manganese effect upon algae growth', Mikrobiologiya, 6, N 3, 292-307 (in Russian).

Hager, J.L., 1980, 'Sorption of manganese and silica by clay and carbonate', Mar. Chem., 9, N 3, 199-209.

Hajash, A., 1975, 'Hydrothermal processes along midocean ridges: an experimental investigation', Contrib. Mineral. Petrol., 53, 205-226.

Hajash, A. and Archer, P., 1980, 'Experimental seawater/basalt interactions: effects of cooling', Contribs. Mineral and Petrol., 75, N 1, 1-13.

Halbach, P., Giovanoli, R. and Borstel, D., 1982, 'Geochemical processes controlling the relationship between Co, Mn and Fe in early diagenetic deep-sea nodules', Earth and Planet. Sci. Lett., 60, 226-236.

Halbach, P. and Manheim, F.T., 1984, 'Potential of cobalt and other metals in ferromanganese crusts on seamounts of the Central Pacific basin', Mar. Mining, 4, N 4, 319-336.

Halbach, P. and Ozkara, M., 1979, 'Morphological and geochemical classification of deep-sea ferromanganese nodules and its genetical interpretation', in La genèse des nodules de manganèse. Edit. CNRS. N 289, Paris, pp. 77-88.

Halbach, P., Ozkara, M. and Hense, I., 1975, 'The influence of metal content on the physical and mineralogical properties of pelagic manganese nodules', Mineral. Deposita, 10, N 4, 397-411.

Halbach, P., Puteanus, D. and Manheim, F.T., 1984, 'Platinum concentrations in ferromanganese seamount crusts from the Central Pacific', Naturwiss., B. 71, 577-579.

Halbach, P., Rahn, E. and Marchig, V., 1979, 'Distribution of Si, Mn, Fe, Ni, Cu, Co, Zn, Pb, Mg and Ca in grain-size fractions of sediment samples from a manganese-nodules field in the Central Pacific', Mar. Geol., 29, 237-252.

Halbach, P., Segl, M., Puteanus, D. and Mangini, A., 1983, 'Co-fluxes and growth rates in ferromanganese deposits from Central Pacific seamount areas', Nature, 304, N 5928, 716-719.

Halbach, P., Sherhag, C., Hebisch, V. and Marchig, V., 1981, 'Geochemical and mineralogical control of different genetic types of deep-sea nodules from the Pacific ocean', Miner. Deposita, 16, N 1, 59-84.

Hall, R.A., Zook, E.G. and Meaburn, G.M., 1978, 'National marine fisheries survey of trace elements in the fishing resource', U.S. Dept. Commerce NOAA Tech.

Rept. NMFS SS RF-721, 313 pp.

Harada, K., 1984, 'Formation of manganese nodules and micronodules in oceanic systems', Intern. Geol. Congr. XXVII, Moscow, Abstr., 6, S. 12, p. 130.

Harada, K. and Nishida, S., 1976, 'Biostratigraphy of some marine manganese nodules', Nature, 260, N 5554, 770-771.

Harada, K. and Nishida, S., 1979, 'Biochronology of some Pacific ferromanganese nodules and their growth mechanism', in Lalou, C. (ed.), La genèse de nodules de manganèse, Edit. CNRS., Paris, pp. 211-216.

Harriss, R.C., 1968, 'Mercury content of deep-sea manganese nodules', Nature, 219, N 5149, 54-55.

Harriss, R.C., Crocket, J.H. and Stainton, M., 1968, 'Palladium, iridium and gold in deep-sea manganese nodules', Geochim. et Cosmochim. Acta, 32, N 10, 1049-1056.

Harriss, J.E., Fabris, G.J., Statham, P.J. and Tawfik, F., 1979, 'Biogeochemistry of selected heavy metals in Western Port, Victoria and use of invertebrates as indicators with emphasis on Mytilus edulis planulatus', Austral. J. Mar. Freshwater Res., 30, 159-178.

Harrison, P.R., Rahn, K.A., Dams, R., Robbins, J.A., Winchester, J.W., Brar, S.S. and Nelson, D.M., 1971, 'Areawide trace metal concentrations measured by multielement neutron activation analysis', APCA J., 21, 563-570.

Hart, T.Y. and Currie, R.J., 1960, 'The Benguela current', Discovery Rept., 31, 123-298.

Hartmann, M., 1964, 'Zur Geochemie von Mangan und Eisen in der Ostsee', Meyniana, 14, N 53, 3-20.

Hartmann, M., 1969, 'Investigation of Atlantis II Deep samples taken by the F.S. "Meteor"', in E.T. Degens and D.A. Ross (eds.), Hot brines and Recent heavy metal deposits in the Red Sea, N.Y., Springer, pp. 204-207.

Hartmann, M. and Müller, P., 1974, 'Geochemische Untersuchungen an Sedimenten und Porenwassern', Meerestechnik, 5, N 6, 201-202.

Hartmann, M. and Müller, P.J., 1982, 'Trace metals in interstitial waters from Central Pacific ocean sediments', in The dynamic environment of the ocean floor, Lexington, D.C., Heath and Co . pp. 285-301.

Haynes, B.W., Law, S.L. and Barron, D.C., 1986, 'An elemental description of Pacific manganese nodules', Mar. Mining, 5, N 3, 239-276.

Heath, G.R., 1979, 'Burial rates, growth rates and size distributions of deep sea manganese nodules', Science, 205, N 4409, 903-904.

Heath, G.R. and Dymond, J., 1977, 'Genesis and transformation of metalliferous sediments from East Pacific Rise, Bauer Deep and Central Basin, northwest Nazca plate', Geol. Soc. Amer. Bull., 88, 723-733.

Heidam, N.Z., 1981, 'On the origin of the arctic aerosol: a statistical approach', Atm. Envir., 15, N 8, 1421-1427.

Hekinian, R., Fevrier, M., Bishoff, J.L. et al., 1980, 'Sulfide deposits from the East Pacific Rise near 21°N', Science, 207, N 4438, 1433-1444.

Hekinian, R., Fevrier, M., Avedik, F. et al., 1983, 'East Pacific Rise near 13°N: geology of new hydrothermal fluids', Science, 219, N 4590, 1321-1324.

Hekinian, R. and Hoffert, M., 1975, 'Rate of pelagonitization and manganese coating on basaltic rocks from the rift valley in the Atlantic Ocean near 36°50'N', Mar. Geol., 19, 91-109.

Helgeson, H.C., 1968, 'Geologic and thermodynamic characteristics of the Salton Sea geothermal system', Amer. J. Sci., 266, N 3, 129-166.

Hem, J.D., 1963, 'Chemical equilibria and rates of manganese oxidation', Geol. Surv. Water Supply Pap., 1667-A. U.S. Gov. Print. Off., 63 pp.

Hem, J.D., 1972, 'Chemical factors that influence the availability of iron and manganese in aqueous systems', Bull. Geol. Soc. Am., 83, 443-450.

Herron, M.M., Langway, C.C., Weiss, H.V. and Cragin, J.H., 1977, 'Atmospheric trace metals and sulfate in the Greenland ice sheet', Geochim. et Cosmochim. Acta, 41, N 7, 915-920.

Heye, D., 1975, 'Wachstumverhältnisse von Manganknollen', Geol. Jb., Reihe E, H. 5, pp. 3-122.

Heye, D., 1978a, 'The internal micro-structure of manganese nodules and their relationship to the growth rate', Mar. Geol., 26, N 3/4, M59-M66.

Heye, D., 1978b, 'Changes in the growth rate of manganese nodules from the Central Pacific in the area of seamount as shown by the ionium method', Mar. Geol., 28, N 3/4, M59-M65.

Heye, D., 1978c, 'Growth conditions of manganese nodules: comparative studies of growth rate, magnetization, chemical compositions and internal structure', in Progress in Oceanography, 7, P. 5/6, Oxford, Pergamon Press, 90 pp.

Heye, D., 1979, 'Ionium dating of manganese nodules: Does diffusion of ionium occure within manganese nodules?', in La genèse de nodules de manganèse, Edit. CNRS, Paris, pp. 93-99.

Heye, D. and Beiersdorf, H., 1973, 'Radioactive und magnetische Untersuchungen an Manganknollen zur Ermittlung der Wachstumsgeschwindigkeit bzw. zur Altersbestimmung', Z. Geophys., 39, 703-726.

Heye, D. and Marchig, V., 1977, 'Relationship between the growth rate of manganese nodules from the central Pacific and their chemical constitution', Mar. Geol., 23, N 1/2, M19-M25.

Hignette, M., 1979, 'Composition des concretions minerales contenues dans les reins de 2 mollusques lamellibranches Pinna nobilis L. et Tridacna maxima', C.R. Acad. Sci. Paris, 289, 1069-1082.

Hirose, K. and Sugimura, Y., 1984, Excess ^{228}Th. in the airbone dust: an indicator of continental dust from the East Asian deserts', Earth and Planet. Sci. Lett., 70, N 1, 110-114.

Hoang, C.T. and Servant, J., 1974, 'Exemple d'un apport continental de quelques métaux dans l'aerosol au-dessus de l'Atlantique Nord à la latitude de 40°N', J. de Recherches Atmospheriques, 8, N 3/4, 791-805.

Hodge, V.S., Johnson, S.R. and Goldberg, E.D., 1978, 'Influence of atmospherically transported aerosols on surface ocean water composition', Geochem. J., 12, 7-20.

Hodge, V.F., Stallard, M., Koide, M. and Goldberg, E.D., 1985, 'Platinum and the platinum anomaly in the marine environment', Earth and Planet. Sci. Lett., 72 N 2/3, 158-162.

Hoffman, G.L., Duce, R.A. and Hoffman, E.J., 1972, 'Trace metals in the Hawaiian marine atmosphere', J. Geophys. Res., 72, 77, N 27, 5322-5329.

Hoffman, G.L., Duce, R.A., Walsh, P.R., Hoffman, E.J., Ray, B.J. and Fashing, J.L., 1974, 'Residence time of some particulate trace metals in the oceanic surface microlayer: significans of atmospheric deposition', J. Rech. Atmosph., 8, N 3-4, 745-759.

Holdren, G.R., Bricher, O.P. and Matisoff, G., 1975, 'A model for the control of dissolved manganese in the interstitial waters of Chesapeake Bay', in T.M. Church (ed.), Marine chemistry in the coastal environment, Wash., Am. Chem. Soc., pp. 364-381.

Holeman, J.N., 1968, 'The sediment yield of major rivers of the world', Water Resour. Res., 4, N 4, 737-747.

Holweger, H., 1979, 'Abundances of the elements in the Sun', in Les elements et leurs isotopes dans l'Universe, Univ. Liège, pp. 117-129.

Honjo, S., 1978, 'Sedimentation of materials in the Sargasso Sea at a 5367 m deep station', J. Mar. Res., 36, N 3, 469-492.

Honjo, S., 1980, 'Material fluxes and modes of sedimentation in the mesopelagic and bathypelagic zones', J. Mar. Res., 38, 53-97.

Honjo, S., Manganini, S.J. and Cole, J.J., 1982, 'Sedimentation of biogenic matter in the deep ocean', Deep Sea Res., 29, 609-625.

Honnorez, J., 1981, 'The aging of the oceanic crust at low temperature', in C. Emiliani (ed.), The Sea, 7, The oceanic lithosphere, N.Y., John Wiley, pp. 525-587.

Honnorez, J., Bohlke, J.R. and Honnorez-Guerstein, B.M., 1979, 'Petrographical and geochemical study of the low temperature submarine alteration of basalt from Hole 396B, Leg 46', Init. Rep. DSDP, 46, Wash. D.C., pp. 299-329.

Horn, R.A., 1969, Marine chemistry. The structure of water and the chemistry of the Hydrosphere, New York, Wiley, 568 pp.

Horn, M.K. and Adams, J.A.S., 1966, 'Computer-derived geochemical balances and element abundances', Geochim. et Cosmochim. Acta, 30, N 3, 279-297.

Horn, D.R., Ewing, M., Horn, B.M. and Delach, M.N., 1972, 'World-wide distribution of manganese nodules', Ocean Ind., 7, N 1, 26-29.

Horn, D.R., Horn, B.M. and Delach, M.N., 1972, 'Ferromanganese deposits of the North Pacific Ocean', U.S. Nat. Sci. Found., Tech. Rep. N 1, GX-33616, 78 pp.

Horn, D.R., Horn, B.M. and Delach, M.N., 1973, 'Factors which control the distribution of ferromanganese nodules and proposed research vessels track, North Pacific', Tech. Rep. Off. Int. Decade Ocean Explor., 8, 20 pp.

Horowitz, A., 1970, 'The distribution of Pb, Ag, Sn, Tl and Zn in sediments on active oceanic ridges', Mar. Geol., 9, N 4, 241-259.

Horowitz, A. and Cronan, D.S., 1976, 'The geochemistry of basal sediments from the North Atlantic Ocean', Mar. Geol., 20, 205-228.

Horowitz, A. and Presley, B.J., 1977, 'Trace metal concentrations and partitioning in zooplancton, neuston and benthos from the South Texas outer continental shelf', Arch. Environ. Contamin. Toxic., 5, 241-255.

Huh, C.A. and Ku, T.L., 1984, 'Radiochemical observations on manganese nodules from three sedimentary environments in the north Pacific', Geochim. et Cosmochim. Acta, 48, N 5, 951-963.

Hulse, M., Mahoney, J.S., Schroeder, G.O., Hacker, C.S. and Pier, S.M., 1980, 'Environmentally acquired lead, cadmium and manganese in cattle egret, Bubulcus ibis, and the laughing gull, Laris atricilla', Arch. Environ. Contamin. Toxicol., 9, 65-78.

Humphris, S. and Thompson, G., 1978, 'Trace elements mobility during hydrothermal alteration of oceanic basalts', Geochim. et Cosmochim. Acta, 42, N 1, 127-136.

Humphris, S., Thompson, R.N. and Marriner, G.F., 1979, 'The mineralogy and geochemistry of basalt weathering, Holes 417 A and 418', in T. Donnelly, J. Francheteau et al., Init. Rep. DSDP, 53, Wash. D.C., 2, 1201-1217.

Hunt, C.D., 1983, 'Incorporation and deposition of Mn and other trace metals by flocculent organic matter in a controlled marine ecosystem', Limnol. oceanogr., 28, N 2, 302-308.

Hunter, K.A., 1980, 'Processes affecting particulate trace metals in the sea surface microlayer', Mar. Chem., 9, N 1, 49-70.

Ichikawa, R., 1961, 'On the concentration factors of some important radionuclides in marine food organisms', Bull. Jap. Soc. Sci. Fich., 27, 66-74.

Iimory, S., 1927, 'Formation of the radioactive manganiferous deposits from Tanokami and the source of manganese in deep sea manganese nodules', Sci. Pap. Inst. Phys. Chem. Res., Tokio, 7, 249-252.

Immel, R. and Osmond, J.K., 1976, 'Micromanganese nodules in deep-sea sediments: uranium isotopic evidence for post-depositional origin', Chem. Geol., 18, N 4, 263-272.

Ingri, J., 1985, 'Geochemistry of ferromanganese concretions in the Barents sea', Mar. Geol., 67, 101-119.

Inoue, T., Huang, Z.Y., Imamura, M., Tanaka, S. and Usui, A., 1983, '^{10}Be and ^{10}Be/^{9}Be in manganese nodules', Geochem. J., 17, N 6, 307-312.

Ireland, M.P. and Wootton, R.J., 1977, 'Distribution of lead, zinc, copper and manganese in the marine gastropods Thais lapillus and Littorina littorea around the coast of Whales', Environ. Poll., 12, 27-41.

Isaeva, A.B., 1960, 'Tungsten in the Sea of Okhotsk bottom deposits', Dokl. AN SSSR, 131, N 2, 416-419 (in Russian).

Isaeva, A.B., 1967, 'Chemical composition of ferromanganese nodules of the Indian Ocean', Lithologia i polezn. iskop., N 3, 43-56 (in Russian).

Isaeva, A.B., 1977, 'Peculiarities of tungsten distribution in the Sea of Okhotsk sediments', Geokhimiya, N 2, 246-253 (in Russian).

Isaeva, A.B., 1982, 'Iron and manganese in the Indian ocean sediments', Okeanologiya, 22, is. 4, 603-609 (in Russian).

Ishii, T., Suzuki, H. and Koyanagi, T., 1978, 'Determination of trace elements in marine organisms. I. Factors for variation of concentration of trace elements',

Bull. Jap. Soc. Fish., 44, 155-162.

Iskowitz, J.M., Lee, J.J.H., Zeitlin, H. and Fernando, Q., 1982, 'Determination of thallium in deep-sea ferromanganese nodules', *Mar. Mining*, 3, N 3-4, 285-295.

Jahnke, R., Heggie, D., Emerson, S. and Grundmanis, V., 1982, 'Pore waters of Central Pacific ocean: nutrient results', *Earth and Planet. Sci. Lett.*, 61, N 2, 233-256.

Jedwab, J., 1975, 'A method for extraction and analysis of possible cosmic particles from manganese nodules', *Meteoritics*, 10, 233-240.

Jedwab, J., 1979, 'Copper, zinc and lead minerals suspended in ocean waters', *Geochim. et Cosmochim. Acta*, 43, N 1, 101-110.

Jedwab, J., 1980, 'Rare antropogenic and natural particles suspended in deep ocean waters', *Eart and Planet. Sci. Lett.*, 49, N 2, 551-564.

Jenkins, W.J., Edmond, J.M. and Corliss, J.B., 1978, 'Excess ^{3}He and ^{4}He in Galapagos submarine hydrothermal waters', *Nature*, 272, N 5649, 156-158.

Jickells, T.D., Knop, A.H. and Church, T.M., 1984, 'Trace metals in Bermuda rainwater', *J. Geophys. Res.*, 89, N D1, 1423-1428.

Johnston, J.H. and Glasby, G.P., 1978, 'The secondary iron oxidehydroxide mineralogy of some deep sea and fossil manganese nodules. A Mössbauer and X-ray study', *Geochem. J.*, N 3, 153-164.

Joly, J., 1908, 'On the radium content of deep-sea sediments', *Phil. Mag.*, 16, 190-197.

Jones, L.H.P. and Milne, A.A., 1956, 'Birnessite, a new manganese oxide mineral from Aberdeenshire, Scotland', *Mineral. Mag.*, 31, 283-288.

Jones, C.J. and Murray, J.W., 1985, 'The geochemistry of manganese in the northeast Pacific ocean off Washington', *Geochim. et Cosmochim. Acta*, N 1, 81-92.

Kachanov, N.N., 1980, 'New data on morphometry of manganese nodules in the Indian ocean', *Dokl. AN UkSSR*, B, N 11, 10-13 (in Russian).

Kadko, D. and Burckle, L.H., 1980, 'Manganese nodule growth rates determined by fossil diatom dating', *Nature*, 287, 725-726.

Kalhorn, S. and Emerson, S., 1984, 'The oxidation state of manganese in surface sediments of the deep sea', *Geochim. et Cosmochim. Acta*, 48, 897-902.

Kalinenko, V.O., 1946, 'Role of bacteria in manganese nodule formation', *Dokl. AN SSSR*, 15, N 5, 364-369 (in Russian).

Kalinenko, V.O., 1949, 'Origin of manganese nodules', *Mikrobiologiya*, 18, N 6, 528-532 (in Russian).

Kalinenko, V.O., Belokopytova, O.V. and Nikolaeva, G.G., 1962, 'Bacteriogenic formation of manganese nodules in the Indian ocean', Okeanologiya, 2, N 6, 1050-1059 (in Russian).

Kallemeyn, G.W. and Wasson, J.T., 1982, 'The compositional classification of chondrites: III. Ungrouped carbonaceous chondrites', *Geochim. et Cosmochim. Acta*, 46, N 11, 2217-2228.

Kaneoka, J. and Teraoka, N., 1978, 'Excess ^{129}Xe and high ^{3}He/^{4}He ratios in olivine phenocrysts of Kapuho lava and xenolithic dunites from Hawaii', *Earth and Planet. Sci. Lett.*, 39, N 3, 382-386.

Karas, M. and Greenslate, J., 1979, 'Marine manganese concentrations: Shape and size related growth models', *Deep-Sea Res.*, 26, A, N 7A, 809-824.

Kastner, M., 1979, 'Zeolites', in R. Burns (ed.), *Marine Minerals*, Wash., Mineral. Soc. Am., pp. 111-122.

Kato, T., Endo, M. and Kato, M., 1983, 'Vertical distribution of various elements in sediment cores from the Japan Sea', *Mar. Geol.*, 53, N 4, 277-290.

Keays, R.R. and Scott, R.B., 1976, 'Precious metals in ocean-ridge basalts: implications for basalts as source rock for gold mineralization', *Econ. Geol.*, 71, N 4, 705-720.

Kennett, J.P. and Watkins, N.D., 1975, 'Deep sea erosion and manganese nodule development in the southeast Indian Ocean', *Science*, 188, N 4192, 1011-1013.

Kepkay, P.E. and Nealson, K.H., 1982, 'Surface enhancement of sporulation and manganese oxidation by a marine bacillus', *J. Bacteriol.*, 151, N 2, 1022-1026.

Khitrov, L.M. and Kholina, Yu.B., 1972, 'Forms of manganese occurrence in oceanic water', in Review of modern geochemistry and analytical chemistry, Moscow, Nauka, pp. 493-500 (in Russian).

Kholodkevich, I.V. and Geptner, A.P., 1982, 'Experimental research of hydrothermal transformation of Icelandic basaltoids', Litol. i polezn. iskop., N 4, 68-78 (in Russian).

Kholodny, N., 1926, 'Die Eisenbacterien', Jena, Gustav Fischer, 162 pp.

Kholodny, N.G., 1953, 'Iron bacteria', Moscow, Izd. AN SSSR, 224 pp. (in Russian).

Kholodov, V.N. and Turovsky, D.S., 1985, 'On sedimentation of the Caspian Sea. Report I. Geochemistry of microelements in recent sediments', Litol. i polezn. iskop., N 1, 17-34 (in Russian).

Khristoforova, N.K. and Bogdanova, N.N., 1980, 'Mineral composition of seaweeds from coral islands of the Pacific ocean as a function of environmental conditions', Mar. Ecol. Prog. Ser., 3, 25-29.

Khristoforova, N.K. and Maslova, L.M., 1983, 'Comparative analysis of heavy metal contamination in the near coastal Atlantic and Western Pacific waters based on mineral composition of fucus algae', Biol. morya, N 1, 3-11 (in Russian).

Khristoforova, N.K., Sinkov, N.A., Saenko, G.N. and Koryakova, M.D., 1976, 'Content of microelements Fe, Mn, Ni, Cr and Zn in proteins of marine algae', Biol. morya, N 2, 69-72 (in Russian).

Khrustalev, Yu.P., 1978, 'Regularities in recent sedimentation in the northern Caspian Sea', Rostov Univ., 207 pp. (in Russian).

Khrustalev, Yu.P., Reznikov, S.A. and Turovsky, D.S., 1977, 'Lithology and geochemistry of the Aral Sea bottom sediments', Rostov Univ., 160 pp. (in Russian).

Khrustalev, Yu.P. and Shcherbakov, F.A., 1974, 'Upper Quaternary deposits of the Azov Sea and Conditions of their accumulation', Rostov Univ., 149 pp. (in Russian).

King, D.E.C. and Pasho, D.W., 1979, 'A generalized estimating model for the Kennecott Joint Venture, manganese nodule processing facility', Ottawa, Dept. Energy, Mines and Resources, 34 pp.

Kirchner, S.J., Oona, H., Fernando, Q. et al., 1973, 'Identification of elemental sulfur in deep-sea ferromanganese nodules', Anal. Chem., 50, N 12, 1701-1702.

Kizevetter, I.V., 1960, 'Industry and processing of marine plants in Primorye', Vladivostok, Primorskoye knizhnoe izd-vo, 120 pp. (in Russian).

Kizevetter, I.V., 1973, 'Biochemistry of row food materials of marine origin', Moscow, Pischevaya promyshlennost, 424 pp. (in Russian).

Klenova, M.V., 1948, 'Marine geology', Moscow, Uchpedgiz, 500 pp. (in Russian).

Klenova, M.V. and Pakhomova, A.S., 1940, 'Manganese in sediments of the northern seas', Dokl. AN SSSR, 18, N 1, 87-89 (in Russian).

Klinkhammer, G.P., 1980, 'Early diagenesis in sediments from the eastern equatorial Pacific. II. Pore water metal results', Earth and Planet. Sci. Lett., 49, N 1, 81-101.

Klinkhammer, G.P., 1980, 'Observations on the distribution of manganese over the East Pacific Rise', Chem. Geol., 29, N 3-4, 211-226.

Klinkhammer, G.P. and Bender, M.L., 1980, 'The distribution of manganese in the Pacific ocean', Earth and Planet. Sci. Lett., 46, N 3, 361-384.

Klinkhammer, G., Bender, M. and Weiss, R.F., 1977, 'Hydrothermal manganese in the Galapagos Rift', Nature, 269, N 5626, 319-320.

Klinkhammer, G.P., Heggie, D.T. and Graham, D.W., 1982, 'Metal diagenesis in oxic marine sediments', Earth and Planet. Sci. Lett., 61, N 2, 211-219.

Klinkhammer, G., Rona, P., Greaves, M. and Elderfield, H., 1985, 'Hydrothermal manganese plumes in the Mid-Atlantic Ridge rift valley', Nature, 314, 727-731.

Knauer, G.A., 1970, 'The determination of magnesium, manganese, iron, copper and zinc in marine shrimp', Analyst, 93, 476-480.

Kondratyev, K.Ya., Moskalenko, N.I. and Pozdnyakov, D.V., 1983, 'Atmospheric aerosol', Leningrad, Gidrometeoizdat, 224 pp. (in Russian).

Konovalov, G.S., Ivanova, A.A. and Kolesnikova, T.Kh., 1968, 'Trace and rare elements in dissolved and suspended matter of the major rivers of the USSR', in Geochemistry of sedimentary rocks and ores, Moscow, Nauka, pp. 72-87 (in Russian).

Kononov, V.I., 1983, 'Geochemistry of thermal water in areas of Recent volcanism', Tr. Geol. Inst. AN SSSR, N 379, 215 pp. (in Russian).

Kontorovich, A.E., 1968, 'Forms of migration of elements in rivers of humic zone: data on western Siberia and other regions', in Geochemistry of sedimentary rocks and ores, Moscow, Nauka, pp. 88-101 (in Russian).

Koszy, F., 1949, 'Thorium in sea-water and marine sediments', Geol. Fören Stockh. Förh., 71, 238-242.

Kovalskiy, V.V., Gribovskaya, I.F., Chekunova, V.I. and Rezaeva, L.G., 1974, 'Concentration of microelements by ascidia of the Sea of Okhotsk', Tr. biogeokhim. lab., 13, 217-223 (in Russian).

Kraemer, T. and Schornick, J.C., 1974, 'Comparison of elemental accumulation rates between ferromanganese deposits and sediments in the South Pacific Ocean', Chem. Geol., 13, N 3, 187-196.

Krasintseva, V.V., Grichuk, D.V., Romanova, G.I. and Kadukin, A.I., 1982, 'Migration processes and forms of chemical elements in pore water of bottom deposits from Ivankovo reservoir', Geokhimiya, N 9, 1342-1354 (in Russian).

Kremling, K., 1983a, 'Trace metal fronts in European shelf waters', Nature, 303, N 5419, 225-227.

Kremling, K., 1983b, 'The behavior of Zn, Cd, Cu, Ni, Co, Fe and Mn in anoxic Baltic waters', Mar. Chem., 13, N 2, 87-108.

Kremling, K. and Peterson, H., 1978, 'The distribution of Mn, Fe, Zn, Cd and Cu in Baltic Sea water; a study on the basis of anchor station', Mar. Chem., 6, N 2, 155-170.

Krinsley, D., 1960, 'Trace elements in the tests of planctonic foraminifera', Micropaleontology, 6, 297-300.

Krishnaswami, S. and Cochran, J.K., 1978, 'Uranium and thorium series nuclides in oriented ferromanganese nodules: growth rates, turnover times and nuclide behavior', Earth and Planet. Sci. Lett., 40, N 1, 45-62.

Krishnaswami, S., Cochran, J.K., Turekian, K.K. and Sarin, M.M., 1979, 'Time scales of deep-sea ferromanganese nodule growth', in La genèse de nodules de manganèse. Edit. CNRS, Paris, 251-260.

Krishnaswami, S. and Lal, D., 1972, 'Manganese nodules and budget of trace solubles in oceans', in D. Dyrssen and D. Jagner (eds.), The changing chemistry of the Oceans, Proc. Nobel Symp. 20-th., N.Y., Interscience, pp. 307-320.

Krishnaswami, S., Mangini, A., Thomes, J.H., Sharma, P., Cochran, J.R., Turekian, K.K. and Parker, P.D., 1982, 'Be and Th isotopes in manganese nodules and adjacent sediments: nodule growth histories and nuclide behaviour', Eart and Planet. Sci. Lett., 59, N 2, 217-234.

Krishnaswami, S., Somayajulu, B.L.K. and Moore, W.S., 1972, 'Dating of manganese nodules using beryllium-10', in D.R. Horn (ed.), Ferromanganese deposits on the ocean floor, Nat. Sci. Found., Wash., pp. 117-122.

Krom, M.D. and Sholkowitz, E.R., 1978, 'On the association of iron and manganese with organic matter in anoxic marine sediments', Geochim. et Cosmochim. Acta, 42, N 6, 607-611.

Krumbein, W.E., 1971, 'Manganese oxidizing fungi and bacteria in recent shelf sediments of the Bay of Biscay and the North Sea', Nature, 58, 56-57.

Krylov, A.Y., Mamyrin, B.A., Silin, Y.I. and Khabarin, L.V., 1973, 'Helium isotopes in sediments of the oceans', Geokhimiya, N 2, 284-288 (in Russian).

Krylov, O.T., Tikhomirov, V.N., Gromov, V.V. and Spitsin, V.I., 1979, 'Determination of manganese species in sea water', Dokl. AN SSSR, 249, N 5, 1209-1212 (in Russian).

Krylov, O.T., Tikhomirov, V.N., Gromov, V.V. and Yashkevich, V.I., 1981, 'Laboratory modelling of manganese sorption from sea water by nodules and estimates of nodule growth rates', Okeanologiya, 21, N 6, 1047-1052 (in Russian).

Ku, T.L., 1977, 'Rates of accretion', In Glasby, G.P. (ed.), Marine Manganese Deposits, Elsevier, N.Y., pp. 249-267.

Ku, T.L. and Broecker, W.S., 1967, 'Uranium, thorium and protactinium in a manganese nodule', Earth and Planet, Sci. Lett., 2, N 4, 317-320.

Ku, T.L. and Broecker, W.S., 1969, 'Radiochemical studies on manganese nodules of deep-sea origin', Deep-Sea Res., 16, N 6, 625-637.

Ku, T.L. and Glasby, G.P., 1972, 'Radiometric evidence for the rapid growth rate of shallow-water, continental margin manganese nodules', Geochim. et Cosmochim. Acta, 36, N 6, 699-703.

Ku, T.L. and Knauss, K.G., 1979, 'Radioactive disequilibrium in fissure-filling material and its implication in dating of manganese nodules', in La genèse de nodules de manganèse, Ed. CNRS, Paris, pp. 289-293.

Ku, T.L., Kusakabe, M., Nelson, D.E., Southon, J.R., Korteling, R.G., Vogel, G. and Nowikov, I., 1982, 'Constancy of oceanic deposition of ^{10}Be as recorded in manganese crusts', Nature, 299, N 5880, 240-242.

Ku, T.L., Omura, A. and Chen, P.S., 1979, '^{10}Be and U-series isotopes in manganese nodules from the Central North Pacific', in Bischoff, J.L., Piper, D.Z. (eds.), Marine geology and oceanography of the Pacific Manganese Nodule Province. Plenum, N.Y., a, London, 791-814.

Kulesza-Owsikowska, G., 1979, 'Iron concretions from the southern Baltic', Bull. Acad. Pol. Sci., Ser. sci. terra, 7, N 3-4, 137-141.

Kulikov, N.N., 1961, 'Sedimentation in the Kara Sea', in Recent marine and oceanic sediments, Moscow, Izd. AN SSSR, pp. 437-447 (in Russian).

Kulm, L.V.D. et al., 1972, Initial Report of the Deep-Sea Drilling Project. Wash., Gov. Print. Off., v. XVIII, 1077 pp.

Kunzendorf, H. and Friedrich, G., 1976a, 'Uranium and thorium in deep-sea manganese nodules from the central Pacific', Trans. Inst. Min. Metall., Sect.B, 85, 283-288.

Kunzendorf, H. and Friedrich, G.H.W., 1976b, 'The distribution of U and Th in growth zones of manganese nodules', Geochim. et Cosmochim. Acta, 40, N 7, 849-852.

Kunzendorf, H. and Friedrich, G., 1978, 'Die Verteilung von Uran in Mangan-Knollen in Abhängigkeit von der Knollenfazies und der Morphologie des Meersbodens', Erzmetall, 30, N 12, 590-592.

Kunzendorf, H., Glasby, G.P., Plüger, W.L. and Friedrich, G., 1978, 'The distribution of uranium in some Pacific manganese nodules and crusts', Uranium, 1, 19-36.

Kunzendorf, H., Plüger, W.L. and Friedrich, G.H., 1983, 'Uranium in Pacific deep-sea sediments and manganese nodules', J. Geochem. Explor., 19, 147-162.

Kunzendorf, H., Walter, P., Stoffers, P. and Gwozdz, R., 1984, 'Metal variations in divergent plate-boundary sediments from the Pacific', Chem. Geol., 47, N 1-2, 133-133.

Kurbatov, L.M., 1936, 'Radioactivity of bottom sediments in lakes and seas within the USSR territory', Arktika, N 4, pp. 95 (in Russian).

Kurbatov, L.M., 1937, 'On the radioactivity of bottom sediments', Am. J. Sci., 33, 147-153.

Kurnosov, B.V., 1984, 'Hydrothermal changes in oceanic crust basalts', Tikhookeanskaya geologiya, N 6, 90-94 (in Russian).

Kusakabe, M. and Ku, T.L., 1984, 'Incorporation of Be isotopes and other trace metals into marine ferromanganese deposits', Geochim. et Cosmochim. Acta, 48, N 11, 2187-2193.

Kuznetsov, Y.V., 1976, Radiochronology of the Ocean, Moscow, Atomizdat, 280 pp. (in Russian).

Kuznetsov, Y.V., Legin, V.K., Lisitsin, A.P. and Simonyak, Z.I., 1967, 'Radioactivity of oceanic suspensions. 2. Uranium in oceanic suspensions', Radiochimiya, 9, N 4, 489-500 (in Russian).

Kvenvolden, K.A. and Blunt, D.J., 1979, 'Animo acid dating of bone nuclei in manganese nodules from the North Pacific ocean', in Bischoff, J.L. and Piper, D.Z. (eds.), Marine geology and oceanogr. of the Pacific Manganese Nodule province, Plenum, N.Y., a. London, pp. 763-773.

Laevastu, T. and Mellis, O., 1955, 'Extraterrestrial material in deep-sea deposits', Trans. Amer. Geophys. Un., 36, N 3, 385-389.

Lakin, H.W., Thompson, C.E. and Davidson, D.F., 1963, 'Tellurium content of marine manganese oxides and other manganese oxides', Science, 142, N 3599, 1568-1569.

Lal, D., 1977, 'The oceanic microcosm of particles', Science, 198, 997-1009.

Lal, D. and Lerman, A., 1973, 'Dissolution and behaviour of particulate biogenic matter in the ocean: some theoretical considerations', J. Geophys. Res., 78, N 24, 7100-7111.

Lallier-Verdes, E. and Clinard, C., 1983, 'Ultra-thin section study of the mineralogy and geochemistry of manganese micronodules from the South Pacific', Marine Geology, 52, N 3-4, 267-280.

Lalou, C. and Brichet, E., 1972, 'Signification des mesures radiochimiques dans l'évaluation de la vitesse de croissance de nodules de manganèse', C.R. Acad. Sci., Paris, t.275, pp. 815-818.

Lalou, C. and Brichet, E., 1976, 'On some relationships between the oxide layers and the cores of deep-sea manganese nodules', Miner. deposita, 11, N 3, 267-277.

Lalou, C. and Brichet, E., 1978, 'Replies on the comments of G.P. Glasby and P. Halbach on the publication by C. Lalou and E. Brichet: on some relationships between the oxide layers and the cores of deep sea manganese nodules', Mineral. Deposita, 13, N 1, 139-144.

Lalou, C. and Brichet, E., 1980, 'Anomalously high uranium contents in the sediment under Galapagos hydrothermal mounds', Nature, 284, N 5753, 251-253.

Lalou, C., Brichet, E. and Bonte, P., 1980, 'Some new data on the genesis of manganese nodules', in Geology and geochemistry of manganese, Akad. Kiado, Budapest, 3, 31-90.

Lalou, C., Brichet, E. and Le Cressus, C., 1973a, 'Etude d'un nodule de manganèse au microscope electronique à balayage et par microanalyse X: Implications dans le mode de formation des nodules', Anales Inst. Océan. N.S., 49, 5-17.

Lalou, C., Brichet, E., Poupeau, G. et al., 1979a, 'Growth rates and possible age of a North Pacific manganese nodule', in Bishoff, J.L., Piper, D.Z. (eds.) Marine Geology and Oceanography of the Pacific manganese nodule province, Plenum, N.Y., a. London, pp. 815-834.

Lalou, C., Brichet, E. and Ranque, D., 1973b, 'Certains nodules de manganèse trouvés en surface des sédiments sont-ils des formations contemporaines de la sédimentation?', C.R. Acad. Sci., Paris, t. 276, 1661-1664.

Lalou, C., Delibrias, G., Brichet, E. and Labeyrie, J., 1973b, 'Existence de carbon - 14 au centre de deux nodules de manganèse du Pacifique, âges ^{14}C et ^{230}Th de ces nodules', C.R. Acad. Sci., Paris, t. 276, N 23, 3013-3015.

Lalou, C., Ku, T.L., Brichet, E., Poupeau, G. and Pomary, P., 1979c, 'TECHNO enrustation. Part I: Radiometric studies', in La genèse de nodules de manganèse, Edit. CNRS, Paris, pp. 261-269.

Lambert, C.E., Bishop, J.K.B., Biscaye, P.E. and Chesselet, R., 1984, 'Particulate aluminium, iron and manganese chemistry at the deep Atlantic boundary layer', Earth and Planet. Sci. Lett., 70, N 2, 237-248.

Landergren, S., 1964, 'On the geochemistry of deep sea sediments', Rep. Swed. Deep-Sea Exped., X, N 5, 61-148.

Landing, W.M. and Bruland, K.W., 1980, 'Manganese in the North Pacific', Earth and Planet. Sci. Lett., 49, N 1, 45-56.

Landing, W. and Bruland, K., 1982, 'The biochemistry of Fe and Mn at VERTEX-II', EOS, 63, N 45, 960.

Landmesser, C.W., Kroenke, L.W., Glasby, G.P., Sawtell, G.H., Kingan, S., Utanga, E., Utanga, A. and Cowan, G., 1976, 'Manganese nodules from the South Penrhyn Basin, Southwest Pacific', South Pacific mar. geol. notes, 1, N 3, 17-40.

Langston, W.J., 1982, 'The distribution of mercury in British estuarine sediments and its availability to deposit-feeding bivalves', J. Mar. Bol. Assoc. U.K., 62, N 3, 667-684.

Lantzy, R.J. and Mackensie, F.T., 1979, 'Atmospheric trace metals: global cycles and assessement of man's impact', Geochim. et Cosmochim. Acta, 43, N 4, 511-525.

Larson, L.T., 1962, 'Zinc-bearing todorokite from Phillipsburg, Montana', Am. Mineral., 47, N 1, 59-66.

Lazur, Yu.M., Varentsov, I.M. and Yermilov, V.V., 1984, 'Volcanic copper and zink minerals in early-cretaceous deposits in the central Nort-Western Pacific', Naturwiss., B. 43, H 2, S. 129-137.

Lazurenko, V.I., 1982, 'Some aspects of biological hypothesis of manganese nodule formation', in Geology and geochemistry of manganese, Moscow, Nauka, pp. 259-262 (in Russian).

Lazurenko, V.M. and Khoruzhyi, V.Ya., 1980, 'New data on biological nature of manganese nodules', Dokl. AN SSSR, ser. B, N 1, 19-21 (in Russian).

Lebedev, L.I. and Maev, E.G., 1974, 'Upper Khvalyn sediments (Mangyshlak and Dagestan beds) in the deep parts of the Caspian Sea', in Comprehensive study of the Caspian Sea, Moscow, Izd-vo MGU, N 4, 30-46 (in Russian).

Lebedev, L.I., Maev, E.G., Bordovskiy, O.G. and Kulakova, L.S., 1973, 'Sediments of the Caspian Sea', Moscow, Nauka, 118 pp. (in Russian).

Lebedev, L.M., 1975, 'Recent metalliferous hydrotherms', Moscow, Nedra, 261 pp. (in Russian).

Lebedev, L.M. (Ed.), 1977, 'Recent hydrotherms and mineral formation', Moscow, Nauka, 202 pp. (in Russian).

Lebedev, L.M. and Nikitina, I.B., 1977, 'Compositional peculiarities and metal bearing potential of hydrotherm of volcanoes constructions: Mendeleev and Golovnin volcanoes', in Recent hydrotherms and mineral formation, Moscow, Nauka, pp. 5-25 (in Russian).

Legin, V.K., Kuznetsov, Yu.V. and Lobanov, Yu.N., 1974, 'Forms of manganese-54 in sea water', in Forms of elements and radionuclides in sea water, Moscow, Nauka, pp. 71-75 (in Russian).

Lerman, A., 1981, 'Controls of river water composition and the mass balance of river systems', in River inputs to ocean systems, U.N., Sci. Com. of Oceanic Res., N.Y., pp. 1-4.

Levashov, G.B., Barsukov, V.L., Sushchevskaya, T.M. and Malkov, I.I., 1975, 'Molybdenum and tungsten in sediments of some Pacific regions', Geokhimiya, N 2, 207-216 (in Russian).

Levitan, M.A. and Gordeev, V.V., 1981, 'Morphology and chemical composition of manganese nodules from the Central Indian ocean', Litol. i polezn. iskop, N 5, 27-37 (in Russian).

Li, Y.H., 1981a, 'Ultimate removal mechanisms of elements from the ocean', Geochim. et Cosmochim. Acta, 45, N 10, 1659-1664.

Li, Y.H., 1981b, 'Geochemical cycles of elements and human perturbation', Geochim. et Cosmochim. Acta, 45, N 12, 2073-2084.

Li, Y.H., 1982, 'A brief discussion on the mean oceanic residence time of elements', Geochim. et Cosmochim. Acta, 46, N 12, 2671-2675.

Li, Y.H., Bischoff, J.L. and Mathieu, G., 1969, 'The migration of manganese in the Arctic Basin sediments', Earth and Planet. Sci. Lett., 7, N 3, 265-270.

Li, Y.H., Teraoka, H., Yang, T.S. and Chen, J.S., 1984, 'The element composition of suspended particles from the Yellow and Yangtze Rivers', Geochim. et Cosmochim. Acta, 48, N 7, 1561-1564.

Lindberg, S.E., 1982, 'Factors influencing trace metals sulfate and hydrogen ion concentrations in rain', Atm. Envir., 16, N 7, 1701-1709.

Lindberg, S.E. and Harris, R.C., 1983, 'Water and acid soluble trace metals in atmospheric particles', J. Geophys. Res., 88, N C9, 5091-5100.

Linnik, P.N. and Nabivanetz, B.I., 1978, 'Dynamics of various manganese forms in the Dnieper River zone', Gidrobiol. zhurn., 14, N 1, 104-110 (in Russian).

Lisitsin, A.P., 1959, 'Bottom deposits of the Bering Sea', Tr. Inst. okeanologii AN SSSR, 29, 65-173 (in Russian).

Lisitsin, A.P., 1961a, 'Processes of recent sedimentation in the southern and central Indian ocean', in Recent sediments of seas and oceans, Moscow, Izd. AN SSSR, pp. 124-174 (in Russian).

Lisitsin, A.P., 1961b, 'Distribution and composition of suspended matter in seas and oceans', in ibid., pp. 175-231 (in Russian).

Lisitsin, A.P., 1964, 'Distribution and chemical composition of suspension in the Indian ocean', Okeanologicheskie issledovaniya, N 10, 136 pp. (in Russian).

Lisitsin, A.P., 1966, 'Processes of Recent sedimentation in the Bering Sea', Moscow, Nauka, 574 pp. (in Russian).

Lisitsin, A.P., 1974, 'Oceanic sedimentation', Moscow, Nauka, 438 pp. (in Russian).

Lisitsin, A.P., 1977, 'Terrigenous sedimentation, climatic zoning and proportions of biogenic and terrigenous matter in the ocean', Litol. i polezn. iskop., N 6, 3-22 (in Russian).

Lisitsin, A.P., 1978, 'Processes of oceanic sedimentation', Moscow, Nauka, 392 pp. (in Russian).

Lisitsin, A.P., 1981, 'Endogenic matter input to oceanic sedimentation', in Lithology at a new stage of geological knowledge, Moscow, Nauka, pp. 20-45 (in Russian)

Lisitsin, A.P., 1983, 'Bio-inert system of oceanic hydrotherms: endogenic matter supply', in Biogeochemistry of the ocean, Moscow, Nauka, pp. 60-72 (in Russian).

Lisitsin, A.P., Bogdanov, Yu.A., Murdmaa, I.O. et al., 1976, 'Metalliferous sediments and their genesis', in Geological-geophysical study of the southeastern Pacific, Moscow, Nauka, pp. 289-379 (in Russian).

Lisitsin, A.P. and Gordeev, V.V., 1974, 'On chemical composition of suspension and water in seas and oceans', Litol. i polezn. iskop., N 3, 38-57 (in Russian).

Lisitsin, A.P., Gordeev, V.V., Demina, L.L. and Lukashin, V.N., 1985, 'Marine geochemistry of manganese', Izv. AN SSSR, ser. geol., N 3, 3-29 (in Russian).

Lisitsina, N.A. and Butuzova, G.Yu., 1979, 'Lithologofacial types of bottom sediments', in Lithology and geochemistry of the Pacific sediments: transoceanic profile, Moscow, Nauka, pp. 13-42 (in Russian).

Liss, P.S., Slinn, W.G.N. (eds.), 1983, 'Air-Sea exchanges of gases and particulates', D. Reidel Publ. Co., Dordrecht, Holland, 561 pp.

Listova, L.P., 1961, 'Physico-chemical study of oxide and carbonate manganese ore formation', Moscow, Izd. AN SSSR, 120 pp. (in Russian).

Livingstone, D.A., 1963, 'Chemical composition of rivers and lakes: data of geochemistry', U.S. Geol. Surv. Prof. Pap., 440, 64 pp.

Livingstone, H.D. and Thompson, G., 1971, 'Trace elements concentrations in some modern corals', Limnol. Oceanogr., 16, N 5, 786-795.

Logvinenko, N.V., 1974, 'Petrography of sedimentary rocks', 2d edition, Moscow, Vysshaya shkola, 398 pp. (in Russian).

Logvinenko, N.V., Volkov, I.I. and Sokolova, E.G., 1972, 'Rhodochrosite in the Pacific deep-oceanic sediments', Dokl. AN SSSR, 203, N 1, 204-207 (in Russian).

Lonsdale, P., Burns, V.M. and Fisk, M., 1980, 'Nodules of hydrothermal birnessite in the caldera of a young seamount', J. Geol., 88, N 5, 611-618.

Lopatin, G.V., 1952, 'River sedimentary load in the USSR', Moscow, Gidrometeoizdat, 366 pp. (in Russian).

Lubchenko, I.Yu. and Belova, I.V., 1973, 'Migration of elements in river water', Litol. i polezn. iskop., N 2, 23-29 (in Russian).

Luck, J.M. and Turekian, K.K., 1983, 'Osmium-187/Osmium-186 in manganese nodules and the Cretaceous-Tertiary boundary', Science, 222, N 4624, 613-615.

Lukashin, V.N. and Lisitsin, A.P., 1980, 'Gallium', in Geochemistry of hydrolizating elements, Moscow, Nauka, pp. 50-70 (in Russian).

Lukashin, V.N., Lisitsin, A.P. and Emelyanov, E.M., 1980, 'Zirconium', in ibid, pp. 150-180 (in Russian).

Lunde, G., 1970, 'Analysis of trace elements in sea weeds', J. Sci. Food Agric., 21, 416-418.

Lupton, J.E., Klinkhammer, G.P., Normark, W.R., Haymon, R., MacDonald, K.C., Weiss, R.F. and Craig, H., 1980, 'Helium-3 and manganese at the 21°N East Pacific Rise hydrothermal site', Earth and Planet. Sci. Lett., 50, N 1, 115-127.

Lupton, J.E. and Craig, H., 1981, 'A major Helium-3 source at 15°S on the East Pacific Rise', Science, 214, N 4516, 13-18.

Lupton, J.E., Weiss, R.J. and Craig, H., 1977, 'Mantle helium in hydrothermal plumes in the Galapagos Rift', Nature, 267, N 5612, 603-604.

Lvovich, M.I., 1974, 'World water resources and their future', Moscow, Mysl, 448 pp. (in Russian).

Lyle, M., 1976, 'Estimation of hydrothermal manganese input to the oceans', Geology, 4, N 12, 733-736.

Lyle, M., 1981a, 'Formation and growth of ferromanganese oxides on the Nazca plate', Geol. Soc. Amer. Memoir., 154, 269-293.

Lyle, M., 1981b, 'Estimating growth rate of ferromanganese nodules from chemical compositions: implications for nodule forming processes', Geochim. et Cosmochim. Acta, 46, N 11, 2301-2306.

Lyle, M., 1983, 'The brown-green color transition in marine sediments: a marker of the Fe (III) -Fe (II) redox boundary', Limnol. Oceanogr., 28, 1026-1033.

Lyle, M., Dymond, J. and Heath, G.R., 1977, 'Copper-nickel enriched ferromanganese nodules and associated crusts from the Bauer Deep, Northwest Nazca Plate', Earth and Planet. Sci. Lett., 35, 55-64.

Lynn, D.C. and Bonatti, E., 1965, 'Mobility of manganese in diagenesis of deep-sea sediments', Mar. geol., 3, N 6, 457-474.

Lyons, W.B. and Fitzgerald, W.F., 1983, 'Trace metal speciation in nearshore anoxic and suboxic pore waters', in C.S. Wong, et al.(eds.), Trace metals in sea water, N.Y., Plenum Press, pp. 621-641.

Macdougall, J.D., 1979, 'The distribution of total alpha radioactivity in selected manganese nodules from the North Pacific: implications for growth processes', in Bishoff, J.L. and Piper, D.Z. (eds.), Marine geology and oceanography of the Pacific Manganese Nodule Province, N.Y., a. London, Plenum, pp. 775-789.

Maenhaut, W., Raemdonck, H., Selen, A., Van Grieken, R. and Winchester, J.W., 1983, 'Characterization of the atmospheric aerosol over the Eastern Equatorial Pacific', J. Geophys. Res., 88, N C9, 5353-5364.

Maenhaut, W. and Zoller, W.H., 'Determination of the chemical composition of the South pole aerosol by instrumental neutron activation analysis', J. of Radioanalytical Chem., 37, N 2, 637-650.

Maenhaut, W., Zoller, H.W., Duce, R.A. and Hoffman, G.L., 1979, 'Concentration and size distribution of particulate trace elements in the South Polar atmosphere', J. Geophys. Res., 84, N C5, 2421-2431.

Maev, E.G. and Lebedev, L.I., 1970, 'Accumulation peculiarities of calcium carbonate, iron and manganese in Novo-Caspian off shelf deposits of the Caspian Sea', in Comprehensive study of the Caspian Sea, Moscow, Izd. MGU, N 1, pp. 97-110 (in Russian).

Mamyrin, B.A. and Tolstikhin, I.N., 1984, Helium isotopes in nature, Amsterdam, Elsevier, 274 pp.

Manganese Nodules: Dimensions and Perspectives, 1979, United Nations Ocean Economics and Technology Office, D. Reidel Publ. Co., Dordrecht, Holland, 194 pp.

Manheim, F.T., 1965, 'Manganese-iron accumulations in the shallow marine environment', in Symp. Mar. Geochem. Occas. Publ. Narrangasett Mar. Labor., Univ. Rhode Island, 3, pp. 217-276.

Manheim, F.T., 1976, 'Interstitial waters of marine sediments', in Chemical Oceanogrpahy, 6, N.Y., Academic Press, pp. 115-185.

Manheim, F.T., Aruscavage, P.J., Simon, F.O. and Woo, C.C., 1980, 'Composition and mineralogy of western Atlantic ferromanganese nodules with special emphasis on platinum metals', Int. Geol. Congr., 26th, Paris, Resumes, 3, N 26, p. 966.

Manheim, F.T., Popenoe, P., Siapno, W. and Lane, C., 1982, 'Manganese-phosphorite deposits of the Blake Plateau (Western North Atlantic Ocean)', in Marine mineral deposits - new research results and economic prospects, Proc. Clausthaler Workshop, Sept. 1982, Essen, pp. 9-44.

Manskaya, S.M. and Drozdova, T.V., 1964, 'Geochemistry of organic matter', Moscow, Nauka, 315 pp. (in Russian).

Marchig, V. and Gundlach, H., 1976, 'Zur Geochemie von Manganknollen aus dem Zentralpazific und ihrer Sedimentunterlage. 1. Manganknollen', Geol. Jb., D 16, S. 59-77.

Marchig, V. and Gundlach, H., 1977, 'Zur Geochemie von Manganknollen aus dem Zentralpazific und ihrer Sedimentunterlage. 3. Sedimente', Geol. Jb., D 23, S. 91-104.

Marchig, V., Gundlach, H., Möller, P. and Schley, F., 1982, 'Some geochemical
    indicators for discriminating between diagenetic and hydrothermal metalliferous
    sediments', Mar. Geol., 50, N 3, 241-256.
Marchig, V. and Reyss, J.L., 1984, 'Diagenetic mobilization of manganese in Peru
    Basin sediments', Geochim. et Cosmochim. Acta, 48, 1349-1352.
Margolis, S.V., 1973, 'Manganese deposits encountered during Deep-Sea Drilling
    Progect Leg 29 in subantarctic waters', in M. Morgenstein (ed.), The origin
    and distribution of manganese nodules in the Pacific and prospects of
    exploration, Honolulu, Hawaii Inst. Geophys., pp. 109-113.
Margolis, S.V. and Glasby, G.P., 1973, 'Microlaminations in marine manganese
    nodules as revealed by scanning electron microscopy', Bull. Geol. Soc. Am.,
    84, N 11, 3601-3610.
Markhinin, E.K., 1967, 'Role of volcanism in the formation of the Earth's crust',
    Moscow, Nauka, 255 pp. (in Russian).
Markhinin, E.K. and Sapozhnikova, A.M., 1962, 'On Ni, Co, Cr, V and Cu content
    in volcanic rocks of Kamchatka and Kuril Islands', Geokhimiya, N 4, 372-376
    (in Russian).
Markhinin, E.K. and Stratula, D.S., 1977, 'Hydrotherms of Kuril Islands', Moscow,
    Nauka, 212 pp. (in Russian).
Marshall, K.C., 1979, 'Biogeochemistry of manganese minerals', in Biogeochem.
    Cycling Miner. Form. Elem. Amsterdam e.a., Elsevier, pp. 253-292.
Marshall, K.C., Stout, R. and Mitchell, R., 1971, 'Mechanism of the initial events
    in the sorption of marine bacteria to surfaces', J. Gen. Microbiol., 68, 337-348.
Martens, C.S. and Harriss, R.C., 1973, 'Chemistry of aerosols, cloud droplets and
    rain in Puerto Rican marine atmosphere', J. Geophys. Res., 78, N 6, 949-957.
Martin, J.H., Elliot, P.D., Anderlini, V.C., Girvin, D., Jacobs, S.A., Risen-
    brough, R.W., Delong, R.L. and Gilmartin, W.G., 1976, 'Mercury-selenium-
    bromine imbalance in premature parturient California sea lions', Mar. Biol.,
    35, 91-104.
Martin, J.H. and Gordeev, V.V., 1984, 'River input to ocean system: a
    reassessment', in Estuarine processes: an application to the Tagus estuary
    (UNESCO-CNA, SCOR Workshop, Lisbon, 1982), Lisbon, pp. 75-99.
Martin, J.H. and Knauer, G.A., 1973, 'The elemental composition of plancton',
    Geochim. et Cosmochim. Acta, 37, 1639-1654.
Martin, J.H. and Knauer, G.A., 1980, 'Manganese cycling in northeast Pacific
    waters', Earth and Planet. Sci. Lett., 51, N 2, 266-274.
Martin, J.H. and Knauer, G.A., 1983, 'VERTEX: manganese transport with CaCO3',
    Deep-Sea Res., 30, 411-425.
Martin, J.H. and Knauer, G.A., 1984, 'VERTEX: manganese transport through
    oxygen minima', Earth and Planet. Sci. Lett., 67, N 1, 35-47.
Martin, J.H. and Knauer, G.A., 1985, 'The lateral transport of manganese within
    the north-east Pacific Gyre oxigen minimum', Nature, 314, 6011, 524-526.
Martin, J.H. and Knauer, G.A., 'VERTEX: Distribution and fluxes of Ag, Al, Ba,
    Cd, Co, Cr, Cu, Fe, Mn, Mo, Ni, Pb, V and Zn in sub-oxic waters off
    Mexico', (submitted for publication).
Martin, J.H., Kulbicki, G. and de Groot, A.J., 1973, 'Terrigenous supply of radio-
    active and stable elements to the ocean', in E. Ingerson (ed.), Proceed. Symp.
    Hydrogeochem. biogeochem., Washington, The Clarke Co, pp. 463-483.
Martin, J.H. and Meybeck, M., 1978, 'The content of major elements in the
    dissolved and particulate load of rivers', in E.D. Goldberg (ed.),
    Biogeochemistry of estuarine sediments, Paris, UNESCO, pp. 95-110.
Martin, J.H. and Meybeck, M., 1979, 'Elemental mass-balance of material carried
    by major world rivers', Mar. Chem., 7, N 2, 173-206.
Marushkin, A.I., Korsakov, O.D., Vishtalyuk, S.D. and Kruglyakov, V.V., 1985,
    'Gases in pelagic ferromanganese nodules', Doklady AN SSSR, 283, N 4, 997-
    999 (in Russian).
Marvin, U.B., Einaudi, M.T. 1967, Black, magnetic spherules from Pleistocene and
    recent beach sands', Geochim. et Cosmochim. Acta, 31, N 10, 1871-1884.

Mason, B., 1979, 'Cosmochemistry. Part. 1. Meteorites', in M. Fleischer (ed.),
    Data of Geochemistry, 6th. edn., Geol. Survey Prof. Pap., U.S. Gov. Print.
    Off., 440-B-1, 132 pp.
Matthews, C.M.E., 1954, 'An investigation of the thorium content of manganese
    nodules using nuclear plates', Sci. Proc. Roy. Dublin Soc., 26, 275-288.
Maury, R.C., Le Guen, M. and Picot, P., 1980, 'Présence du cuivre natif dans une
    andésite de l'arc volcanique des Petites Antilles (Isles de Sainte-Lucie)',
    Bull. miner., 103, N 5, 503-506.
McKelvey, V.E. and Wang, F.F., 1969, 'Preliminary maps of world subsea mineral
    resources', Wash., U. Geol. Surv.
McKelvey, V.E., Wright, N.A. and Bowen, R.W., 1983, 'Analysis of the World
    distribution of metal-rich subsea manganese nodules', U.S. Geol. Surv. Circ.,
    N 886, 55 pp.
McKelvey, V.E., Wright, N.A. and Rowland, R.W., 1979, 'Manganese nodule
    resources in the Northeastern equatorial Pacific', in J.L. Bischoff and
    D.Z. Piper (ed.), Marine Geology and oceanography of the Pacific manganese
    nodule province, N.Y., Plenum, pp. 747-762.
Melo, U., Guazelli, W. and Ataide, C.M.P., 1978, 'Nodules polimetalicos com nucleo
    de fosforita no plato de Pernambuco', Reconh. Glob. Mar. Contin. Brasil,
    proj. REMAC, 3, Rio de Janeiro, 15-32.
Menard, H.W., 1964, The marine geology of the Pacific, N.Y., Mc Graw Hill, 271 pp.
Menyailov, I.A., Nikitina, L.P., Shapar, V.N. and Miklishansky, A.Z., 1982,
    'The role of active volcanism in enrichment of the atmosphere in chalcophile
    elements', J. Geophys. Res., 87, N C 13, 11113-11118.
Mero, J., 1965, The mineral resources of the sea, Amsterdam-London-New York,
    Elsevier, 312 pp.
Merrihue, G.M., 1964, 'Rare gases as a proof of maintenance of cosmic dust in red
    clay', Ann. N.Y. Acad. Sci., 119, 351-357.
Meybeck, M. and Carbonnel, J.P., 1975, 'Chemical transport by the Mekong River',
    Nature, 255, N 5504, 134-136.
Meyer, J.P., 1979, 'The significance of the carbonaceous chondrite abundunce',
    in Les Elements et leurs Isotopes dans l'Universe, Univ. Liège, pp. 153-188.
Meyer, K., 1973, 'Surface sediment and manganese nodule facies encountered on
    R/V Valdivia cruises 1972/73', in M. Morgenstein (ed.), The origin and
    distribution of manganese nodules in the Pacific and prospects for exploration,
    Hawaii Inst. Geophys., pp. 125-130.
Meylan, M.A., 1968, 'The mineralogy and geochemistry of manganese nodules from
    the Southern Ocean', Cont. Sediment. Res. Lab. Fla. St. Univ., N 22, 177 pp.
Meylan, M.A., 1974, 'Field description and classification of manganese nodules',
    in Ferromanganese deposits on the Ocean floor. Hawaii Inst. Geophys. Rept.
    HJG-74-9. Honolulu, pp. 158-168.
Meylan, M.A., Glasby, G.P., Knedler, K.E. and Johnston, J.H., 1981, 'Metallife-
    rous deep-sea sediments', in K.H. Wolf (ed.), Handbook of Strata-Bound and
    Stratiform ore deposits, Amsterdam, Elsevier, pp. 77-178.
Michard, G., Albarède, F., Michard, A., Minster, J.F., Charlou, J.L. and
    Tan, N., 1984, 'Chemistry of solution from the 13°N East Pacific Rise
    hydrothermal site', Earth and Planet. Sci. Lett., 67, N 3, 297-307.
Michard, G., Grimaud, D. and Lavergne, D., 1974, 'Concentration de manganese
    dissous dans les eaux interstitielles des sediments du Pacific equatorial',
    C.R. Acad. Sci., 278, Ser. D, 3157-3160.
Migdisov, A.A., Bogdanov, Yu.A., Lisitsin, A.P., Gurvich, E.G., Lebedev, A.I.,
    Lukashin, V.N., Gordeev, V.V., Girin, Yu.P. and Sokolova, E.G., 1979,
    'Geochemistry of metalliferous sediments', in Metalliferous sediments of the
    south-eastern Pacific, Moscow, Nauka, pp. 122-200 (in Russian).
Miklishansky, A.Z., 1983, 'Bio-inert system of the atmosphere', in Biogeochemistry
    of the ocean, Moscow, Nauka, pp. 72-89 (in Russian).
Miklishansky, A.Z., Pavlotskaya, F.I., Saveliev, B.V. and Yakovlev, Yu.V., 1977,
    'Concentrations and forms of microelements in the lower air stratum and in
    atmospheric precipitation', Geokhimiya, N 11, 1673-1682 (in Russian).

Miklishansky, A.Z., Yakovlev, Yu.V., Menyailov, I.A., Nikitina, L.P. and
    Saveliev, B.V., 1979, 'On geochemical role of chemical element supply along
    with volatiles of active volcanism', Geokhimiya, N 11, 1652-1661 (in Russian).
Miller, A.R., Densmore, C.D., Degens, E.T., Hathaway, Y.C., Manheim, F.T.,
    McFarlin, P.F., Pocklington, R. and Jokela, A., 1966, 'Hot brines and recent
    iron deposits in deeps of the Red Sea', Geochim. et Cosmochim. Acta, 30,
    N 3, 341-343.
Milliman, J.D., 1981, 'Transfer of river-borne particulate material to the ocean',
    in River inputs to ocean systems, Switzerland, UNEP and UNESCO, pp. 5-12.
Milliman, J.D. and Meade, R.H., 1983, 'World-wide delivery of river sediment to the
    ocean', J. Geol., 91, N 1, 1-21.
Mitropolsky, A.Yu., Bezborodov, A.A. and Ovsyanyi, E.I., 1982, 'Geochemistry of
    the Black Sea', Kiev, Naukova Dumka, 144 pp. (in Russian).
Mizuno, A., 1981, 'Deep-sea mineral resources investigation in the northern part of
    Central Pacific Basin', Geol. Surv. Japan, cruise GH, 79-1, 309 pp.
Mizuno, A. and Chujo, J., 1975, 'Deep-sea mineral resources investigation in the
    Eastern Central Pacific Basin', Geol. Surv. Japan, cruise GH, 74-5, 103 pp.
Mizuno, A. and Moritani, T., 1977, 'Deep-sea mineral resources investigation in
    the central-eastern part of Central Pacific Basin', Geol. Surv. Japan, cruise
    GH 76-1, 217 pp.
Mokievskaya, V.V., 1961, 'Manganese in the Black Sea water', Dokl. AN SSSR, 137,
    N 6, 1445-1447 (in Russian).
Monget, J.M., Murray, J.W. and Mascle, J., 1976, 'A world-wide compilation of
    published multicomponent analysis of ferromanganese concretions', NSF-IDOE
    Tech. rep. 12, 127 pp.
Moorby, S.A. and Cronan, D.S., 1981, 'The distribution of elements between
    co-existing phases in some marine ferromanganese-oxide deposits', Geochim. et
    Cosmochim. Acta, 45, N 10, 1855-1877.
Moorby, S.A., Cronan, D.S. and Glasby, G.P., 1984, 'Chemistry of hydrothermal
    Mn-oxide deposits from the South-West Pacific island arc', Geochim. et Cosmo-
    chim. Acta, 48, N 3, 433-441.
Moore, J.G., 1966, 'Rate of palagonitization of submarine basalt adjacent to Hawaii',
    U.S. Geol. Surv. prof. pap., 550 D, 163-171.
Moore, W.S., 1984, 'Thorium and radium isotopic relationships in manganese nodules
    and sediments at MANOP Site S', Geochim. et Cosmochim. Acta, 48, N 5,
    987-992.
Moore, R.M., Burton, J.D., Le Williams, P.J. and Young, M.L., 1979, 'The
    behaviour of dissolved organic material, iron and manganese in estuarine
    mixing', Geochim. et Cosmochim. Acta, 43, N 6, 919-926.
Moore, W.S., Ku, T.L., Macdougall, J.D., Burns, V.M., Burns, R., Dymond, J.,
    Lyle, M.W. and Piper, D.Z., 1981, 'Fluxes of metals to a manganese nodule:
    radiochemical, chemical, structural and mineralogical studies', Earth and Planet.
    Sci. Lett., 52, N 1, 151-171.
Morgan, J.J., 1964, 'Chemistry of aqueous manganese II and IV', Ph.D. diss.
    Harvard Univ., Cambridge, MA, 244 pp.
Morgenstein, M. and Riley, J.J., 1975, 'Hydration-rind and dating of basaltic glass:
    a new method for archaeological chronologies', Asian Perspectives, 17, 145-159.
Moritani, T., Maruyama, S.. Nohara, M. et al.,1977, 'Description and classification
    of manganese nodules', Geol. Surv. Japan., Cruise Rept., N 8, 136-158.
Morozov, A.A., 1985, 'On fixation mechanism of Mn and Fe at the surface of
    manganese nodules', Dokl. AN SSSR, 282, N 3, 688-692 (in Russian).
Morozov, N.P., 1968, 'On geochemistry of alkaline elements in marine sediments',
    Litol. i polezn. iskop., N 6, 3-16 (in Russian).
Morozov, N.P., 1969, 'Geochemistry of lithium and rubidium in the sea', Ph.D.
    Dissertation. Moscow, Institute of Oceanology, Academy of Sciences of the
    USSR, (in Russian).
Morozov, N.P., 1983, 'Chemical elements in biota and in food chains', in
    Biogeochemistry of the ocean, Moscow, Nauka, 127-165 (in Russian).
Morozov, N.P., Baturin, G.N., Gordeev, V.V. and Gurvich, E.G., 1974, 'On
    suspension and sediment composition in mouth areas of the Northern Dvina,

Mezen, Pechora and Ob rivers', Gidrokhim. materialy, 60, 60-73 (in Russian).

Morozov, N.P., Patin, S.A. and Nikolenko, E.M., 1976, 'Microelements in water, suspended matter and marine biota of the Black Sea', Geochimiya, N 9, 1391-1399 (in Russian).

Morozov, N.P., Petukhov, S.A., Petrov, A.A. and Tikhomirova, A.A., 1979, 'On concentration ability of biotic components of the Indian ocean ecosystem', Geokhimiya, N 7, 1112-1117 (in Russian).

Morten, L., Landimi, F., Bocchi, G., Mottana, A. and Brunfelt, A.O., 1980, 'Fe-Mn crusts from the southern Tyrrhenian Sea', Chem. Geol., 28, 261-278.

Mottle, M.J., Holland, H.D. and Carr, R.F., 1979, 'Chemical exchange during hydrothermal alteration of basalt by seawater. II. Experimental results for Fe-Mn and sulfur species', Geochim. et Cosmochim. Acta, 43, N 6, 869-884.

Mroz, E. and Zoller, W.H., 1975, 'Composition of atmospheric particulate matter from the eruption of Heimaey, Iceland', Science, 190, N 4213, 461-464.

Müller, D., 1979, 'Sulfide inclusions in manganese nodules of the Northern Pacific', Miner. Dep., 14, N 3, 375-380.

Mullin, J.B. and Riley, J.P., 1956, 'On the occurence of cadmium in sea-water and in marine organisms and sediments', J. Mar. Res., 15, N 2, 103-122.

Munda, J.M., 1978, 'Trace metal concentrations in some Iceland seaweeds', Botan. Marina, 21, 261-263.

Murdmaa, I.O., Bazilevskaya, E.S., Gordeev, V.V., Emelyanov, E.M., Kuzmina, T.G. and Turanskaya, N.V., 1979, 'Geochemical peculiarities of sedimentary formations', in Geological formations of the north-western Atlantic, Moscow, Nauka, pp. 96-147 (in Russian).

Murphy, K. and Dymond, J., 1984, 'Rare earth element fluxes and geochemical budget in the eastern equatorial Pacific', Nature, 307, N 5950, 444-447.

Murray, J., 1876, 'Preliminary report on speciments of the sea bottom', Proc. Roy. Soc., London, 24, 471-532.

Murray, J. and Irvine, R., 1893, 'On the chemical changes which take place in the composition of the sea-water associated with blue muds on the floor of the ocean', Trans. Roy. Soc. Edinburgh, 37, 481-507.

Murray, J. and Irvine, R., 1895, 'On the manganese oxides and manganese nodules in marine deposits', Trans. Roy. Soc. Edinburgh, 37, 721-742.

Murray, J. and Lee, G.V., 1909, 'The depth and marine deposits of the Pacific', Mem. Mus. Comp. Zool., Cambridge, 38, N 1, 7-169.

Murray, J. and Renard, A.F., 1884, 'On the microscopic characters of volcanic ashes and cosmic dust, and their distribution in the deep sea deposits', Proc. Roy. Soc. Edinb., 12, 474-495.

Murray, J. and Renard, A.F., 1891, 'Deep-sea deposits', Rep. Sci. Res. Explor. Vayage Challenger. London. Eyre and Stottiswoode, 525 pp.

Murray, J.W., 1979, Iron oxides', in R.G. Burns (ed.), Marine minerals, Wash., Mineral. Soc. Amer., pp. 47-98.

Murray, J.W., Balistrieri, L.S. and Paul, B., 1984, 'The oxidation state of manganese in marine sediments and ferromanganese nodules', Geochim. et Cosmochim. Acta, 48, 1237-1247.

Murray, J.W. and Brewer, P.G., 1977, 'Mechanisms of removal of manganese, iron and other trace metals from sea water', in G.P. Glasby (ed.), Marine manganese deposits, Amsterdam-London-New York, Elsevier, pp. 291-325.

Murray, J.W. and Dillard, J.G., 1979, 'The oxidation of cobalt (II) adsorbed on manganese dioxide', Geochim. et Cosmochim. Acta, 43, 781-787.

Murray, J.W., Spell, B. and Paul, B., 1983, 'The contrasting geochemistry of manganese and chromium in the eastern tropical Pacific Ocean', in C.S. Wong et al. (eds.), Trace metals in sea water, N.Y., Plenum Press, pp. 643-669.

Naboko, S.I., 1980, 'Metalliferous capacity of recent hydrotherms in regions of tectonic-magmatic activity', Moscow, Nauka, 200 pp. (in Russian).

Nazarova, T.N., Rybakov, A.K., Vasyukova, Z.V. and Vasiliev, Yu.D., 1975, 'Meteoritic matter by cosmic observational data', Kosmich. issled., 14, N 3, 435-444 (in Russian).

Naumov, G.B., Ryzhenko, B.N. and Khodakovsky, I.L., 1971, 'Handbook of thermodynamic values', Moscow, Atomizdat, 239 pp. (in Russian).

Nealson, K.H., 1978, 'The isolation and characterization of marine bacteria which catalyze manganese oxidation', in W.E. Krumbein (ed.), Environmental Biogeochemistry and geomicrobiology, Michigan, Ann Arbor Sci. Publ., Vol. 3, pp. 847-858.

Nealson, K.H., 1983, 'The microbial manganese cycle', in W.E. Krumbein (ed.), Microbial geochemistry: studies in microbiology, Oxford, Blackwell Press, pp. 191-221.

Nealson, K.H. and Cohen, H., 1977, 'Control of manganese oxidation by marine bacteria (abstr.)', 3d Intl. Symp. Environ. Beogeochem., Univ. Oldenburg, Environ. Lab. Occ. Publ., 1, 97-98.

Neil, S.T., 1980, 'Chemical analysis of USGS manganese nodule reference samples', Geostand. Newsletter, 4, N 2, 205-212.

Nesterova, I.L., 1960, 'Forms of migration of elements in the Ob River', Geokhimiya, N 4, 355-362 (in Russian).

Nevessky, E.N., Medvedev, V.S. and Kalinenko, V.V., 1977, 'The White Sea. Holocene sedimentation and evolution', Moscow, Nauka, 236 pp. (in Russian).

Niffeler, U.P., Li, Y.H. and Santschi, P.H., 1984, 'A kinetic approach to describe trace elements distribution between particles and solution in natural aqueous systems', Geochim. et Cosmochim. Acta, 48, 1513-1522.

Nikolaev, D.S. and Efimova, E.I., 1963, 'On age-dating of manganese nodules in the Indian and Pacific oceans', Geokhimiya, N 7, 678-688 (in Russian).

Nikolaev, D.S. and Efimova, E.I., 1965, 'On forms of radioactive elements in manganese nodules', Radiokhimiya, N 5, 614-620 (in Russian).

Nishiizumi, K., 1983, 'Measurement of ^{53}Mn in deep-sea iron and stony spherules', Earth and Planet. Sci. Lett., 63, N 2, 223-228.

Nissenbaum, A. and Swaine, D.J., 1976, 'Organic matter-metal interactions in Recent sediments: the role of humic substances', Geochim. et Cosmochim. Acta, 40, N 7, 809-816.

Nohara, M. and Nasu, N., 1977, 'Mineralogical and geochemical characteristics of manganese nodules from the Suiko Seamount, North-western Pacific ocean', Bull. Geol. Surv. Japan, 28, N 9, 615-621.

Noro, T., 1978, 'Effect of manganese on the growth of a marine green alga, Dunaliella tertiolecta', Jap. J. Phycol., 26, 69-72.

Novgorodova, M.I., 1979, 'Findings of native aluminium in quartz veins', Dokl. AN SSSR, 248, N 4, 965-968 (in Russian).

Novgorodova, M.I., 1983, 'Native metals in hydrothermal ores', Moscow, Nauka, 288 pp. (in Russian).

Nozaki, Y. and Yang, H.S., 1985, 'Non-destructive alpha-spectrometry and radiochemical studies of deep-sea manganese nodules', Geochim. et Cosmochim. Acta, 49, N 8, 1765-1774.

Nozdryukhina, L.P., 1977, 'Biological role of microelements in organisms and plants', Moscow, Nauka, 184 pp. (in Russian).

Nriagu, J.O., 1979, 'Global inventory of natural and anthropogenic emissions of trace metals to the atmosphere', Nature, 279, N 5712, 409-411.

Nyffeler, U.P., Li, Y.H. and Santschi, P.H., 1984, 'A kinetic approach describe trace element distribution between particles and solution in natural aquatic systems', Geochim. et Cosmochim. Acta, 48, 1513-1522.

Okada, A., Okada, T. and Shima, M., 1972, 'Study on the manganese nodules. IV. Some aspects of the chemical form of iron in the manganese nodule', Int. Pacific. Congr., Sci. Papers, N 66, 178-183.

Olaffson, J. and Riley, J.P., 1972, 'Some data on the marine geochemistry of rhenium', Chem. Geol., 9, N 3, 227-232.

Oleinikov, B.V., Okrugin, A.V. and Leskova, N.V., 1978, 'Petrological value of native aluminium findings in basites', Dokl. AN SSSR, 243, N 1, 191-194 (in Russian).

O'Nions, R.K., Carter, S.R., Cohen, R.S., Evensen, N.M. and Hamilton, P.J., 1978, 'Pb, Nd and Sr isotopes in oceanic ferromanganese deposits and ocean floor basalts', Nature, 273, N 5662, 435-438.

Oreshkin, V.N., 1977, 'Cadmium distribution in the surficial layer of the Pacific bottom sediments', Okeanologiya, 17, N 4, 666-671 (in Russian).

Oreshkin, V.N., Ashchyan, T.O., Vnukovskaya, G.L. and Belyaev, Yu.I., 1985, 'Mercury in geochemical reference samples', Geokhimiya, N 10, 1526-1529 (in Russian).

Orlando, E. and Mauri, M., 1978, 'Accumulation of manganese in Donax trunculus L (Bivalvia)', J. Etud. Pollut., 4, 297-299.

Ostroumov, E.A., 1954, 'Manganese in the Sea of Okhotsk bottom deposits', Dokl. AN SSSR, 97, N 2, 285-288 (in Russian).

Ostroumov, E.A., 1955, 'Distribution of manganese in the Sea of Okhotsk bottom deposits', Izv. AN SSSR, ser. geol., N 5, 83-88 (in Russian).

Ostwald, J., 1983, 'Possible authigenic sulphide inclusions in a central Pacific manganese nodule', Austral. Miner., N 78, 230-232.

Ostwald, J. and Frazer, F.W., 1973, 'Chemical and mineralogical investigations on deep-sea manganese nodules from the Southern Ocean', Miner. Deposita, 8, N 4, 303-311.

Ovsyanyi, E.I. and Eremeeva, L.V., 1980, 'Distribution of microelements in the north-western Pacific and river discharge', in Complex hydrophysical and hydrochemical study of the Black Sea, Sevastopol, MGI AN UkSSR, pp. 119-129 (in Russian).

Ozima, M., Takayanagi, M., Zashu, S. and Amari, S., 1984, 'High ^{3}He/^{4}He ratio in oceanic sediments', Nature, 311, N 5985, 448-450.

Ozima, M. and Zashu, S., 1983, 'Primitive helium in diamonds', Science, 219, N 4588, 1067-1068.

Pachadzhanov, D.N., Bandurkin, G.A., Migdisov, A.A. and Girin, Yu.P., 1963, 'Some data on geochemistry of the Indian ocean manganese nodules', Geokhimiya N 5, 493-499 (in Russian).

Pak, C.K., Yang, K.R. and Lee, J.K., 1977, 'Trace metals in several edible marine algae of Korea', J. Oceanol. Soc. Korea, 12, 41-47.

Pakhomova, A.S., 1948, 'Manganese in marine sediments', Tr. Gos. okeanogr. Inst. 5, 101-150 (in Russian).

Palme, H., Suess, H.E. and Zeh, H.D., 1981, 'Abundances of the elements in the solar sytem', in K.H. Hellwege (ed.), Landolt-Börnstein, Group VI: Astronomy, Astrophysics and Space Res., vol. 2, Astronomy and Astrophysics, Extension and suppl. to vol. 1, subvol. a, Berlin, Springer, pp. 257-272.

Papanastassiou, D.A., Brownlee, D.E. and Wasserburg, G.L., 1982, 'Chemical and isotopic studies of deep sea spherules', Abstr. Pap 13th. Lunar Planet Sci. Conf., March 15-19, Houston, Tex., 1982, Pt. 2, pp. 613-614.

Patel, B., Valanju, P.G., Mulay, C.D., Balani, M.C. and Patel, S., 1973, 'Radio-ecology of certain molluscs in Indian coastal waters', in Radioactive contamination of the marine environment, Int. Atom. Energ. Agen, Vienna, Austria, pp. 307-330.

Paul, A.Z., 1976, 'Deep-sea bottom photographs show that benthic organisms remove sediment cover from manganese nodules', Nature, 263, 50-51.

Pautot, G. and Melguen, M., 1979, 'Influence of deep-water circulation and seafloor morphology on the abundance and grade of central South Pacific manganese nodules', in J.L. Bischoff and D.Z. Piper (eds.), Marine geology and oceanography of the Pacific manganese nodule province, N.Y., Plenum, pp. 621-649.

Pavlova, L.G.,1982a, 'Geochemical analysis of pore-water in the Norwegian and Barentz Seas for April-June, 1981', in Comprehensive study of the northern seas, Apatity, pp. 30-35 (in Russian).

Pavlova, L.G., 1982b, 'Biogenic matter, manganese and iron in pore water of the Kara Sea', in ibid, pp. 36-46 (in Russian).

Pearcy, W.G. and Osterberg, 1968, 'Zink-65 and Mn-54 in albacore Thunnus alalunga from the west coast of North America', Limnol. Oceanogr., 13, 490-498.

Peirson, D.H., Cawse, P.A., Salmon, L. and Cambray, R.S., 1973, 'Trace elements in the atmospheric environment', Nature, 241, N 5387, 252-256.

Peirson, D.H., Cawse, P.A. and Cambray, R.S., 1974, 'Chemical uniformity of airbone particulate material and a maritime effect', Nature, 251, N 5477, 675-679.

Pentreath, R.J., 1973, 'The accumulation and retention of ^{65}Zn and ^{54}Mn by the plaice, Pleuronectes platessa L', J. Exper. Mar. Biol. Ecol., 12, 1-18.

Pentreath, R.J., 1976, 'Some further studies on the accumulation and retention of Zn-65 and Mn-54 by the plaice, Pleuronectes platessa L', J. Exp. Mar. Biol. Ecol., 21, 179-189.

Perelman, A.I., 1961, 'Geochemistry of landscape', Moscow, Geografgiz, 496 pp. (in Russian).

Perelman, A.I., 1968, 'Geochemistry of epigenetic processes', Moscow, Nedra, 331 pp. (in Russian).

Perfilyev, B.V., 1926, 'New data on the role of microbes in ore formation', Izv. Geol. kom., 45, N 7, 795-819 (in Russian).

Perfilyev, B.V. and Gabe, D.R., 1964, 'Study of bacteria accumulating manganese and iron in bottom deposits by microbic landscape technique', in The role of microorganisms in iron-manganese ore formation, Moscow-Leningrad, Izd. AN SSSR, pp. 16-53 (in Russian).

Peschevitsky, B.I., Anoshin, G.N. and Erenburg, A.M., 1970, 'Chemical forms of gold and problems of oxidation-reduction potential of the sea water, in Chemical resources of seas and oceans, Moscow, Nauka, pp. 141-144 (in Russian).

Petelin, V.P. and Ostroumov, E.A., 1961, 'Geochemistry of the Sea of Okhotsk bottom sediments', in Recent marine and oceanic sediments, Moscow, Izd. AN SSSR, pp. 380-403 (in Russian).

Peterson, J.T. and Yunge, C.E., 1971, 'Sources of particulate matter in the atmosphere', in W.H. Matthews, W.W. Kellog and G.D. Robinson (eds.), Man's Impact on the Climate, MIT Press, pp. 310-320.

Peterson, M.N.A., Edgar, N.T. and von der Borch, C.C., 1970, 'Cruise leg summary and discussion'. Init. Rep. of the Deep-Sea Drilling Project, v. II, Wash., Nat. Sci. Found, 413-427.

Petkevich, T.A., 1967, 'Elementary composition of the flesh of plankton eating and bentic fishes from the north-western Black Sea', Dokl. AN UkSSR, ser. B, 29, N 2, 142-146 (in Russian).

Pettersson, H., 1943, 'Manganese nodules and the chronology of the Sea floor, Medd. Oceanogr. Inst. Göteborg, 6, 1-39.

Pettersson, H., 1945, 'Iron and manganese on the ocean floor', Medd. Oceanogr. Inst. Geteborg, Ser. B, N 3, 1-37.

Pettersson, H., 1955, 'Manganese nodules and oceanic radium', Deep-Sea Res., 3 (Suppl.), 335-345.

Pettersson, H., 1959, 'Manganese and nickel on the ocean floor', Geochim. et Cosmochim. Acta, 17, N 3/4, 209-213.

Pettersson, H. and Rotschi, H., 1952, 'The nickel content of deep-sea deposits', Geochim. et Cosmochim. Acta, 2, N 2, 81-90.

Pfeiffer, G., Förstner, U. and Stoffers, P., 1982, 'Speciation of reducible metal compounds in pelagic sediments by chemical extraction', Senckenberg. Marit., B. 14, N 1/2, 23-38.

Philippi, E., 1910, 'Die Grundproben der Deutschen Südpolar Expedition 1901-1903'. Deutsche Südpolar Expedition, B.2, H 6, 411-416.

Phillips, D.J.H., 1978, 'The common mussel Mytilus edulis as an indicator for trace metals in Scandinavian waters. II. Lead, iron and manganese', Mar. Biol., 46, 147-156.

Piotrowicz, S.R., Ray, B.J., Hoffman, G.L. and Duce, R.A., 1972, 'Trace metal enrichment in the sea surface microlayer', J. Geophys. Res., 77, N 27, 5243-5254.

Piper, D.Z., 1971, 'Origin of metalliferous sediments from the East Pacific Rise', Earth and Planet. Sci. Lett., 19, N 1.

Piper, D.Z., 1972, 'Rare earth elements in manganese nodules from the Pacific Ocean', in Horn, D.R. (ed.), Ferromanganese deposits on the Ocean floor, Nat. Sci. Found., Wash. D.C., pp. 123-130.

Piper, D.Z., 1974a, 'Rare earth elements in ferromanganese nodules and other marine phases', Geochim. et Cosmochim. Acta, 38, N 7, 1007-1022.

Piper, D.Z., 1974b, 'Rare elements in the sedimentary cycle: a summary', Chem. Geol., 14, 285-304.

Piper, D.Z., Basler, J.R. and Bischoff, J.L., 1984, 'Oxidation state of marine
    manganese nodules', Geochim. et Cosmochim. Acta, 48, N 11, 2347-2355.
Piper, D.Z. and Blueford, J.R., 1982, 'Distribution, mineralogy and texture of
    manganese nodules and their relation to sedimentation of DOMES Site A in the
    equatorial North Pacific', Deep-sea Res., 29, N 8 A, 927-951.
Piper, D.Z. and Fowler, B., 1980, 'New constraint on the maintenance of manganese
    nodules on the sediment surface', Nature, 286, N 5776, 880-882.
Piper, D.Z. and Graef, P.A., 1974, 'Gold and rare earth elements in sediments
    from the East Pacific Rise', Mar. Geol., 17, N 5, 287-297.
Piper, D.Z., Leong, K. and Cannon, W.F., 1979, 'Manganese nodule and surface
    sediment compositions: DOMES Sites A, B and C', in J. Bischoff and D.Z.
    Piper (eds.), Marine Geol. and oceanogr. of the Pacific manganese nodule
    province, N.Y., Plenum, pp. 437-473.
Piper, D.Z. and Williamson, M.E., 1977, 'Composition of Pacific ocean ferromangane-
    se nodules', Mar. Geol., 23, N 4, 285-303.
Piper, D.Z. and Williamson, M.E., 1981, 'Mineralogy and composition of concentric
    layers within a manganese nodule from the North Pacific ocean', Mar. Geol.,
    40, N 3-4, 255-268.
Plaskett, D. and Potter, I.C., 1979, 'Heavy metal concentrations in the muscle
    tissue of 12 species of teleost from Cockburn Sound, Western Australia',
    Austral. J. Mar. Freshwat. Res., 30, 607-616.
Plüger, W.L., Kunzendorf, H. and Friedrich, G., 1984, 'Rare earth elements in
    manganese nodules from the Southwest Pacific', Abstr. 27th. Intern. Geol.
    Congr., Moscow, 4-14 Aug. 1984, vol. 3, Sect. 06, 07, p. 65.
Pogrebnyak, Yu.F. and Krendelev, F.P., 1976, 'Microelementary composition of pore
    water in the south-eastern Pacific bottom sediments', in Geological-geophisical
    study of the south-easter Pacific. Moscow, Nauka, pp. 270-288 (in Russian).
Poreda, R. and Craig, H., 1979, 'Helium and neon in oceanic volcanic rocks',
    EOS, 60, 969.
Potter, R.M. and Rossman, G.R., 1979, 'A magnesium analogue of chalcophanite in
    manganese-rich concretion from Baja California', Am. Miner., 64, 1227-1229.
Pratt, R.M., 1971, 'Lithology and rocks dredged from the Blake Plateau',
    Southeastern Geology, 13, N 1, 19-38.
Pratt, R.M. and Manheim, F.T., 1967, 'The relation of manganese to phosphorite
    concretions on the Blake Plateau', Trans. Amer. Geophys. Un., 48, N 1,
    144-145.
Pratt, R.M. and McFarlin, P.E., 1966, 'Manganese pavement on the Blake Plateau',
    Science, 151, N 3714, 1080-1082.
Presley, B.J., Brooks, R.R. and Kaplan, I.R., 1967, 'Manganese and related
    elements in the interstitial water of marine sediments', Science, 158, N 3803,
    906-910.
Presley, B.J., Kolodny, Y., Nissenbaum, A. and Kaplan, J.R., 1972, 'Early
    diagenesis in a reducing fjord, Saanich Inlet, British Columbia. II. Trace
    element distribution in interstitial water and sediment', Geochim. et Cosmochim.
    Acta, 36, N 10, 1073-1090.
Preston, A., Jeffris, D.F., Dutton, J.W., Harvey, B.R. and Steel, A.K., 1972,
    'British isles coastal waters: the concentrations of selected heavy metals in
    seawater, suspended matter and biological indicators - a pilot survey',
    Environ. Pollut., 3, 69-82.
Pritchard, R.G., 1979, 'Alteration of basalts from DSDP legs 51, 52 and 53, Holes
    417 A and 418 A', in T. Donnelly and J. Francheteau (eds.), Init. Rep. DSDP
    51, 52, 53, P. 2, Wash. D.C., pp. 1185-1199.
Prospero, J.M., 1979, 'Mineral and sea-salt aerosol concentrations in various oceanic
    regions', J. Geophys. Res., 84, N C2, 725-731.
Prospero, J.M., 1981, 'Eolian transport to the World Ocean', in C. Emiliani (ed.),
    The Oceanic Lithosphere, Vol. 7, The Sea, N.Y., John Wiley, pp. 801-874.
Prospero, J.M. and Bonatti, E., 1969, 'Continental dust in the atmosphere of the
    eastern Equatorial Pacific', J. Geophys. Res., 74, N 13, 3362-3371.

Pushkina, Z.V., 1967, 'Iron, manganese, silicon, phosphorus, boron, aluminium in sea water near the Santorini volcano (Aegean Sea)', Litol. i polezn. iskop., N 2, 87-96 (in Russian).

Pushkina, Z.V., 1971, 'Distribution of trace elements in the Indian ocean sediments south of Ceylon', Litol. i polezn. iskop., N 4, 34-45 (in Russian).

Pushkina, Z.V., 1980, 'Interstitial waters of sediments on the transoceanic profile and their diagenetic changes', in Geochemistry of diagenesis of the Pacific Ocean sediments (transoceanic profile), Moscow, Nauka, pp. 70-98 (in Russian).

Pushkina, Z.V., Stepanetz, M.I. and Cherkasova, E.V., 1977, 'Iron, manganese, copper, nickel and vanadium in pore water of the north-eastern Pacific sediments', Lithol. i polezn. iskop., N 3, 3-16 (in Russian).

Pushkina, Z.V., Stepanetz, M.I., Orlova, L.P. and Sinani, T.I., 1981, 'Fe, Mn, Cu, Co, Ni, Zn, Pb, Cd, B and Si in pore water of the Red Sea metalliferous sediments (Atlantis II and Discovery deeps)', Litol. i polezn. iskop., N 6, 62-69 (in Russian).

Quinby-Hunt, M.S. and Turekian, K.K., 1983, 'Distribution of elements in sea water', EOS, 64, N 14, 130-131.

Raab, W., 1972, 'Physical and chemical features of Pacific deep sea nodules and their implications to the genesis of nodules', in Ferromanganese deposits on the ocean floor, Wash. D.C., Natl. Sci. Found., pp. 31-49.

Raab, W.J. and Meylan, M.A., 1977, 'Morphology', in G.P. Glasby (ed.), Marine manganese deposits, Amsterdam, Elsevier, pp. 109-146.

Raevsky, B.N., Shestakov, Yu.M. and Terekhov, B.A., 1981, 'Application of diffusion technique to study physico-chemical forms of ^{60}Co and ^{54}Mn in the systems simulating sea water', Okeanologiya, 21, is. 3, pp. 473-475 (in Russian).

Rahn, K.A., 1976, 'The chemical composition of the atmospheric aerosol', Tech. Rep. Univ. Rhode Island, School of Oceanography Kingstone R.I., 203 pp.

Rancitelli, L.A. and Perkins, R.W., 1970, 'Trace element concentrations in the troposphere and lower stratosphere', J. Geophys. Res., 75, N 15, 3055-3064.

Rankama, K.K., 1948, 'Geochemistry of columbium', Ann. Acad. Sci. fenn., Geol. Geogr., 3, N 13, 1-57.

Rankin, P.C. and Glasby, G.P., 1979, 'Regional distribution of rare earth and minor elements in manganese nodules and associated sediments in the South-west Pacific and other localities', in J.L. Bischoff and D.Z. Piper (eds.), Marine geology and oceanography of the Pacific manganese nodule province, Plenum, New York a. London, pp. 681-697.

Rawson, M.D. and Ryan, W.B.F., 1978, 'Ocean floor sediment and polymetallic nodules maps', Lamont-Doherty Geol. Observ., Columbia Univ., N.Y.

Reimers, C.E. and Suess, E., 1983, 'The partitioning of organic carbon fluxes and sedimentary organic carbon decomposition rates in the ocean', Mar. Chem., 13, N 2, 141-168.

Reinolds, P.H. and Dasch, E.J., 1971, 'Lead isotopes in marine manganese nodules and the ore-lead growth curve', J. Geophys. Res., 76, N 21, 5124-5129.

Revelle, R.R., 1944, 'Scientific results of the cruise VII of the "Carnegie"', Publ. Carnegie Inst., N 556, 1-180.

Reyss, J.L. and Lalou, C., 1981, 'Nodules and associated sediments in the Madagascar Basin', Chem. Geol., 34, N 1/2, 31-41.

Reyss, J.L., Lemaitre, N., Ku, T.L., Marchig, V., Southon, J.R., Nelson, D.E. and Vogel, J.S., 1985, 'Growth of a manganese nodule from Peru basin: A radiochemical anomaly', Geochim. et Cosmochim. Acta, 49, N 11, 2401-2408.

Reyss, J.L., Marchig, V. and Ku, T.L., 1982, 'Rapid growth of a deep-sea manganese nodule', Nature, 295, N 5844, 401-403.

Reyss, J.L. and Yokoyama, Y., 1976, 'Aluminium-26 in a manganese nodule', Nature, 262, N 5565, 203-204.

Ridley, W.P., Dizikes, L.J. and Wood, J.M., 1977, 'Biomethylation of toxic elements in the environment', Science, 197, N 4301, 329-332.

Ridout, P.S., 1984, 'A large volume squeezing system for the analysis of rare elements in deep oceanic pore waters', Mar. Chem., 15, N 3, 193-201.

Riley, J.P. and Roth, J., 1971, 'The distribution of trace elements in some species of phytoplancton grown in culture', J. Mar. Biol. Assn. U.K., 51, N 1, 63-72.

Riley, J.P. and Segar, D.A., 1970, 'The distribution of the major and some minor elements in marine animals. 1. Echinoderms and coelenterates', J. Mar. Biol. Assn. U.K., 50, 721-730.

Riley, J.P. and Sinhaseni, P., 1958, 'Chemical composition of three manganese nodules from the Pacific Ocean', J. Mar. Res., 17, N 1, 466-482.

Robinson, E. and Robbins, R.C., 1971, 'Emissions, concentrations and fate of particulate atmospheric pollutants', Am. Petrol. Inst. Pub., N 76.

Romankevich, E.A., 1977, 'Geochemistry of organic matter in the ocean', Moscow, Nauka, 256 pp. (in Russian).

Rona, P.A., 1984, 'Hydrothermal mineralization at seafloor spreading centers', Earth-Sci. Rev., 20, 104 pp.

Rona, P.A., Bostrom, K., Laubier, L. and Smith K.L. (eds.), 1983, 'Hydrothermal processes at seafloor spreading centers (NATO Conf. Ser., IV Marine Sciences)', N.Y., Plenum, 796 pp.

Ronov, A.B. and Yaroshevsky, A.A., 1967, 'Chemical composition of the Earth's crust', Geokhimiya, N 11, 1285-1309 (in Russian).

Rosson, R.A. and Nielson, K.H., 1982, 'Mn binding and oxidation by spores of marine bacillus', J. Bacteriol., 151, N 2, 1022-1026.

Rozanov, A.G. and Sokolov, V.S., 1984, 'Oxidation-reduction processes in the Arabian basin bottom deposits (the Indian ocean)', Litol. i polezn. iskop., N 6, 41-56 (in Russian).

Rozanov, A.G., Sokolov, V.S. and Volkov, I.I., 1972, 'Iron and manganese forms in the northwestern Pacific sediments', Litol. i polezn. iskop., N 4, 26-39 (in Russian).

Rozanov, A.G., Volkov, I.I. and Sokolov, V.S., 1980, 'Oxidation-reduction processes. Iron and manganese forms in sediments and their variations', in Geochemistry of diagenesis of the Pacific sediments, Moscow, Nauka, pp. 22-50 (in Russian).

Rozanov, A.G., Volkov, I.I., Sokolov, V.S., Pushkina, Z.V. and Pilipchuk, M.F., 1976, 'Oxidation-reduction processes in sediments of the Californian Bay and the adjacent Pacific area (iron and manganese species)', in Biogeochemistry of oceanic sediments diagenesis, Moscow, Nauka, pp. 96-135 (in Russian).

Rucker, J.B. and Valentine, J.W., 1961, 'Salinity response of trace element concentration in Crassostrea virginica', Nature, 190, 1099-1100.

Rydell, H., Kraemer, T., Bostrom, K. and Joensuu, O., 1974, 'Postdepositional injection of uranium-rich solution into East Pacific Rise', Mar. Geol., 17, N 3, 151-164.

Sadasivan, S., 1980, 'Trace constituents in cloud water, rain water and aerosol samples collected near the west coast of India during the southwest monsoon', Atm. Envir., 14, N 1, 33-38.

Sadasivan, S., Anand, S.J.S. and Vohra, K.G., 1974, 'Trace constituents in the monsoon rains near Bombay', Jour. Rech. Atmosph., 8, N 3-4, 873-882.

Saenko, G.N., 1968, 'Formation of modern concepts on photosynthetic reduction of carbon dioxide', in Problems of plant physiology, Moscow, Nauka, pp. 220-263 (in Russian).

Saenko, G.N., Koryakova, M.D., Mokienko, V.F. and Dobrosmyslova, I.G., 1976, 'Concentration of polyvalent metals by seaweeds in Vostok Bay, Sea of Japan', Mar. Biol., 34, N 2, 169-176.

Saenko, G.N., Radkevich, R.O. and Belcheva, N.N., 1977, 'Distribution of microelements in marine plants and sediments as related to land geology of the adjacent coastal regions', in Marine geology and geological structure of discharge areas: the Sea of Japan and Okhotsk Sea, Vladivostok, DVNC AN SSSR, pp. 138-163 (in Russian).

Samoilov, Ya.V. and Gorshkova, T.I., 1924, 'Sediments of the Barentz and Kara Seas', Tr. Morsk. nauchn. instituta, 1, is. 1, 1-40 (in Russian).

Samoilov, Ya.V. and Titov, A.G., 1922, 'Iron-manganese nodules from the bottom of the Black, Baltic and Barentz Seas', Tr. Geol. i miner. muzeya Ross. AN., 3, is. 2, 25-112 (in Russian).

Sanders, J.G., 1978a, 'Enrichment of estuarine phytoplancton by the addition of
    dissolved manganese', Mar. Environ. Res., 1, 59-66.
Sanders, J.G., 1978b, 'The sources of dissolved Mn to Calico Creek, North Carolina',
    Est. Coast. Mar. Sci., 6, 231.
Sanderson, B., 1985, 'Haw bioturbation supports manganese nodules at the sediment
    water interface', Deep-Sea Res., A 32, N 10, 1281-1285.
Sano, Y., Toyoda, K. and Wakita, H., 1985, '3He/4He ratios of marine ferro-
    manganese nodules', Nature, 317, N 6037, 518-520.
Savenko, V.S., 1981, 'Stability of oxic manganese compounds in sea water and some
    problems of manganese nodule genesis', Dokl. AN SSSR, 257, N 5, 1217-1220
    (in Russian).
Savenko, V.S., 1984, 'Biogenic sedimentation, diagenesis and genesis of pelagic
    manganese nodules', Dokl. AN SSSR, 276, N 2, 431-434 (in Russian).
Savenko, V.S., 1985, 'On solubility of manganese dioxide in water solutions',
    Geokhimiya, N 3, 416-419 (in Russian).
Savenko, V.S., Anikeev, V.V. and Kolesov, G.M., 1975, 'Influx of toxic elements
    to sea water out of the atmosphere!' In Problems of mixing of waste water and
    self-purification of water bodies, Moscow, Nauka, pp. 135-139 (in Russian).
Savenko, V.S. and Baturin, G.N., 1981, 'Problems of modelling manganese
    precipitation from sea water and manganese nodule genesis', Litol. i polezn.
    iskop., N 5, 64-70 (in Russian).
Savenko, V.S., Gordeev, V.V., Zhivago, V.N., Korzh, V.D., Serzhenko, T.P.
    and Shinkar, G.G., 1978, 'On concentration of microelements in the atmosphere
    over the ocean, coastal regions of the inner sea basin and over the continent',
    Geokhimiya, N 3, 433-436 (in Russian).
Sawlan, J.J. and Murray, J.W., 1983, 'Trace metal remobilization in the interstitial
    waters of red clay and hemipelagic marine sediments', Earth and Planet. Sci.
    Lett., 64, N 2, 213-230.
Schindler, P.W., 1975, 'Removal of trace metals from the oceans: a zero order
    model', Thalassia Jugosl., 11, 101-111.
Schmitt, R.A., Goles, G.G., Smith, R.H. and Osborn, T.W., 1972, 'Elemental
    abundances in stone meteorites', Meteoritics, 7, N 2, 131-213.
Schnier, C., Gundlach, H. and Marchig, V., 1978, 'Trace elements in pore water
    and sea water in the radiolarian ooze of the Central Pacific as related to the
    genesis of manganese nodules', Environ. Biogeochem., and Geomicrobiol. Proc.
    3d Int. Symp., Wolfenbüttel. vol. 3. Ann. Arbor. Mich., pp. 859-867.
Schuett, C., 1982, 'Role of bacteria in the formation of manganese nodules', in
    Coll. Intern. Bacteriol. Mar. Marseille, Centre Nat. Res. Sci., p. 8.
Schuett, C. and Ottow, J.C.G., 1977, 'Mesophilic and psychrophilic manganese-
    precipitating bacteria in manganese nodules of the Pacific ocean', Zeitschr.
    Allg. Microbiol., B. 17, H. 8, S. 611-616.
Schutz, L., Jaenicke, R. and Pietrek, H., 1980, 'Saharan dust transport over the
    North Atlantic Ocean', Geol. Soc. Amer. Spec. Pap., 186, 87-100.
Schutz, L. and Rahn, K.A., 1982, 'Trace element concentrations in erodible soils',
    Atm. Envir., 16, N 1, 171-176.
Schvoerer, M., Dautant, A. and Bechtel, F., 1979, 'Datation par thermolumines-
    cence des nodules polymetalliques', in La genèse de nodules de manganèse,
    Edit. CNRS, Paris, pp. 295-304.
Schweisfurth, R.D., Eleftheriadis, D., Gundlach, H., Jacobs, M and Jung, W.,
    1978, 'Microbiology and precipitation of manganese', in W.E. Krumbein (ed.),
    Environmental biochemistry and geomicrobiology, v.3. Methods metals and
    assessement. Michigan, Ann. Arbor. Sci. Publ., pp. 923-928.
Segar, D.A., Collins, J.D. and Riley, J.P., 1971, 'The distribution of the major
    and some minor elements in marine animals. Pt. II. Molluscs', J. Mar. Biol.
    Assn. U.K., 51, 131-136.
Segl, M., Mangini, A., Bonani, G., Hoffmann, H.J., Nessi, M., Suter, M.,
    Wölfli, W., Friedrich, G., Plüger, W.L., Wiechowski, A. and Beer, J., 1984a,
    '10Be-dating of a manganese crust from Central North Pacific and implications
    for ocean paleocirculation', Nature, 309, N 5968, 540-543.

Segl, M., Mangini, A., Bonani, G., Hoffmann, H.J., Morenzoni, E., Nessi, M.,
    Suter, M. and Wölfli, W., 1984b, '^{10}Be dating of the inner structure of
    Mn-incrustations applying the Zürich tandem accelerator', Nucl. Instrum. and
    Meth. Phys. Res., 233, 5 B, N 2, 359-364.
Semenov, A.D., Zaletov, V.G., Fuksman, A.L. et al., 1968, 'Experience in
    determining migration forms of material dissolved in natural water', Gidrokhim.
    materialy, 47, 194-202 (in Russian).
Sevastyanov, V.F. and Volkov, I.I., 1965, 'Re-distribution of iron and manganese
    at oxidation-reduction processes in the deposits of the oxic Black Sea zone',
    Litol. i polezn. iskop., N 4, 72-84 (in Russian).
Sevastyanov, V.F. and Volkov, I.I., 1967, 'Re-distribution of chemical elements in
    oxidized sedimentary layer during formation of manganese nodules in the Black
    Sea', Tr. Inst. okeanologii AN SSSR, 83, 135-157 (in Russian).
Sevastyanova, E.S., 1982, 'On correlation of phosphorus and tetravalent manganese
    in pelagic oxidized clays', Okeanologiya, 22, is. 6, 970-974 (in Russian).
Seyfried, W.E. and Bischoff, J.L., 1977, 'Hydrothermal transport of heavy metals
    by seawater: the role of seawater/basalt ratio', Earth and Planet. Sci. Lett.,
    34, N 1, 71-77.
Seyfried, W.E. and Bischoff, J.L., 1979, 'Low temperature basalt alteration by
    seawater: an experimental study at 70°C and 150°C', Geochim. et Cosmochim.
    Acta, 43, N 12, 1937-1947.
Seyfried, W.E. and Bischoff, J.L., 1981, 'Experimental seawater-basalt interaction
    at 300°C and 500 bars: chemical exchange, secondary mineral formation and
    implications for the transport of heavy metals', Geochim. et Cosmochim. Acta,
    45, N 2, 135-147.
Seyfried, W.E. and Mottle, M.J., 1977, 'Origin of submarine metal-rich hydrothermal
    solutions: experimental basalt-seawater interaction in a seawater dominated
    system at 300°C, 500 bars', Proc. 2nd. Int. Symp. Water-Rock interaction,
    Strasbourg, Sec. 4, pp. 173-180.
Seyfried, W.E. and Mottle, M.J., 1982, 'Hydrothermal alteration of basalt by sea-
    water under seawater-dominated conditions', Geochim. et Cosmochim. Acta,
    46, N 6, 985-1002.
Shamov, G.I., 1954, 'River suspension load', Leningrad, Gidrometeoizdat, 346 pp.
    (in Russian).
Sharma, P. and Somayajulu, B.L.K., 1979, 'Growth rates and composition of two
    ferromanganese nodules from the central North Pacific', in La genèse de nodule
    de manganèse. Edit, CNRS, Paris, pp. 281-288.
Sharma, P. and Somayajulu, B.L.K., 1982, '^{10}Be dating of large manganese nodules
    from world oceans', Earth and Planet. Sci. Lett., 59, N 2, 235-244.
Sharma, P. and Somayajulu, B.L.K., 1983, 'Growth rates and composition of Indian
    Ocean ferromanganese nodules', Indian J. Mar. Sci., 12, N 3, 174-176.
Shilo, N.A., Razin, L.V., Khomenko, G.A. and Agaltsov, G.I., 1977, 'On forms of
    gold and palladium in deep oceanic manganese nodules', Dokl. AN SSSR, 232,
    N 2, 466-470 (in Russian).
Shima, M. and Okada, A., 1968, 'Study on the manganese nodule. 1. Manganese
    nodules collected from a long deep-sea core on the Mid-Pacific ocean floor',
    Proc. Soc. Petrol. Miner. Deposits, 60, 47-56.
Shimkus, K.M., 1981, 'Late Quaternary sedimentation in the Mediterranean',
    Moscow, Nauka, 240 pp. (in Russian).
Shiozawa, T., Akira, H., Osamu, T. and Terumi, T., 1982, 'Seasonal cycle of
    manganese in seawater in Harima Sound', J. Oceanogr. Soc. Jap., 38, N 1,
    15-20.
Shishkina, O.V., 1972, 'Chemistry of sea and oceanic pore water', Moscow, Nauka,
    228 pp. (in Russian).
Shishkina, O.V., Baturin, G.N. and Gordeev, V.V., 1982, 'Peculiarity in iron-
    manganese distribution and accumulation in pore water of oceanic sediments',
    Abstracts of the reports to the VI Symposium of IAGOD, Tbilisi, pp. 316-317
    (in Russian).
Shishkina, O.V., Gordeev, V.V., Blazhchishin, A.I. and Mitropolsky, A.Yu., 1981,
    'Microelements in the Baltic Sea pore water', in Sedimentation in the Baltic Sea,
    Moscow, Nauka, pp. 207-215 (in Russian).

Shishkina, O.V., Gordeev, V.V., Tsvetkov, V.A. and Girin, Yu.P., 1979, 'Some data on microelements in pore water of metalliferous sediments from the south-eastern Pacific', in Metalliferous sediments of the south-eastern Pacific, Moscow, Nauka, pp. 217-223 (in Russian).

Shnyukov, E.F. and Orlovsky, G.N., 1980, 'Manganese nodules of the Indian ocean' Geol. Zhurnal., 40, N 2, 46-64 (in Russian).

Shterenberg, L.E., 1978, 'Major manganese minerals of oceanic manganese nodules', Litol. i polezn. iskop., N 1, 32-49 (in Russian).

Shterenberg, L.E., Aleksandrova, V.A., Vasilieva, G.L. et al., 1980, 'Products of volcanic activity in sediments at Site 665 (the north-eastern Pacific)', Litol. i polezn. iskop., N 2, 17-32 (in Russian).

Shterenberg, L.E., Aleksandrova, V.A., Ilichev, L.V., Sivtsov, A.V. and Stepanova, K.N., 1986, 'Post-sedimentation transformations of oceanic manganese nodules and crusts', Izv. AN SSSR, ser. geol., 1, 80-88 (in Russian).

Shterenberg, L.E., Bazilevskaya, E.S. and Chigireva, T.A., 1966, 'Iron and manganese carbonates in bottom sediments of the Punnus-Yarvi Lake', Dokl. AN SSSR, 170, 691-694 (in Russian).

Shterenberg, L.E., Dubinina, G.A. and Stepanova, K.N., 1975, 'Formation of tabular Fe-Mn nodules', in Problems of lithology and geochemistry of sedimentary rocks and ores, Moscow, Nauka, pp. 166-179 (in Russian).

Shterenberg, L.E., Lavrushin, Yu.A., Golubev, Yu.K., Sivtsov, A.V., Spiridonov, M.A. and Rybalko, A.E., 1985, 'Manganese nodules of the White Sea strait', Litol. i. polezn. iskop., N 5, 66-75 (in Russian).

Shterenberg, L.E. and Shurina, G.N., 1974, 'Clay minerals in the Pacific manganese nodules', Dokl. AN SSSR, 218, N 4, 908-910 (in Russian).

Shterenberg, L.E. and Vasilieva, G.L., 1979, 'Native metals and intermetallic compounds in the north-eastern Pacific sediments', Litol. i polezn. iskop., N 2, 133-139 (in Russian).

Shterenberg, L.E., Vasilieva, G.L., Voronin, V.I. and Korina, E.A., 1981, 'Gold and silver minerals in the Pacific metalliferous sediments', Izv. AN SSSR, ser. geol., N 3, 151-154 (in Russian).

Shterenberg, L.E., Voznesensky, A.I. and Eliseeva, T.G., 1977, 'Clay minerals in the Pacific manganese nodules (the eastern part of the Trans-Pacific profile)', Dokl. AN SSSR, 234, N 3, 693-695 (in Russian).

Shulkin, V.M. and Bogdanova, N.N., 1984, 'Geochemistry of iron, manganese, zinc and copper in coastal sedimentation', Litol. i polezn. iskop., N 6, 105-117 (in Russian).

Shumilin, E.N. and Tikhomirov, V.N., 1982, 'Study of the plankton effect upon [54]Mn, [57]Co, [144]Ce, [95]Zr, [95]Nb radionuclides in sea water by selective leaching', Radiokhimiya, N 2, 250-251 (in Russian).

Siegel, M.D. and Turner, S., 1983, 'Crystalline todorokite associated with biogenic debris in manganese nodules', Science, 219, N 4581, 172-174.

Siesser, W.G., 1978, 'Native copper in DSDP Leg 40 sediments', Init. Rep. DSDP, Wash. D.C., 38-41 Suppl., 761-765.

Sievering, H, Eastman, J. and Schmidt, J.A., 1982, 'Air-sea particle exchange at a nearshore oceanic site', J. Geophys. Res., 87, N C13, 1127-1137.

Sillen, L.G., 1961, 'The physical chemistry of sea water', in M. Sears (ed.), Oceanography, Am. Assoc. Adv. Sci. Publ., 67, 549-581.

Simpson, R.D., 1979, 'Uptake and loss of zinc and lead by mussels (Mytilus edulis) and relationship with body weight and reproductive cycle', Mar. Pol. Bull., 10, 74-78.

Sivalingam, P.M., 1978, 'Effects of high concentration stress of trace metals on their biodeposition modes in Ulva reticulata Forksal', Japan. J. Phycol., 26, 157-160.

Skopintsev, B.A. and Popova, T.P., 1963, 'On manganese accumulation in waters of H_2S-basins: the Black Sea', Tr. Geol. Inst. AN SSSR, is. 97, 24-43 (in Russian).

Skornyakova, N.S., 1961, 'Bottom deposits of the north-eastern Pacific', Tr. Inst. okeanologii AN SSSR, 45, 22-64 (in Russian).

Skornyakova, N.S., 1964, 'Dispersed iron and manganese in the Pacific sediments', Litol. i polezn. iskop., N 5, 3-20 (in Russian).

Skornyakova, N.S., 1970, 'Dispersed iron and manganese in the Pacific sediments', in The Pacific. Sedimentation in the Pacific, Moscow, Nauka, book 2, pp. 159-202 (in Russian).

Skornyakova, N.S., 1976a, 'Dispersed Fe, Mn, Ti and some minor elements in sediments enclosing ferromanganese nodules', in Ferromanganese nodules of the Pacific Ocean, Moscow, Nauka, pp. 168-189 (in Russian).

Skornyakova, N.S., 1976b, 'Chemical composition of ferromanganese nodules of the Pacific', ibid., 190-240.

Skornyakova, N.S., 1979, 'Zonal regulaties in occurence, morphology and chemistry of manganese nodules of the Pacific ocean', in J. Bischoff and D.Z. Piper (eds.), Marine geology and oceanography of the Pacific manganese nodule province, N.Y., Plenum, pp. 699-728.

Skornyakova, N.S., 1984, 'Morphogenetic types of manganese nodules from the Pacific radiolarian belt', Litol. i polezn. iskop., N 6, 67-83 (in Russian).

Skornyakova, N.S., 1986, 'Local variations of manganese nodule fields', in Manganese nodules of the Central Pacific, Moscow, Nauka, pp. 109-184 (in Russian).

Skornyakova, N.S. and Andruschenko, P.F., 1968, 'Manganese nodules of the South Pacific', Okeanologiya, 8, is. 5, 865-877 (in Russian).

Skornyakova, N.S. and Andruschenko, P.F., 1970, 'Manganese nodules of the Pacific', in Sedimentation in the Pacific, Moscow, Nauka, book 2, pp. 203-268 (in Russian).

Skornyakova, N.S. and Andruschenko, P.F., 1971, 'On some morphological and internal features of manganese nodules of the Pacific', Litol. i polezn. iskop., N 1, 3-14 (in Russian).

Skornyakova, N.S. and Andruschenko, P.F., 1976, 'Morphology and internal structure of manganese nodules', in Manganese nodules of the Pacific, Moscow, Nauka, pp. 91-122 (in Russian).

Skornyakova, N.S. and Vanshtein, B.G., 1983, 'Ferromanganese nodules of the Indian Ocean', Lithologiya i polezn. iskop., N 2, 86-98 (in Russian).

Skornyakova, N.S., Baturin, G.N. and Murdmaa, I.O., 1984, 'Manganese nodules from the near equatorial Pacific radiolarian zone', Moscow, Nauka, Tr. 27 Mezhd. Geol. Kongr., Geologia Mirovogo okeana, Sec C.06, Doklady, 6, pt. I, 19-27 (in Russian).

Skornyakova, N.S., Baturin, G.N. and Zaikin, V.N., 1986, 'Molybdenum in the Pacific manganese nodules', Geokhimiya, N 12, 1799-1805. (in Russian).

Skornyakova, N.S., Bazilevskaya, E.S. and Gordeev, V.V., 1975, 'Some problems of mineralogy and geochemistry of the Pacific manganese nodules', Geokhimiya, N 7, 1064-1076 (in Russian).

Skornyakova, N.S., Bezrukov, P.L., Bazilevskaya, E.S. and Gordeev, V.V., 1979, 'Manganese nodules of the eastern Indian ocean: zonation and local variability', Litol. i polezn. iskop., N 3, 3-18 (in Russian).

Skornyakova, N.S., Bezrukov, P.O. and Murdmaa, I.O., 1981, 'Major distribution patterns and composition of oceanic manganese nodule fields', Litol. i polezn iskop., N 5, 51-64 (in Russian).

Skornyakova, N.S., Gordeev, V.V., Anikeeva, L.I., Chudaev, O.V. and Kholodkevich, I.V., 1985, 'Local variations of nodules in the Clarion-Clipperton ore province', Okeanologiya, 25, N 4, 630-637 (in Russian).

Skornyakova, N.S. and Zenkevich, N.L., 1976, 'Regularities in spatial distribution of manganese nodules', in Manganese nodules of the Pacific, Moscow, Nauka, pp. 37-81 (in Russian).

Slowey, D.F., Hayes, D., Dixon, B. and Hood, D.W., 1965, 'Distribution of gamma-emitting radionuclides in the Gulf of Mexico', Occl. Publ. Narrangansett Mar. Lab., Univ. Rhode Island, Kingstone R.I., 3, 109-129.

Slowey, D.F. and Hood, D.W., 1966, 'Investigation of the distribution of copper, manganese and zinc in oceans by neutron activation analysis', in 2nd. Intern. Oceanogr. Congr. Abstr., Moscow, Nauka, p. 353.

Slowey, D.F. and Hood, D.W., 1971, 'Cu, Mn and Zn concentrations in Gulf of Mexico waters', Geochim. Coamochim. Acta, 35, N 2, 121-138.

Smith, D.G. and Westgate, J.A., 1969, 'Electron probe technique for characterising pyroclastic deposits', Earth and Planet. Sci. Lett., 5, 313-319.

Smith, J.D. and Burton, J.D., 1972, 'The occurence and distribution of tin with particular reference to marine environment', Geochim. et Cosmochim. Acta, 36, N 5, 621-629.

Sobotovich, E.V., 1976, 'Cosmic matter in the Earth's crust', Moscow, Atomizdat, 159 pp. (in Russian).

Sobotovich, E.V., 1982, 'Cosmic matter at the Earth', Kiev, Naukova Dumka, 168 pp. (in Russian).

Sokolova, E.G. and Pilipchuk, M.F., 1973, 'On geochemistry of selenium in the north-western Pacific deposits', Geokhimiya, N 10, 1537-1546 (in Russian).

Somayajulu, B.L.K., 1967, 'Berillium-10 in manganese nodule', Science, 156, N 3779, 1219-1220.

Somayajulu, B.L.K., Heath, G.R., Moore, T.C. and Cronan, D.S., 1971, 'Rates of accumulation of manganese nodules and related sediment from the equatorial Pacific', Geochim. et Cosmochim. Acta, 35, N 6, 621-624.

Sorem, R.K., 1967, 'Manganese nodules: nature and significance of internal structure', Econ. Geol., 62, N 1, 141-147.

Sorem, R.K., 1973, 'Manganese nodules: as indicators of long-term variations in sea floor environment', in M. Morgenstein (ed.), The origin and distribution of manganese nodules in the Pacific and prospects for exploration, Hawaii Inst. Geophys., pp. 151-164.

Sorem, R.K. and Fewkes, R.H., 1977, 'Internal characteristics', in G.P. Glasby (ed.), Marine manganese deposits, Amsterdam, Elsevier, pp. 147-183.

Sorem, R.K. and Fewkes, R.H., 1979, 'Manganese nodules. Research data and methods of investigation', N.Y., Plenum, 723 pp.

Sorem, R.K. and Fewkes, R.H., 1980, 'Distribution of todorokite and birnessite in manganese nodules from the "Horn Region", Eastern North Pacific', in J.M. Varentsov and G.Y. Grasselly (eds.), Geology and Geochemistry of manganese, Budapest, Akad. Kiado, pp. 203-229.

Sorem, R.K., Fewkes, R.H., McFarland. W.A. and Reinhart, W.R., 1979, 'Physical aspects of the growth environment of manganese nodules in the "Horn Region", East Equatorial Pacific Ocean', in C. Lalou (ed.), Sur la genèse des nodules de manganèse, Proc. Coll. Intern. du CNRS, Paris, 61-76.

Sorem, R.K. and Foster, A.R., 1972, 'Marine manganese nodules: importance of structural analysis', 24th. Intern. Geol. Congr., Sec. 8, pp. 192-200.

Sorem, R.K., Reinhart, W.R., Fewkes, R.H. and McFarland, W.D., 1979, 'Occurence and character of manganese nodules in DOMES Sites A, B and C, Equatorial Pacific Ocean', in Mar. Geol. and oceanogr. of the Pacific manganese nodule province, N.Y., a. London, Plenum, pp. 475-527.

Sorokin, Yu.I., 1971, 'On microflora of manganese nodules at the ocean bottom', Mikrobiologiya, 40, N 3, 563-566 (in Russian).

Sorokin, Yu.I., 1972, 'On the role of biogenic factors in iron, manganese and cobalt sedimentation and in nodule formation', Okeanologiya, 12, N 1, 3-14 (in Russian).

Spencer, D.W. and Brewer, P.G., 1971, 'Vertical advection, diffusion and redox potentials as controls on the distribution of manganese and other trace metals dissolved in waters of the Black Sea', J. Geophys. Res., 76, N 24, 5877-5892.

Spencer, D.W., Brewer, P.G., Fleer, A., Honjo, S., Krishnaswamy, S. and Nozaki, J., 1978, 'Chemical fluxes from a sediment trap experiment in the deep Sargasso Sea', J. Mar. Res., 36, N 3, 493-523.

Spencer, D.W., Brewer, P.G. and Sachs, P.L., 1972, 'Aspects of the distribution and trace element composition of suspended matter in the Black Sea', Geochim. et Cosmochim. Acta, 36, N 1, 71-86.

Spirn, R.V., 1965, 'Rare earth distributions in the marine environment', Ph. D. Thesis, Massachusetts Institute of Technology, Cambridge, Mass., 165 pp.

Stackelberg, U. von, 1984, 'Significance of benthic organisms for the growth and movement of manganese nodules, equatorial North Atlantic', Geo-Mar. Lett., 4, N 1, 37-42.

Stallard, R.F. and Edmond, J.M., 1983, 'Geochemistry of the Amazon. 2. The influence of geology and weathering environment on the dissolved load', J. Geophys. Res., 88, N C14, 9671-9688.

Stanier, R.Y., Adelberg, E.A. and Ingraham, J.L., 1976, The microbial world, Englewood Cliffs, Prentice-Hall, Vol. 1-3.

Staudigel, H. and Hart, S.R., 1983, 'Alteration of basaltic glass: mechanisms and significance for the oceanic crust-seawater budget', Geochim. et Cosmochim. Acta, 47, N 3, 337-350.

Stevenson, J.S. and Stevenson, L.S., 1970, 'Manganese nodules from the Challenger Expedition at Redpath Museum', Can. Mineral., 10, 599-615.

Stevenson, R.A. and Ufret, S.L., 1966, 'Iron, manganese and nickel in skeletons and food for the sea urchins Tripneustes esculentus and Echinometra locunter', Limnol. Oceanogr., 11, 11-17.

Stoffers, P., Schmitz, W. and Glasby, G.P., 1983, 'Geochemistry of deep-sea sediments from the Indian ocean collected during DSDP legs 22, 24, 26 and 27', Chem. Erde, B. 42, N 1, 15-30.

Straczek, J.A., Horen, A., Ross, M. and Warshaw, C.M., 1960, 'Studies of the manganese oxides. IV. Todorokite', Am. Mineral., 45, 1174-1184.

Strakhov, N.M., 1954, 'Sedimentation in the Caspian Sea', in Sedimentation in recent water bodies, Moscow, Izd-vo AN SSSR, pp. 137-179 (in Russian).

Strakhov, N.M., 1960, 'Fundamental problems of lithogenesis theory', Moscow, Izd. AN SSSR, 2, 572 pp. (in Russian).

Strakhov, N.M., 1963, 'Lithogenic types and their evolution through the Earth's history', Moscow, Gosgeoltechizdat, 535 pp. (in Russian).

Strakhov, N.M., 1976, 'Geochemical problems of recent oceanic lithogenesis', Moscow, Nauka, 300 pp. (in Russian).

Strakhov, N.M. and Nesterova, I.L., 1968, 'On volcanism effect upon geochemistry of marine deposits: the Okhotsk Sea', in Geochemistry of sedimentary rocks and ores, Moscow, Nauka, 438-483 (in Russian).

Strakhov, N.M., Shterenberg, L.E., Kalinenko, V.V. and Tikhomirova, Z.S., 1968, 'Geochemistry of manganese ore sedimentation', Tr. Geol. Inst. AN SSSR, 185, 495 pp. (in Russian).

Stumm, W., Hohl, H. and Dalong, F., 1976, 'Interaction of metal ions with hydrous oxide surfaces', Croatica Chem. Acta, 48, N 4, 491-504.

Stumm, W. and Morgan, J.J., 1970, Aquatic Chemistry, N.Y., Wiley, 583 pp.

Stumm, W. and Morgan, J.J., 1981, Aquatic Chemistry, 2nd. ed. N.Y., Wiley, 780 pp.

Suess, E., 1979, 'Mineral phases formed in anoxic sediment by microbial decomposition of organic matter', Geochim. et Cosmochim. Acta, 43, N 3, 339-352.

Suess, E., 1980, 'Particulate organic carbon flux in the ocean - surface productivity and oxygen utilization', Nature, 288, N 5788, 260-263.

Sugimura, Y., Miyake, Y. and Yanagawa, H., 1975, 'Chemical composition and the rate of accumulation of ferromanganese nodules in the western North Pacific', Pap. Met. Geophys., 26, N 2, 47-54.

Summerhayes, C. and Willis, J., 1975, 'Geochemistry of manganese deposits in relation to environment of sea floor around Southern Africa', Mar. Geol., 18, N 3, 159-173.

Sunda, W.G., Huntsman, S.A. and Harvey, G.R., 1983, 'Photoreduction of manganese oxides in seawater and its geochemical and biological implications', Nature, 301, N 5897, 234-236.

Sundby, B. and Silverberg, N., 1985, 'Manganese fluxes in the benthic boundary layer', Limnol. Oceanogr., 30, N 2, 372-381.

Sundby, B., Silverberg, N. and Chesselet, R., 1981, 'Pathways of manganese in an open estuarine system', Geochim. et Cosmochim. Acta, 45, N 3, 293-307.

Sung, W. and Morgan, J.J., 1981, 'Oxidative removal of Mn (II) from solution catalyzed by the γ-FeOOH (lepidocrocite) surface', Geochim. et Cosmochim. Acta, 45, 2377-2384.

Sung, F.C., Nevissi, A.E. and Dewalle, F.B., 1982, 'Water-soluble constituents of mount St. Helens ash', J. Environ. Sci. and Health, A. 17, N 1, 45-55.

Supatashvili, G.D., Goliadze, N.S. and Gvelisiani, L.T., 1984, 'Manganese distribution in suspension and bottom sediments in Georgia water bodies', Litol. i polezn. iskop., N 6, 97-104 (in Russian).

Svalnov, V.N., 1983, 'Quaternary sedimentation in the eastern Indian ocean', Moscow, Nauka, 192 pp. (in Russian).

Swanson, V.E., Palocas, J.G. and 'Love Alonza A., 1967. Geochemistry of deep-sea sediment along the 160° W meridian in the North Pacific Ocean', U.S. Geol. Surv. Prof. Pap., 575-B, 137-144.

Tatsumoto, M. and Goldberg, E.D., 1959, 'Some aspects of the marine geochemistry of uranium', Geochim. et Cosmochim. Acta, 17, N 3/4, 201-208.

Taylor, S.R., 1964, 'Abundance of chemical elements in the continental crust: a new table', Geochim. et Cosmochim. Acta, 28, N 8, 1273-1286.

Tennant, D.A. and Forster, W.D., 1969, 'Seasonal variation and distribution of ^{65}Zn, ^{54}Mn and ^{51}Cr in tissues of the crab Cancer magister Dana', Health Phys., 18, 649-659.

Teraoka, H. and Kobayashi, J., 1980, 'Concentration of 21 metals in the suspended solids collected from the principal 166 rivers and 3 lakes in Japan', Geochem.J., 14, N 5, 203-225.

Terashima, S., Nakao, S. and Mochizuki, T., 1982, 'Sulfur and carbon contents of manganese nodules from the Central Pacific, GH-80-1 cruise', Bull. Geol. Soc. Jap., 33, N 3, 111-123.

Thiel, G.A., 1925, 'Manganese precipitated by microorganisms', Econ. Geol., 20, 301-310.

Thiel, H., 1978, 'The faunal environment of manganese nodules and aspects of deep sea time scales', in W.E. Krumbein (ed.), Environmental Biogeochemistry and Geomicrobiology, Vol. 3. Methods, Metals and Assessement. Ann. Arbor, Mich., Ann Arbor Sci. Publ., pp. 887-896.

Thiel, K., Peters, J., Vogt, S. and Herr, W., 1983, 'On the nature of the metallic and glassy spherules in Antarctic ice', Meteoritics, 18, N 4, 408.

Thompson, G., 1983, 'Hydrothermal fluxes in the Ocean', in Chemical Oceanography, Vol. 8, London, Acad. Press., pp. 272-337.

Thompson, G. and Humphris, S.E., 1977, 'Sea water-rock interactions in the oceanic basement', Proc. 2nd. Int. Symp. Water-Rock Interact., Strasbourg, Sec. 2-3, pp. III/13-III/18.

Tikhomirov, V.N., 1962, 'Study of manganese sorption by oceanic bottom sediments and estimates of manganese nodule growth rates from manganese sorption rates', Geol. Zhurnal, N 2, 56-63 (in Russian).

Tikhomirov, V.N., 1983, 'Study of hydrolitical properties of ^{54}Mn in sea water by ion exchange technique', Radiokhimiya, N 1, 85-89 (in Russian).

Tikhomirov, V.N., 1984, 'Study of metal sorption by sediments from oceanic water using radiotracers, Litol. i polezn. iskop., N 3, 99-106 (in Russian).

Tikhomirov, V.N., 1986, 'Study of the state and sorption behaviour of metals at the water / bottom interface using radiotracers', in Manganese nodules of the Central Pacific, Moscow, Nauka, 270-283 (in Russian).

Tikhomirov, V.N. and Gromov, V.V., 1971, 'Study of the state of ^{99}Tc and ^{54}Mn in distilled and sea waters', Radiokhimiya, 13, N 2, 316-318 (in Russian).

Tikhomirov, V.N. and Gromov, V.V., 1983, 'Behaviour of radioactive nuclides of manganese and technetium in the ocean', Ocean Sci. a. Eng., 8, N 3, 299-328.

Tikhomirov, V.N., Gromov, V.V., Bernovskaya, R.N. and Spitsin, V.I., 1970, 'The role of plankton in ^{99}Tc and ^{54}Mn behaviour in oceanic water', Dokl. AN SSSR, 194, N 2, 445-447 (in Russian).

Tikhomirov, V.N., Gromov, V.V. and Konnov, V.A., 1979, 'Sorption of abiogenic and biogenic species of manganese and nickel by calcareous sediments of the Indian ocean', Okeanologiya, 19, N 4, 632-637 (in Russian).

Tikhomirov, V.M., Gromov, V.V., Krylov, O.T., 1982, 'Desorption technique to release elements from bottom sediments to oceanic water and to study forms of manganese occurrence', Okeanologiya, 22, N 1, 53-56 (in Russian).

Tikhomirov, V.M., Gromov, V.V., Krylov, O.T. and Spitsin, V.I., 1978, 'Study of regularities in manganese sorption by bottom sediments from oceanic water', Dokl. AN SSSR, 241, N 5, 1193-1196 (in Russian).

Tikhomirov, V.M. and Lukashin, V.N., 1983, 'Study of biogeochemical behaviour of microelements in the ocean using radiotracers', in Biogeochemistry of the ocean, Moscow, Nauka, pp. 302-312 (in Russian).

Tikhomirov, V.N. and Shakhova, N.F., 1987, 'Study of the state of some metals in near-bottom and pore waters of the Pacific and formation of manganese nodules',

Okeanologiya, 27, N 1, 63-68 (in Russian).

Tischenko, P.Ya. and Gramm-Osipov, L.M., 1985, 'Formation of manganese solid phases in the ocean', Dokl. AN SSSR, 280, N 1, 231-232 (in Russian).

Tlig, S., 1982, 'Distribution des terres rares dans les fractions de sediments et nodules de Fe et Mn associés en l'ocean Indien', Mar. Geol., 50, N 3, 257-274.

Toth, J.R., 1980, 'Deposition of submarine crusts rich in manganese and iron', Bull. Geol. Soc. Amer., 91, N 1, 44-54.

Trefrey, J.H. and Presley, B.J., 1976, 'Heavy metal transport from the Mississippi River to the Gulf of Mexico', in H.L. Windom and R.A. Duce (eds.), Marine Pollutant Transfer, Lexington Books, Mass., Toronto, pp. 39-76.

Trefrey, J.H. and Presley, B.J., 1982, 'Manganese fluxes from Mississippi Delta sediments', Geochim. et Cosmochim. Acta, 46, N 10, 1715-1726.

Trimble, R.B. and Ehrlich, H.L., 1968, 'Bacteriology of manganese nodules. III. Reduction of MnO_2 by two strains of nodule bacteria', Appl. Microbiol., 16, 695-702.

Trimble, R.B. and Ehrlich, H.L., 1970, 'Bacteriology of manganese nodules. IV. Induction of an MnO_2-reductase system in a marine bacillus', Appl. Microbiol., 19, 966-972.

Trukhin, Yu.P. and Shuvalov, R.A., 1977, 'Recent hydrothermal process and evolution of volcanism', Moscow, Nauka, 351 pp. (in Russian).

Tsunogai, S. and Kondo, T., 1982, 'Sporadic transport and deposition of continental aerosols to the Pacific Ocean', J. Geophys. Res., 87, N c 11, 8870-8874.

Tsunogai, S. and Kusakabe, M., 1982, 'Migration of manganese in the deep-sea sediments', in K.A. Fanning and F. Manheim (eds.), The dynamic environment of the ocean floor, Toronto, Lexington Books, pp. 257-273.

Tsunogai, S., Uematsu, M., Noriki, S., Tanaka, N. and Yamada, M., 1982, 'Sediment trap experiment in the western North Pacific', Geochem. J., 16, N 3, 129-147.

Turekian, K.K., 1965, 'Some aspects of the geochemistry of marine sediments', in Chemical Oceanography, vol. 2, London, Academic Press, pp. 81-126.

Turekian, K.K., 1967, 'Estimates of the average Pacific deep-sea clay accumulation rate from material balance calculations', in Progress in Oceanography, 4, 227-244.

Turekian, K.K., 1968, 'Deep-sea deposition of barium, cobalt and silver', Geochim. et Cosmochim. Acta, 32, N 6, 603-612.

Turekian, K.K., 1971, 'Rivers, tributaries and estuaries', in D.W. Hood (ed.), Impingement of man on the oceans, N.Y., Wiley, N 4, pp. 9-73.

Turekian, K.K. and Cochran, J.K., 1981a, '[210]Pb in surface air at Enewetak and the asian dust flux to the Pacific', Nature, 292, N 5823, 522-524.

Turekian, K.K. and Cochran, J.K., 1981b, 'A correction to the letter ^{210}Pb in surface air at Enewetak and the Asian dust flux to the Pacific', Nature, 294, N 5842, p. 670.

Turekian, K.K. and Imbrie, J., 1966, 'The distribution of trace elements in deep-sea sediments of the Atlantic Ocean', Earth and Planet. Sci. Lett., 1, 161-168.

Turekian, K.K. Nozaki, Y., and Benninger, L.K., 1977, 'Geochemistry of atmospheric radon and radon products', Annu. Rev. Earth Planet. Sci., 5, 227-255.

Turekian, K.K. and Scott, M.R., 1967, 'Concentrations of Cu, Ag, Mo, Ni, Co and Mn in suspended material in streams', Environmental Sci. and Technol., 1, N 9, 940-942.

Turekian, K.K. and Wedepohl, K.H., 1961, 'Distribution of the elements in some major units of the earth's crust', Bull. Geol. Soc. Amer., 72, N 1, 175-192.

Turner, S. and Buseck, P.R., 1981, 'Todorokites: a new family of naturally accurring manganese oxides', Science, 212, N 4498, 1024-1027.

Turner, S., Siegel, M.D. and Buseck, P.R., 1982, 'Structural features of todorokite intergrowths in manganese nodules', Nature, 296, N 5860, 841-842.

Udelnova, T.M., Pusheva, M.A., Laktionov, N.V. and Koryakin, A.V., 1974, 'Content of some polyvalent metals in blue-green algae', Mikrobiologiya, 43, N 6, 1064-1068 (in Russian).

Uematsu, M., Duce, R.A. Prospero, J.M., Chen, L., Merrill, J.T. and McDonald, R.L., 1983, 'Transport of mineral aerosol from Asia over the North Pacific

Ocean', <u>J. Geophys. Res.</u>, <u>88</u>, N C9, 5343-5352.

Uematsu, M. and Tsunogai, S., 1983, 'Recycling of manganese in the coastal sea, Funka Bay, Japan', <u>Mar. Chem.</u>, <u>13</u>, N 1, 1-14.

Usui, A., 1979a, 'Nickel and copper accumulations as essential elements in 10 Å manganite of deep-sea manganese nodules', <u>Nature</u>, <u>279</u>, N 5712, 411-413.

Usui, A., 1979b, 'Minerals, metal contents and mechanism of formation of manganese nodules from the Central Pacific Basin (GH 76-1 and GH 77-1 areas)', in J.L. Bischoff and D.Z. Piper (eds.), <u>Marine geology and oceanography of the Pacific manganese nodule province</u>, N.Y., Plenum, pp. 651-679.

Usui, A., 1983, 'Regional variations of manganese nodule facies on the Wabe-Tahiti transect: morphological, chemical and mineralogical study', <u>Mar. Geol.</u>, <u>54</u>, N 1, 27-51.

Vakhrushev, V.A. and Prokoptsev, N.G., 1969, 'Primary magmatic sulphide formations in oceanic crust basalts and in ultramorfic rock inclusions', <u>Geologia rud. mestorozhd.</u>, N 6, 14-26 (in Russian).

Van As, D., Fourie, H.O. and Vleggaar, C.M., 1975, 'Trace elements concentrations in marine organisms from the Cape West coast', <u>S. Afr. J. Sci.</u>, <u>71</u>, 151-154.

Van As, D., Fourie, H.O. and Vleggaar, C.M., 1973, 'Accumulation of certain elements in the marine organisms from the sea around the Cape of Good Hope', in <u>Radioactive contamination of the Marine Environment</u>, Int. Atom. Ener. Agen., Vienna, Austria, pp. 615-624.

Van der Veijden, C.H. and Kruissink, E.C., 1977, 'Some geochemical controls on lead and barium concentrations in ferromanganese deposits', <u>Marine Chemistry</u>, <u>5</u>, N 2, 93-112.

Varentsov, I.M. and Blazhchishin, A.I., 1976, 'Iron-manganese formations', in <u>Geology of the Baltic Sea</u>, Vilnius, Mokslas, pp. 307-348. (in Russian).

Varentsov, I.M., Dikov, Y.P. and Bakova, N.V., 1978, 'On ore formation modelling in recent basins', <u>Geokhimiya</u>, N 8, 1198-1209 (in Russian).

Varentsov, I.M. and Pronina, N.V., 1972, 'On the study of mechanisms of iron-manganese ore formation in recent basins: the experimental data on nickel and cobalt', <u>Miner. Depos.</u>, <u>8</u>, 161-178.

Vassiliou, A.H. and Blount, A.M., 1977, 'Crystallographic data for authigenic phillipsite from manganese nodules', <u>Miner. Depos.</u>, <u>12</u>, N 2, 171-174.

Vernadskiy, V.I., 1954, 'Collection of papers', <u>Moscow Izd. AN SSSR</u>, <u>1</u>, 696 pp. (in Russian).

Vershinin, A.V., Petrii, O.A. and Rozanov, A.G., 1982, 'On nature of platinum electrode potential in oxygen-rich marine environment', <u>Okeanologiya</u>, N 6, 1018-1022 (in Russian).

Vilenskii, V.D. and Miklishanskii, A.Z., 1976, 'Chemical composition of snow sheet of the Eastern Antarctic', <u>Geokhimiya</u>, N 11, 1683-1690 (in Russian).

Vinogradov, A.P., 1935-1944, 'Chemical element composition of marine organisms', <u>Tr. biogeokhim. lab. AN SSSR</u>, part I, 1935, pp. 63-278; part II, 1937, pp. 5-225; part III, 1944, pp. 373 (in Russian).

Vinogradov, A., 1953, 'The elementary chemical composition of marine organisms', <u>Sears Found. Mar. Res. Mem. 2</u>, New Haven, 647 pp.

Vinogradov, A.P., 1957, 'Geochemistry of microelements in soils', <u>Moscow, Izd. AN SSSR</u>, 238 pp. (in Russian).

Vinogradov, A.P., 1959, 'Chemical evolution of the Earth', Moscow, <u>Izd. AN SSSR</u>, 44 pp. (in Russian).

Vinogradov, A.P., 1962, 'Average concentration of chemical elements in major types of igneous crustal rocks', <u>Geokhimiya</u>, N 7, 555-571 (in Russian).

Vinogradov, A.P., 1967, 'Introduction to the geochemistry of the ocean', Moscow, Nauka, 215 pp. (in Russian).

Vinogradsky, S.N., 1952, <u>Microbiology of the soil</u>, Moscow, Academy of sci. USSR, 792 pp. (in Russian).

Vinogradova, Z.A. and Kovalskii, V.V., 1962, 'On analysis of elemental composition of the Black Sea plankton', <u>Dokl. AN SSSR</u>, <u>147</u>, N 6, 1458-1460 (in Russian).

Vizhenskii, V.A., Shnykin, B.A., 1984, 'General concentration of microelements in suspension of the Indian and Pacific oceans', in <u>Monitoring of general pollution of natural environment</u>, Leningrad, Gidrometeoizdat, N 2, pp. 168-177 (in Russian).

Voinar, A.I., 1960, Biological role of microelements in organism of animals and men, Moscow, Vys'shaya Shkola, 544 pp. (in Russian).

Volkov, I.I., 1973, 'Main regularities in the distribution of chemical elements in the Black Sea deep sediments', Litologia i polezn. iskop., N 2, 3-22 (in Russian).

Volkov, I.I., 1975, 'Chemical elements in river discharge and forms of their supply to the sea (rivers of the Black Sea basin)', in Problems of lithology and geochemistry of sedimentary rocks and ores, Moscow, Nauka, pp. 85-113 (in Russian).

Volkov, I.I., 1979, 'Manganese nodules', in Chemistry of the ocean. Geochemistry of bottom sediments, Moscow, Nauka, 2, 414-467 (in Russian).

Volkov, I.I., 1980, 'Re-distribution of chemical elements in diagenesis of sediments', in Geochemistry of diagenesis of the Pacific sediments, Moscow, Nauka, pp. 144-168 (in Russian).

Volkov, I.I., 1981, 'Manganese balance in oceanic sedimentation cycle', Litoligia i polezn. iskop., N 3, 25-34 (in Russian).

Volkov, I.I., 1984, 'Geochemistry of sulphur in oceanic sediments', Moscow, Nauka, 272 pp. (in Russian).

Volkov, I.I. and Fomina, L.S., 1967, 'Rare-earth elements in sediments and manganese nodules of the ocean', Litologia i polezn. iskop., N 5, 66-85 (in Russian).

Volkov, I.I. and Fomina, L.S., 1973, 'New data on geochemistry of rare-earth elements in the Pacific sediments', Geokhimiya, N 11, 1603-1614 (in Russian).

Volkov, I.I., Fomina, L.S. and Yagodinskaya, T.A., 1976, 'Chemical composition of manganese nodules of the Pacific along the Wake Atoll - Mexico coast profile', in Biogeochemistry of diagenesis of oceanic sediments, Moscow, Nauka, pp. 186-204 (in Russian).

Volkov, I.I. and Sevastyanov, V.F., 1968, 'Re-distribution of chemical elements in diagenesis of the Black Sea sediments', in Geochemistry of sedimentary rocks and ores, Moscow, Nauka, pp. 134-182 (in Russian).

Volkov, I.I., Shterenberg, L.E. and Fomina, L.S., 1980, 'Manganese nodules', in Geochemistry of diagenesis of the Pacific sediments, Moscow, Nauka, pp. 169-223 (in Russian).

Volkov, I.I. and Sokolov, V.S., 1970, 'Germanium in manganese nodules of Recent sediments', Litologia i polezn. iskop., N 6, 24-29 (in Russian).

Volkov, I.I., Sokolov, V.S., Sokolova, E.G. and Pilipchuk, M.F., 1974, 'Rare and dispersed elements in the northwestern Pacific sediments', Litologia i polezn. iskop., N 2, 3-21 (in Russian).

Volkov, I.I., Sokolov, V.S. and Fomina, L.S., 1980, 'Reactive forms of elements in sediments along the northeastern Pacific profile', in Geochemistry of diagenesis of the Pacific sediments, Moscow, Nauka, pp. 6-21 (in Russian).

Volkov, I.I. and Sokolova, E.G., 1984, 'Thallium in oceanic manganese nodules', in Geology of oceans and seas, Abstracts of the reports to the 6th. All-Union seminar on marine geology, Moscow, Nauka, 3, 33-35 (in Russian).

Vologdin, A.G., 1946, 'On microbiological origin of some economic ore fields in Kazakhstan', Izv. Kaz. fil. AN SSSR, ser. geol., N 8 (26), 61-66 (in Russian).

Vologdin, A.G., 1947, 'Geological activity of microorganisms', Izv. AN SSSR, ser. geol., N 3, 19-36 (in Russian).

Von Buttlar, H. and Houtermans, G., 1950, 'Fotographische Bestimmung der Aktivitätsverteilung in einer Manganknolle der Tiefsee', Naturwiss., 37, 400-401.

Von Stackelberg, U., 1979, 'Sedimentation, hiatuses and development of manganese nodules, Valdivia site VA-13/2, Northern Central Pacific', in J.L. Bischoff and D.Z. Piper (eds.), Marine geology and oceanography of the Pacific manganese nodule province, N.Y., Plenum Press, pp. 559-586.

Von Stackelberg, U., 1985, 'Significance of benthic organisms for the growth and movement of manganese nodules in the equatorial North Pacific', Geo-Marine Lett., N 4, 37-42.

Wallace, G.T. and Duce, R.A., 1975, 'Concentration of particulate trace metals and particulate organic carbon in marine surface waters by a bubble flotation mechanism', Mar. Chem., 3, N 2, 157-181.

Wallace, G.T., Hoffman, G.L. and Duce, R.A., 1977, 'The influence of organic matter and atmospheric deposition on the particulate trace metal concentration of northwest Atlantic surface seawater', Mar. Chem., 5, N 2, 143-170.

Wallast, R., Billen, G. and Duinker, J.C., 1979, 'Behaviour of manganese in the
    Rhine and Scheldt estuaries. Pt. 1: Physico-chemical behaviour', Estuarine
    Coast. Mar. Sci., 9, N 2, 161-169.
Wangersky, P. and Gordon, D., 1965, 'Particulate carbonate, organic carbon and
    Mn^{2+} in the open ocean', Limnol. Oceanogr., 10, 544-550.
Wangersky, P. and Joensuu, O., 1964, 'Strontium, magnesium and manganese in
    fossil foraminiferal carbonates', J. Geol., 72, 477-483.
Watkins, N.D. and Kennett, J.P., 1976, 'Erosion of deep-sea sediments in the
    Southern Ocean between longitudes 70°E and 190°E and contrasts in manganese
    nodule environment', Mar. Geol., 23, N 1, 103-111.
Watling, H.R. and Watling, R.J., 1976, 'Trace metals in Choromytilus meridionalis',
    Mar. Poll. Bull., 7, 91-94.
Wedepohl, K.H., 1960, 'Spurenanalytische Untersuchungen an Tiefseetonen aus dem
    Atlantic', Geochim. et Cosmochim. Acta, 18, N 3-4, 200-231.
Wedepohl, K.H., 1980, 'Potential sources for manganese oxide precipitation in the
    oceans', in Geology and geochemistry of manganese, Vol. 3, Budapest, Akad.
    Kiado, pp. 13-22.
Weiss, H., Bertine, K., Koide, M. and Goldberg, E.D., 1975, 'The chemical compo-
    sition of a Greenland glacier', Geochim. et Cosmochim. Acta, 39, N 1, 1-10.
Weiss, H.V., Herron, M.M. and Langway, C.C., 1978, 'Natural enrichment of ele-
    ments in snow', Nature, 274, N 5669, 352-353.
Weiss, R.F., 1977, 'Hydrothermal manganese in the deep-sea: Scavenging residence
    times and manganese/helium-3 relationships', Earth Planet. Sci. Lett., 37,
    257-262.
Wendt, J., 1974, 'Encrusting organisms in deep-sea manganese nodules', Spec. Publ.
    Int. Assoc. Sedim., 1, 437-447.
White, W.M., Patchett, J. and Ben Othman, D., 1986, 'Hf isotope ratios of marine
    sediments and Mn nodules: evidence for a mantle source of Hf in seawater',
    Earth and Planet. Sci. Lett., 79, N. 1/2, 46-54.
Whitfield, M. and Turner, D.R., 1982, 'Discussion on the paper: "Ultimate removal
    mechanisms of elements from the ocean - a comment"', Geochim. et Cosmochim.
    Acta, 46, N 10, 1989-1995.
Wildeman, T.R. and Haskin, L., 1965, 'Rare-earth elements in ocean sediments',
    J. Geophys. Res., 70, N 12, 2905-2910.
Wilkniss, P.E., Bressan, D.J., Carr, R.A. and Larson, R.E., 1974, 'Chemistry of
    marine aerosols and meteorological influences', J. Rech. Atmosph., 8, N 3-4,
    883-893.
Williams, R.B. and Murdock, M.B., 1969, 'The potential importance of Spartina al-
    terniflora in conveying zinc, manganese and iron into estuarine food chains',
    Proc. 2nd. Nat. Symp. Radioecol. US AEC Conf., pp. 431-440.
Willis, J.P. and Ahrens, L.H., 1962, 'Some investigations on the composition of
    manganese nodules with particular reference to certain trace elements', Geochim.
    et Cosmochim. Acta, 26, 751-764.
Wilson, T.R.S., 1978, 'Evidence for denitrification in aerobic pelagic sediments',
    Nature, 274, N 5669, 354-356.
Winchester, J.W., Weixin, L., Lixin, R., Mingxing, W. and Maenhaut, W., 1981,
    'Fine and coarse aerosol composition from a rural area in North China',
    Atm. Envir., 15, N 6, 933-937.
Windom, H.L., 1969, 'Atmospheric dust records in permanent snowfields: implications
    to marine sedimentation', Bull. Geol. Soc. Amer., 80, 761-782.
Windom, H.L., 1970, 'Contribution of atmospherically transported metals to South
    Pacific sediments', Geochim. et Cosmochim. Acta, 34, N 4, 509-514.
Windom, H.L., Beck, K.O. and Smith, R., 1971, 'Transport of trace metals to the
    Atlantic ocean by three southeastern rivers', Southeast. Geol., 12, N 2, 169-181.
Windom, H.L. and Smith, R.G., 1972, 'Distribution of iron, manganese, copper, zinc
    and silver in oysters along the Georgia coast', J. Fish. Res. Bd. Canada, 29,
    450-452.
Winogradsky, S., 1888, 'Ueber Eisenbacterien', Botan. Zeit., B. 46, 261-270.

Winterer, E.W. and Ewing, J.J. et al., 1972, 'Initial Report of the Deep-Sea Drilling Project', Wash., Covt. Print. Off., v.XVII. 930 pp.

Wolfe, D.A., Cross, F.A. and Jennings, C.D., 1973, 'The flux of Mn, Fe and Zn in an estuarine ecosystem', in Radioactive Contamination of the Marine Environment, Int. Atom. En. Agen. Vienna, Austria, pp. 159-175.

Wong, M.H., Chan, K.Y., Kwan, S.H. and Mo, C.F., 1979, 'Metal contents of the two marine algae found on iron ore tailings', Mar. Poll. Bull., 10, 56-59.

Woo, C.C., 1973, 'Scanning electron micrographs of marine manganese micronodules, marine pebble - sized nodules and fresh water manganese nodules', in Morgenstein (ed.), The origin and distribution of manganese nodules and prospects for exploration, Hawaii Inst. Geophys., pp. 165-171.

Wood, J.M., 1974, 'Biological cycles for toxic elements in the environment', Science, 183, N 4129, 1049-1052.

Wooster, W.S. and Reid, J.L., 1963, 'Eastern boundary currents', in M.N. Hill (ed.), The sea, N.Y., Wiley, pp. 253-280.

World Register of Rivers discharging into the Ocean. 1978, Paris, UNESCO, 21 pp.

Xavier, A., 1976, 'Sedimentologische und structurelle Untersuchungen zur Genese der marinen Eisen-Mangan-Akkretionen ("Manganknollen")', Senckenberg. Marit., 8, N 4/6, 271-309.

Xavier, A. and Klemm, D.D., 1976, 'The ultrastructures of the pelagic manganese nodules from the Pacific Ocean and implications for their genesis', Abstr., 25th. Intl. Geol. Congr., 2, 359-360.

Xavier, A. and Klemm, D.D., 1979, 'Authigenic gypsum in deep-sea manganese nodules', Sedimenthology, 26, N 2, 307-310.

Xia Ming, 1984, Rate of accretion of manganese crust from South China Sea basin and its geochemical characteristics; 27th. Intern. Geol. Congr., Moscow, 4-14 Aug., 1984. Abstr., Vol. 5, Sect. 10-11, pp. 440-441.

Yabuki, H. and Shima, M., 1973, 'Uranium and other heavy elements in deep sea sediments coexisting with manganese nodules', Sci. pap. Inst. Phys. and Chem. Res., 67, N 3, 155-156.

Yano, N., Katsuragawa, H. and Maebashi, K., 1974, 'Assessment of sources of maritime aerosols by neutron activation analysis', J. Rech. Atmosph., 8, N 3-4, 807-817.

Yasushi, K., Masahiro, S. and Matsumoto, E., 1980, 'Partitioning of heavy metals into mineral and organic fractions in a sediment core from Tokio Bay', Geochim. et Cosmochim. Acta, 44, N 9, 1279-1285.

Yeats, P.A., Sundby, B. and Bewers, J.M., 1979, 'Manganese recycling in coastal waters', Mar. Chem., 8, N 1, 43-55.

Yokoyama, Y. and Nguyen, H.V., 1979, 'Non destructive radiometric studies of manganese nodules by gamma-ray spectroscopy with Ge Li detector', in La genèse de nodules de manganèse, Edit. CNRS, Paris, pp. 327-332.

Yokoyama, Y. and Nguyen, V., 1979, 'Determination des vitesses de sedimentation marine et celles de la croissance des nodules de manganèse par la spectrométrie gamma, non destructive à haute résolution', C. r. Acad. Sci., D. 289, N 3, 229-232.

Yoshimura, T., 1934, 'Todorokite, a new manganese mineral from the Todoroki Mine, Hokkaido, Japan', J. Fac. Sci., Hokkaido, Univ., Sapporo, Ser. 4, N 2, pp. 289-297.

Yoshimura, A., Tada, H., Sakai, M., Harada, T. and Oishi, K., 1976, 'Distribution of inorganic constituents of Kombu blade. III. Inorganic constituents of acceleratodly cultured ma-combu-at various growth stages', Bull. Jap. Soc. Sci. Fish., 42, 661-664.

Young, E.G. and Langille, W.M., 1958, 'The occurence of inorganic elements in marine algae of the Atlantic provinces of Canada', Can. J. Botany, 36, 301-310.

Young, J.A. and Silker, W.B., 1980, 'Aerosol deposition velocities to the Pacific and Atlantic oceans calculated from ^{7}Be measurements', Earth and Planet. Sci. Lett., 50, N 1, 92-104.

Young, L.G. and Nelson, L., 1974, 'The effects of heavy metals ions on the motility of sea urchins spermatozoa', Biol. Bull., 147, 236-246.

Yushko-Zakharova, O.E., Ivanov, V.V., Razina, I.S. and Chernyaev, L.A., 1970, 'Geochemistry, mineralogy and methods of determination of platinum group elements', Moscow, Nedra, 200 pp. (in Russian).

Yushko-Zakharova, O.E., Zakharov, V.E., Golovina, M.S., Dubakina, L.S. and Scherbakov, D.K., 1984, 'Native metals in the World Ocean manganese nodules', Dokl. AN SSSR, 275, N 2, 465-467 (in Russian).

Zalogin, B.S. and Rodionov, N.A., 1969, 'River mouths within the USSR territory', Moscow, Mysl, 312 pp. (in Russian).

Zavarzin, G.A., 1964, 'On mechanism of manganese deposition at mollusks' tests', Dokl. AN SSSR, 154, N 4, 944-945 (in Russian).

Zavarzin, G.A., 1979, Lithotrophic microorganisms, Moscow, Nauka, 324 pp. (in Russian).

Zektzer, I.S., Dzhamalov, R.G. and Meskheteli, A.V., 1984, 'Subsurface water exchange of land and sea', Leningrad, Gidrometeoizdat, 208 pp. (in Russian).

Zelenov, K.K., 1964, 'Iron and manganese in exhalations of the Banu - Wahu submarine volcano (Indonesia)', Dokl. AN SSSR, 155, N 6, 1317-1320.

Zelenov, K.K., 1972, 'Volcanoes as source of ore matter supply to sediments', Moscow, Nauka, 213 pp. (in Russian).

Zhigalovskaya, T.H., Malakhov, S.G., Egorov, V.V., Filina, A.I. and Makhonko, E.P., 1973, 'Some characteristics of the global distribution of microelement concentrations in the lower troposphere', Izv. AN SSSR, Fizika okeana i atmosfery, 9, N 7, 775-780.

Zholskyi, S.T., Kleshchenko, S.A., Emelyanov, V.A. et al., 1979, 'Hydrophysical and hydrochemical properties of bottom sediments in the tropical Indian ocean', Kiev, GIN AN UkSSR, Preprint 79.14, 54 pp. (in Russian).

Zingde, M.D., Singbal, S.Y.S., Moraes, C.F. and Reddy, C.V.G., 1976, 'Arsenic, copper, zinc and manganese in the marine flora and fauna of coastal and estuarine waters around Goa', Ind. J. Mar. Sci., 5, 212-217.

Zoller, W.H., Cladney, E.S. and Duce, R.A., 1974, 'Atmospheric concentrations and sources of trace metals at the South Pole', Science, 183, N 4121, 198-200.

- , diagenetic 104, 112, 145, 159, 214, 256
- , diagenetic-hydrogenous 145
- , discoidal 103, 104
- , ellipsoidal 103, 104, 107, 108
- , fields of 98
- , flat 103, 104
- , form of 103-110
- , forms of metals in 155-159
- genesis of 250-266
- , glassy texture of 111, 112
- , globular texture of 111, 112
- , homogenous texture of 111
- host sediments of 160-162
- , hyaline texture of 111
- , hydrogenous 112, 145, 159, 214, 256
- in sediment cores 102
- , insoluble residue of 122, 164, 211
- , internal structure of 108, 112-120
- , irregular 103, 104, 107
- , laminated texture of 111, 112
- , marine 102, 137, 138
- , massive texture of 111
- , megabelts of 98
- , micronodules 40, 74, 77, 78, 103, 108, 151, 152
- , mineralogy of 112, 121-126, 146
- , Mn/Fe ratio in 140-147, 159, 160
- , morphogenetic type of 144, 145
- , mottled texture of 111, 112
- , nuclei of 104, 108, 169, 175, 176, 180, 220
- , organogenic texture of 111
- , origin of 251-267
- , outer shell of 112
- pavement 103, 104
- , platy 104, 107
- , polynodules 103, 104, 107
- , pore water in 164
- , porosity of 104, 107

- reference samples of 129, 130, 164, 173, 175, 180, 183, 192
- , regions of 98
- , reserves of 98
- , rounded 108
- , shape of 104-108
- , shell of 104, 175, 176, 178, 180, 220
- , side part of 145, 147
- , size of 103, 104-108
- , spatial density of 98
- , structure of 104, 108-120
- , surface of 103, 104-110
- , tabular 103, 104
- , texture of 108-111
- , top side of 145, 147, 166, 168, 169, 175, 201, 214, 246, 247
- , variagated 103
- , ultramicroscopic structure of 112, 117-120
- , weight density of 98-101
Manganite 74, 112 114
Manganosite 114, 122
Manihiki Rise 100, 165
MANOP 145, 161
Marcus-Necker 99, 100
Mariana Trough 234
Masquaren Ridge 101
Mass Balance 50
Mass Exchange 9, 15
Massachusetts 152
Matupi Harbour 18, 20
Meandrina 39
Mediterranean Sea 58, 60, 63, 66, 67, 74
- sediments 63, 66, 67, 74
Medusae 40
Mekong River 6
Menard field 98, 100
Mercenaria 41
Mercury 9, 180
Meretrix 41
Metal-organic complexes 27, 28, 36, 37
Metalliferous sediments 23, 24, 67,